Basics of Virtual Reality

Armin Grasnick

Basics of Virtual Reality

From the Discovery of Perspective to VR Glasses

Springer

Armin Grasnick
Moos-Bankholzen, Baden-Württemberg
Germany

ISBN 978-3-662-64200-9 ISBN 978-3-662-64201-6 (eBook)
https://doi.org/10.1007/978-3-662-64201-6

This book is a translation of the original German edition „Grundlagen der virtuellen Realität" by Grasnick, Armin, published by Springer-Verlag GmbH, DE in 2020. The translation was done with the help of artificial intelligence (machine translation by the service DeepL.com). A subsequent human revision was done primarily in terms of content, so that the book will read stylistically differently from a conventional translation. Springer Nature works continuously to further the development of tools for the production of books and on the related technologies to support the authors.

This Springer imprint is published by the registered company Springer-Verlag GmbH, DE, part of Springer Nature.
The registered company address is: Heidelberger Platz 3, 14197 Berlin, Germany

Dedicated to my beloved wife Annett, without whom this book would have had no beginning and no end

Preface

Perspectives from the point of view of giants

It seems to be a basic need of modern marketing to label any further development with the label of innovation. Innovation is constantly equated with invention as a matter of course. Yet this term is used in its original meaning (in the sense of the economist Schumpeter) in a much more general way (from [1], p. 91)

> Technological change in the production of commodities already in use, the opening up of new markets or of new sources of supply, Taylorization of work, improved handling of material, the setting up of new business organizations such as department stores—in short, any 'doing things differently' in the realm of economic life—all these are instances of what we shall refer to by the term of Innovation.

Since innovation does not necessarily mean invention,[1] but merely describes some kind of change and, moreover, now seems a bit old-fashioned,[2] the reference to an innovation is no longer sufficient to motivate consumers to buy. Stronger incentives are needed to achieve this. In order to advertise the absolute novelty of a product, one therefore likes to use the term "world novelty". Those who like it a little more martial even announce a "revolution". Marx had already proclaimed in his *Capital* ("Kapital") the revolutionary idea[3] of the

---

[1] Schumpeter uses the concept of innovation detached from the concept of invention: "Innovation is possible without anything we should identify as invention and invention does not necessarily induce innovation, but produces of itself, . . ., no economically relevant effect at all" (from [1] p. 91) see also [2].

[2] After all, Schumpeter used it already since the 1930s in his "Theory of Innovations" (in [3], p. 87 ff.).

[3] Clearly more revolutionary than *Das Kapital* is the Manifesto of the Communist Party. It culminates in the statement: "They [the Communists] openly declare that their ends can only be achieved by the violent overthrow of all previous social order. Let the ruling classes tremble before a Communist revolution. The proletarians have nothing to lose but their chains. They have a world to gain" (from [4], p. 23).

fundamental decline of profits.[4] The inevitable reduction in the profits of established companies leads to the appearance of new market players with fresh ideas that create media hype and can thus bring about the inevitable ruin of even large companies. This allows start-ups to build their success on the ruins of their failed predecessors. This "creative destruction"[5] is the more elegant and less bloody form of revolution. In the 1990s, Bower and Christensen had used this concept to explain the failure of large companies and called it "disruptive technologies" [6]. However, far from being outdated, the aged concept of disruption[6] is definitely "in vogue". A basic disruptive attitude can be found, for example, in Facebook's mantra "Move fast and break things" [7], and in 2015, the "Frankfurter Allgemeine" considered disruption to be the word of the year among Germany's business people [8].

The example of innovation illustrates one thing quite well. Even though modern products like to be labelled world firsts, revolutions or disruptions, the previous development history remains. In truth, even the most modern technologies have historical antecedents and have rarely appeared completely out of the blue. Sometimes a new technology is merely a contemporary adaptation of an earlier technology that produces better or faster results with the means and processes available today. Every smartphone camera is still based on the idea of the camera obscura, whose basic mode of operation has been state of the art for more than two millennia.

When I started working on this book, I had initially planned to concentrate primarily on the technical description of the imaging representation. However, the more intensively I dealt with the fundamentals of virtual reality during my research, the more important it seemed to me that I should place it in the context of the history of technology.

An image must be seen and for this you need the visual system. Eyes and vision, like man, did not develop suddenly, but are the result of many millions of years of evolution. The technology for generating images as realistically as possible also has an evolutionary history that spans several millennia. It is not a new observation that the constant increase in knowledge is always based on the previous knowledge of earlier generations.

John of Salisbury, in his 1159 work *Metalogicon* [9], already quotes the philosopher Bernard of Chartres with the remarkable sentence.

> We are like dwarfs sitting on the shoulders of giants, so that we can see more and farther than they—not, however, by virtue of our own visual acuity or height, but because we are lifted up and elevated to the heights by giant-like size.[7]

---

[4]"Law of the tendential fall of the rate of profit" (in [5], p. 191 ff.).

[5]"The Process of Creative Destruction" (Chap. VII in pp. 81 ff.).

[6]Schumpeter's *Business Cycles* of 1939 also (already) mentions disruptive innovation ([3], p. 101), but a full description is only given by Christensen some 60 years later.

[7]German translation by Klinghardt [10], in the Latin edition *Metalogicus* [11] of 1610 (III. book, IV. chapter, p.148): "...,ut possimus plura eis & remotiora videre, non utique proprij visus acumine, aut eminentia corporis, sed quia in altum subvehimur & extollimur magnitudine gigantea."

Newton picked up this image half a century later, writing in a letter to Hooke in 1675 "If I have seen further it is by standing on ye shoulders of Giants". And so it is in the field of virtual reality. You can't help but get a little acquainted with the former ideas, and after researching them, you think you've gotten to know the people behind them a bit. On my virtual journey through time and reading about a thousand publications, I came across more than five hundred personalities who had revealed a part of their lives to me through their books and inventions.

It seems appropriate to introduce these people to you one by one. Since I assume that sometimes you would like to know a little more than the name, I have taken the liberty of adding an index of persons at the end of the book, in which I have given each individual at least a short line. For some of them this was quite easy; for others only little information is known; for some of them only the name has been handed down. However, I have made every effort to obtain basic knowledge about each of them and have studied many additional sources.

In this book, I will first look at the basics of virtual reality very specifically from the point of view of its historical development, before going on to look at how the technology is shaping up. So that you can follow me, I have fully disclosed my sources to you. If something seems particularly interesting to you now and you want to know more about exactly this fact, you may have a first clue from which you can start your further studies.

Virtual reality is generated particularly through vision. Every form of representation is based on a technique that reliably enables a reproduction of the object to be represented. Most information is taken in via the eye, which means that knowledge of our surroundings is not only largely visual in nature, but also always perspectival in essence. It is therefore quite natural that the attempt to reproduce reality also relies mainly on the visual component, and the use of perspective makes the copies seem all the more realistic.

Natural vision only works through the perception of light emanating from objects. Without these illuminating rays, observing the environment is simply not possible. Looking at a painting in a dark room is just as useless as trying to use your laptop with the backlight turned off. The illusion of a painting is not based on the lighting itself, but on the interaction of light with the structures and colours of the object through reflection, absorption, refraction, scattering or diffraction. What we see is not really the object, but the change of the light reaching the eye. The illusion of the object is only created by the light and is consequently an illusion of light.

A perfect illusion of light must completely deceive the perception and completely reproduce the external shape of an object. The illusion is perfect when the viewer cannot tell whether he is really seeing the object or only an image of the object. In screen technology, attempts are being made to come closer and closer to this claim with ever higher resolution, better colour reproduction or simply sheer size. Nevertheless, there remains a decisive difference between the artificial image and real reality: the artificial image does not provide the spatial effect of the real scene. A change of one's own position does not cause a change of perspective in the image; the spatial visual impression remains modest.

However, human vision is always spatial. Perception without the impression of space is unnatural. That is why there have been and still are attempts to expand image reproduction to include the illusion of space. In the spatial image or 3D technology, the most diverse technical processes were and are used to make the illusion even more perfect.

After the publication of my book *3D ohne 3D-Brille* [12] a few years ago, the development of visual space representation has taken a path that I described and justified in the chapter "Limits of stereoscopy": The negative effects of spatial imaging technology have led to the rejection of 3D technology and, as a logical consequence, all major manufacturers of consumer electronics are now gradually burying the production of 3D televisions. The media response manifests itself in headlines such as "3D TV is dead" or "The end of 3D". But is this really the case?

The use of "3D" is not limited to the showing of more or less successful films with a plastic effect in the cinema or in the use of virtual reality glasses in computer games, but is taking on an ever greater role in everyday life.

In fact, the acquisition and rendering of three-dimensional data has become increasingly important in recent years.

The reference to 3D space is obvious in 3D printing. For applications in robotics, artificial intelligence or even autonomous driving, the need for a 3D description of space is much less obvious. On closer inspection, however, it quickly becomes clear that spatial orientation is required in the examples mentioned. This spatial orientation does not only refer to the knowledge of one's own position. At the latest in the example of autonomous driving, it becomes clear that an estimation of distances to other objects is also of considerable advantage.

Today, the reality we know is not only virtualized in 3D, but also expanded with additional information and mixed with the real world.

The term "3D" is probably currently perceived a little old-fashioned and not very hip. Much more popular are all names that fit well with the English word reality, such as virtual reality[8], augmented reality[9] or mixed reality.[10] With today's interactive 3D glasses, people simply try to establish the product names (e.g. RIFT[11], Zeiss VR One[12] or Playstation VR[13]). Of course, modern and technologically advanced sounding names are often used in connection with Digital or Lightfield[14] or even with an alleged reference to holography.[15]

[8] Virtual reality, also in the abbreviation "VR".

[9] Augmented reality, also in the abbreviation "AR".

[10] Mixed reality, abbreviated as "MR".

[11] For the 3D glasses from OCULUS.

[12] For the 3D glasses from CARL ZEISS.

[13] For the 3D glasses from SONY.

[14] Or even together, as for example in the "Digital Lightfield" for the 3D display of MAGIC LEAP.

[15] For example, "Holographic Video" for the 3D rendering of OTOY.

Of course, one does not necessarily have to insist on the abbreviation "3D" to describe a three-dimensional space. The transition from the real environment to the virtual world is not even necessarily bound to the existence of a spatial model; the virtual or augmented reality can also be a superimposition of purely two-dimensional data. Normally, however, a localization of the data and objects or at least the evaluation of the viewer's head position is also desired. The image impression should adapt to the perspective of the viewer.

Background Information

In 1999, Cees van Berkel[16] visited me in my laboratory and we discussed the differences between his 3D monitor (Philips 3D[17]) and mine (4D-Vision[18]). In the course of the debate, Cees said: "In principle, your 3D mask is also just a lenticular". Of course I immediately protested and pointed out the differences, but in principle Cees was absolutely right. It all depends on the way you look at it, the perspective.

Since then, I have persistently studied the representation and over the years have researched the most diverse systems for pretending artificial reality. In the process, it has become increasingly clear that despite the frequent emphasis on the differences in the various methods, the similarities outweigh the differences. The different VR techniques are quite comparable with each other.

A perspective describes the perception of an object from a certain point of view, i.e. the way of looking at it. However, perspective is also the observation of a 3D scene from a viewing position. This is the usual procedure when recording a real scene. From several perspectives, the illusion of reality is created by simple observation if the scene is suitably reproduced.

The observation of an object from several perspectives does not seem to deserve the designation 3D. But this is the actual essence of spatial perception: only the difference of what is seen from different perspectives leads to a spatial impression. This vision is neither based on imagination nor hallucination, but is the basic function of spatial vision.

3D is nevertheless only a part of the big perspective picture. From the light field or a hologram of a scene, the recorded space can be reconstructed—not just displayed. You can now move around in this virtual space, measure in it or obtain a specific perspective from it with the highest resolution.

Is perspective, then, the essential thing? A perspective is the projection of space onto a surface, from a three-dimensional space onto a two-dimensional sensor, from a higher dimension to a lower one. The perspective itself can be three-dimensional, for example, if only a 3D view is obtained from a four-dimensional space. A two-dimensional image can also be created from a series of one-dimensional measurements.

Real 3D is by no means static. The arrangement of the objects in space and one's own perspective change continuously during observation. The real world is subject to the

[16] At that time at Philips Research Laboratories in Redhill, UK.

[17] With lenticular.

[18] With parallax barrier.

elements; wind and waves, light and shadow constantly change the external appearance of the object of observation. The changes over time demand to be described by another dimension: time. Thus, our world becomes four-dimensional (4D). Perspective, space, time and vision are the foundations of spatial perception and illusion.

In this book, I want to get to the bottom of things and introduce you to the various methods of representation from antiquity to the present, but also point out in particular the commonalities of the technologies. You will notice that even the most modern technologies are often only current implementations of historical inventions. But all technologies have one thing in common. They always deliver illusions of light.

The perfect illusion is still a vision. A device whose image reproduction can no longer be distinguished from reality has not yet been presented. Nevertheless, the foundations have been laid. The illusion may soon become a reality.

This book is an attempt to describe the development of artificial perspectives as a whole and to develop the foundations of virtual reality from them. But perhaps some of the old things that were considered long ago can serve today as a basis or inspiration for something new. Or, to put it in Walt Disney's words[19]:

> If you can dream it, you can do it. Remember that this whole thing started with a dream and a mouse.

Moos-Bankholzen, Baden-Württemberg, Germany Armin Grasnick
Autumn 2019

Literature

1. Schumpeter JA. Konjunkturzyklen: Eine theoretische, historische und statistische Analyse des kapitalistischen Prozesses. Göttingen: Vandenhoeck & Ruprecht; 1961.
2. Schumpeter JA. Capitalism, socialism and democracy. 2. Aufl. New York: Harper & Brothers Publishers; 1942.
3. Schumpeter JA. Business cycles: a theoretical, historical and statitical analysis of the capitalist process. New York und London: McGraw-Hill Book Company; 1939.
4. Marx K, Engels F. Manifest der Kommunistischen Partei. London: Bildungs-Gesellschaft für Arbeiter; 1848.
5. Marx K. Das Kapital. Engels F, Editor. Hamburg: Otto Meissner; 1894.
6. Bower JL. Christensen CM. Disruptive technologies: catching the wave. Harv Bus Rev. 1995;73(1):43–53.

[19]From a promotional video by the Walt Disney Company for the reopening of "Horizons" at Walt Disney World Resort in Florida in 1983.

7. WIRED. Mark Zuckerberg's Letter to Investors: „The Hacker Way". WIRED. 1. Februar 2012.https://www.wired.com/2012/02/zuck-letter/. Accessed: 22. Dez. 2018.
8. Meck G, Weiguny B. Das Wirtschaftswort des Jahres: Disruption, Baby, Disruption! Frankfurter Allgemeine. 2015.https://www.faz.net/aktuell/wirtschaft/wirtschaftswissen/das-wirtschaftswort-des-jahres-disruption-baby-disruption-13985491-p2.html?printPagedArticle=true#pageIndex_1. Accessed: 22. Dez. 2018.
9. von Salisbury J. The metalogicon of John of Salisbury: a twelfth-century defense of the verbal and logical arts of the trivium. Berkeley: University of California Press; 1955.
10. Klinghardt M. Das älteste Evangelium und die Entstehung der kanonischen Evangelien. Tübingen: Francke; 2015.
11. von Salisbury J. Metalogicus: E codice M.S. Academiaę Cantabrigensis. Paris: Beysl; 1610.
12. Grasnick A. 3D ohne 3D-Brille: Handbuch der Autostereoskopie. Springer; 2016. http://www.springer.com/de/book/9783642305092#aboutBook.

Contents

1 Introduction

Summary

From a technical point of view, all relevant properties of a scene can be reproduced in the best quality today—except for the spatial impression. The reproduction of this so important feature is still unsatisfactory. Frequently, spatial imaging techniques are still tied to 3D glasses, which can be perceived as unpleasant. VR glasses are advanced over the historical stereoscope—but glasses nonetheless. And so a certain rejection of 3D technology has taken root, which could lead to the misconception that 3D imaging is a waste of time. But 3D images have long been an everyday necessity in medicine and must be displayed in 3D again for depth perception during a surgery. Complete 3D data is not only needed for 3D printing, but is also indispensable for interactive visualization. Autonomous systems and robots need 3D information to navigate safely in space. I want to show you that amazing technologies have been developed constantly to enhance the illusion of reality, to record and render reality. The creation of a realistic illusion is not a useless gimmick, but is occupying an ever increasing space in daily life. If you want it, you can already create spatial images and display them without 3D glasses. Familiarize yourself with the technology of illusion. It is time.

1.1 Creation of Illusions

The Technique of the Images

Overview

"Hoping for many peaceful-colored, but also exciting-colored events".

(continued)

A. Grasnick, *Basics of Virtual Reality*,
https://doi.org/10.1007/978-3-662-64201-6_1

(Willy Brandt on the occasion of the launch of German colour television at the 25th Great German Radio Exhibition in Berlin on August 25, 1967 [1]).

About half a century ago, color television was officially introduced in Germany. By this time, not even 40 years had passed since electronic black-and-white television was first introduced. In this relatively short period of time, the general desire for lifelike, colorful presentation has significantly fueled innovation and eventually led to a technology capable of imitating natural colors.

In addition to natural color perception, humans are also naturally capable of perceiving other characteristics of a scene, most notably movement, resolution, and spatial arrangement.

Today's technology offers many possibilities to reproduce a scene true to life. A current TV in portrait format can easily tower over a grown man and display everyday objects in their original size. Even the fastest movements are displayed smoothly with the highest refresh rates, and the resolutions of the devices have long been high enough to display the smallest details.

There is no doubt that man has always wanted to reproduce his surroundings as faithfully as possible and has used perspective since time immemorial. In this way, man reproduces what he has already seen before in reality. The eye of the observer is the instrument that dictates perception and creates the illusion of perspective. Therefore, the eye is the very first object of our perspective observation. In Chap. 2, various forms of perspective representation in different stages of human history are presented—from the Stone Age, through Antiquity, to the Renaissance. A perspective can have surprising properties and, for example, can only be perceived from a very specific point, as is the case with an anamorphic image. A perspective does not even have to occur solitarily. Natural spatial vision already allows the perception of two different perspectives at the same time. Some or several perspectives can be displayed simultaneously as tabula scalata. In this case, different perspectives are displayed simultaneously, but the perceived image remains flat. In principle, both eyes see the same image and thus receive a perception that essentially corresponds to the vision of the one-eyed person. Surprisingly, however, the one-eyed person can also see 3D.[1] A little experiment will show you how surprisingly well this can work.

Seeing an illusion of space requires the existence of a decent rendering. The development of the corresponding techniques is presented in Chap. 3. The observation of the heavenly bodies in antiquity gave rise to the desire for aids to track them more accurately. The first images of a camera obscura were probably those of the sun, and in its glow the imaging power of light was recognized early on. Magic lanterns and fright lamps

[1] and participate in road traffic in Germany, which requires sufficient spatial perception [2] (Annex 6: Visual acuity requirements).

sometimes created an unpleasant illusion by means of this image, which was subsequently enriched with some movement for increased realism and greater horror. The ever-improving illumination of the projection not only made possible more serious investigations through the solar microscope, but also spurred the representation of motion. The apparatus up to this time was entirely mechanical. The discovery of electricity catapulted technology into a new era. The transmission of image and motion was now over the greatest distances, even continents. In a very short time, a multitude of television sets came into being, and as the picture grew larger, so did the depth. A large picture simply demanded a huge piece of furniture. Only the development of flat screens allowed the creation of truly flat picture reproduction devices.

Usually, the image impression was then also flat. But if a 3D impression can be obtained in natural vision with both eyes or even a single eye, why does one always see only a flat image on a screen? The reason lies in the way the perspectives are displayed. If the perspectives are not only duplicated, but also displayed in such a way that together they form a 3D image, this is called stereoscopy. The spatial impression, or rather the illusion of space, can be achieved in various ways. The Chap. 4 tells about this. The means of choice are optical (e.g. mirrors, lenses, prisms) or use the possibility of a targeted masking (e.g. color filters, polarization, shutter, time parallax). With some training, 3D vision can also be achieved without 3D glasses. It is not possible for everyone to perform parallel or cross vision reliably and without effort. For this reason, the desire to achieve 3D vision without the use of "eye gymnastics" or the use of 3D glasses arose more than 100 years ago. This technique is summarized under the generic term "autostereoscopy ". But even in autostereoscopy, wishes remain unfulfilled. For example, the viewing angle is often limited or the resolution of the display is significantly reduced. In practice, these limitations are created by the optical grid, which turns the 2D screen into a 3D screen.

However, the virtual space created by a 3D monitor is always limited by the size of the display system.

Wouldn't it be wonderful to be able to do without the screen and be right in the middle of the action?

That this wish is also achievable is described in Chap. 5. It is understandable that the first virtual rooms were installed in real circular buildings. These walkable panoramas fulfilled the desire for a virtual environment already in the beginning, but did not allow for the representation of real movement. With the advent of large-format cinema projections, things finally started to move. Even more, the movement was not limited to the screen, but extended to the observer. The beginning of interaction with representation started with the development of shooting and flying trainers at the time of the First World War.

But real interaction is by no means tied to warfare. After the end of the Second World War, computer technology began to develop. This new technology now made it possible to create special virtual spaces that matched the viewer's position. A stereoscope strapped to the head provided the viewer with a wonderful illusion for this purpose. These head-mounted displays, in a wide variety of forms, are the most common motif one encounters when intensively studying virtual reality.

Such systems can provide an extraordinary spatial impression, but also force the wearer of the 3D frame to shut himself off from the world. Interaction with the environment is subject to the restriction "look only, do not touch".

Artificial reality should not be bound to the use of glasses, displays or headsets. Further considerations are given to this in Chap. 5. But what does a lifelike reproduction look like, how must one imagine the ultimate 3D display? Well-known visions of the future are the projection of Princess Leia from Star Wars or the "Medical Holographic Emergency Program" from the starship Voyager. A seemingly close approximation is also the "Holodeck" from the television series Star Trek. A visual approximation might succeed at some point, but will one be able to touch and feel the objects then? Haptic projections only succeed in rudiments, there is hardly any question of a tangible interaction with realistic graphics. But a real experience can also take place without real seeing, feeling or hearing—e.g. in a dream. The written vision[2] for this already exists as "phantomatics" and describes the stimulation of the senses or directly of the brain to create an artificial reality. This direct form of virtual truth is the subject of the film "Matrix", in which this extended but completely unwanted reality was dramatically staged.

Even the most sophisticated virtual reality techniques, however, essentially use peripheral virtualization by creating specific, mostly visual stimuli.

The reproduction of matter and thus a real reconstruction of objects has so far only been possible in the 3D printer under defined conditions and far from the illusion of interaction or even real time. A processing or representation of real spatial data in real time is thus not possible. In order to create objects in space that actually exist, around which one can walk and with which one can interact, one makes use of volume representation. Of course, the possibility of touching the virtual objects is then again dispensed with. Besides various voxel systems, holography is the most elegant of the methods. The recording now even takes place without a camera, but coherent light is required for the recording, which in principle does not occur in nature. Therefore, a laser is needed for illumination. However, the reconstruction can be done with normal light, which increases the applicability. To solve the illumination problem, a combination of autostereoscopy and holography is possible, which should solve all problems. A certain part of the hologram now contains the perspective and the imaging optics in the form of an interference pattern at the same time. The resolution of the image can be so high that these autostereoscopic structures lie below the perceptual threshold and are therefore no longer disturbing. However, this creates another problem. The resolution of the screen would have to be so high that no current monitor would be able to reproduce such an interference image.

There seems to be an immense variety of technologies for the representation of virtual reality. Therefore, a short section is added at the end in which an attempt is made to sort and classify the technology.

[2] As early as 1964, Stanislaw Lem described virtual reality in his "Summa technologiae" [3].

All the methods mentioned so far only describe what already exists. There is no doubt that new technologies will emerge that will overcome many of the previous drawbacks. But will the new techniques also use perspectives? Can the use of the time-honored perspective produce anything new at all? Viewing perspective as vision allows for a different point of view. If one sees the world surrounding us as a field of distributed stimuli that is captured by the senses in space, a real scene can also be interpreted as an event field. The recording of this event field from different perspectives in turn allows the reconstruction of the space. If a large number of cameras are used simultaneously during recording, moving scenes can also be recorded and analyzed. Usually, a sensor with the highest possible resolution is used for this purpose. Surprisingly, images can also be recorded if the sensor has only a single pixel. However, a single-pixel camera ultimately only records the intensity of the incident light. If you also look at the phase of the incident light, you come back to holography. Here, a three-dimensional scene is recorded onto a two-dimensional medium and reconstructed as required. This results in a reduction of the required recording medium by one dimension. If you transfer this idea to a larger scale, you could store the entire universe on a huge surface according to the holographic principle.

It becomes clear that the technique for recording and reproducing reality exists in many variants and variations. For each new application, it will be necessary to select the best method for the specific case.

In the following chapters, a comprehensive overview of the different methods is given, which should enable a historical classification, a well-founded selection and a professional application of image technology and spatial imaging techniques.

But this book is intended to be one thing above all: an instigator for your own ideas through knowledge of the basics of virtual reality.

1.2 Light and Shadow

Light as the Basis of Imaging

There is no question that an image cannot be perceived by humans without the illuminating power of light. The optical perception of the environment is based on the sensory detection of light, which makes the objects in observed space visible in the first place. The physiological structure of the eye allows vision only in a limited section of the electromagnetic spectrum, which is consequently also called the "visible spectrum" (see Fig. 1.1).

The propagation of light occurs in a straight line from a luminous point in different spatial directions. The radiated intensity and the wavelength of the light of countless points result in a radiation field of an infinite number of light rays—the light field. The intensity of what is seen is perceived as brightness, a certain wavelength range gives the impression of a certain colour. The light field emanating from the observed objects can be seen from different positions and gives a slightly different visual impression from each new point of view. This impression essentially corresponds to a geometric central projection of the

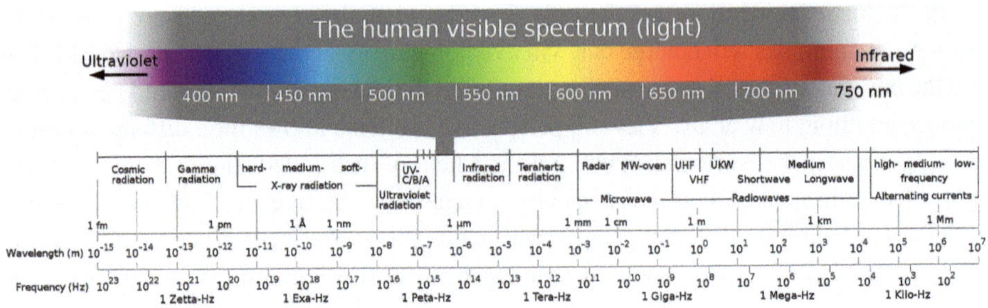

Fig. 1.1 Electromagnetic spectrum, Horst Frank/Phrood/Anony (2008), Wikimedia Commons

scene with the eye as the vanishing point. The perception of the light field from a certain location produces in the eye of the observer the projection of this view as a location-dependent perspective.

The projection of an image represents the reverse process of seeing. The recorded perspective is illuminated and “thrown” onto a suitable surface with a projector. Even the Latin original form “proiectio” denotes the throwing forward or highlighting of something. The German term “Bildwerfer” (image thrower) describes this process excellently.

After the reproduction was only possible as a static image for a long time, the mechanical reproduction of moving images became possible at the end of the nineteenth century. With the use of electric current, an electrification of the display technology also began. Here, too, the demand for illumination applied. Increasingly brighter light bulbs provided better light and allowed larger projections. With a television tube, illumination could be dispensed with altogether. Here, the phosphor hit by the cathode ray glows, providing the perceptible light. A flat screen again usually has a backlight, which is quite adequate for normal indoor lighting conditions. When viewing a laptop screen on a sunny beach, you no longer have this positive impression. However, the brightness does not only have an impact on the readability, but also on the realism of the reproduction.

The brightness of daylight cannot currently be reproduced by any screen. And so even a high-resolution image of a sunny beach is simply recognized for what it is—a beautiful but incomplete picture of reality.

1.3 Holograph and Holoscopy

Possible and Impossible Word Formations in 3D

Overview

A "holograph" could be understood to mean a recording device for holograms or the job title of a manufacturer of holograms. In fact, however, the word "holograph"[3] is not usually used in this context in the German-speaking world, but is occasionally used to refer to a document written in one's own hand and signed by the author.[4] In the English-speaking world, this interpretation is more common [4], and in the Canadian province of Ontario it is even an established legal term (holograph will[5]).

The term "holoscopy" seems to be excellently suited as a generic term for the entire 3D technology, as it is a composition of the prefix "holo"[6] and the suffix "scopy"[7] meaning "to observe the whole completely" and describes the facts in a first-class manner: One observes things completely, in all dimensions, three-dimensionally. An excellent technical term, but hardly used.

These two examples already show that even in 3D technology the terms are not always used as expected, but can deviate in meaning or designation from the usual.

3D Prefixes

In spatial imagery, some prefixes are frequently used in connection with other terms. Knowledge of these prefixes facilitates the understanding of special 3D word creations. In the following table, the usual meaning of the respective words is taken from the Indo-European Etymological Dictionary [6], the Latin Etymological Dictionary [7] and the Ancient Greek Dictionary of Origins [8], respectively, and supplemented with the common meaning where necessary (Table 1.1).

3D Suffixes

Analogous to the prefixes, the suffixes also have a common meaning indicating the probable application of a thing so named. A device denoted by "-meter" will probably be

[3] The word does not exist in the dictionary of the German language in the spelling "Holograph" nor "Holograf".

[4] Then as "holographon" or "holografon" or "holographum" or "holografum".

[5] A handwritten will signed by the author [5].

[6] Derived from the Greek "hólos" meaning whole, complete, entire.

[7] Derived from the Greek "skopein" in the meaning of to observe, aim, examine [6], p. 984.

Table 1.1 3D prefixes

Prefix	Usual meaning	Use
Stereo-	Rigid, firm, hard (also physical → spatial)	Stereoscope, Stereoscopy, Stereophotography
Holo-	Whole, complete, entire	Holography, Hologram
Auto-	Self, alone	Autostereoscopy, Autofocus
Multi-	Many, multiple, numerous	Multiview, Multifocus, Multispectral
Super-	Above, at a higher level	Superresolution, Supermultiview
Hyper-	Above, over . . . beyond, at a much higher level (usually as an intensification of super).	Hyperstereo[a], Hyperview

[a]Stereophotography with enlarged stereo base

used for measuring something, with a "-scope" one will want to observe something (Table 1.2).

The common term 3D (or 3-D[8]) is known to be just short for "three-dimensional".

But did you know that laser is just the acronym for "light amplification by stimulated emission of radiation"? This term was coined by a Columbia University PhD student[9] named Gordon Gould, who first[10] mentioned the name and acronym in his 1957 lab book[11] [14].

1.4 What the Hell. . .?

Autostereoscopy, Holography or Virtual Reality?

In the literature, different designations are used to describe a replica of reality that is as close as possible to reality. The explanation of differences or similarities of the designations is usually not part of a technical education. In fact, the designations are sometimes used in a

[8]The form recommended by the Duden is "3-D", but the spelling "3D" is more common.

[9]With Nobel Prize winner Polykarp Kusch.

[10]It should be mentioned here that Charles Townes developed the MASER (Microwave Amplification by Stimulated Emission of Radiation) at the same university a few years earlier and received the Nobel Prize for it in 1964 together with Nikolai Bassow and Alexander Prochorow ". . .for fundamental work in the field of quantum electronics, which has led to the construction of oscillators and amplifiers based on the maser-laser principle" [13].

[11]Gould writes there: "Some rough calculations on the feasibility of a LASER: Light Amplification by Stimulated Emission of Radiation".

Table 1.2 3D suffixes

Suffix	Usual meaning	Application
-graph	Recording device or job title	Stereograph, Perspectograph
-meter	Measuring device	Photometer, Perspectometer[a], Stereometer[b]
-graphic or –graphy	Recording process	Stereography, Holography
-gram	the recorded	Stereogram
-scope	Observation device	Telescope, Stereoscope, Oscilloscope
-scopy	Observation technique	Microscopy, Stereoscopy, Autostereoscopy

[a]Auxiliary means for the determination of the geodetic orientation of balloon photographs [9]
[b]A stereometer was originally (since the early nineteenth century, see also [10]) an apparatus for determining the volume of powders [11]. In connection with the evaluation of stereo images, the term stereometer was used from about the beginning of the twentieth century for a device with which the distances of identical objects in a stereo image pair could be measured (e.g. by Pulfrich [12])

misleading manner[12] or in a purely striking manner without further explanation. Knowing the meaning of some of the basic terms of 3D representation is beneficial for understanding the following chapters. Therefore, some of the most common word creations will be presented in this section.

Stereoscopy

The term stereoscopy was coined in 1838 by Charles Wheatstone, who in a paper on the physiology of vision also reported "some remarkable and hitherto unobserved phenomena of binocular vision" [15]. Basically, stereoscopy includes all methods of recording and reproducing spatial images. However, the term stereoscopy is usually used to refer only to those methods and techniques that record or reproduce in some way a stereoscopic image pair consisting of a right and a left image.

▶ Stereoscopy refers to the totality of all processes in which a 3D spatial illusion is created by a stereoscopic image pair.

Autostereoscopy

The term autostereoscopy (or, less commonly, auto-stereoscopy) is usually used to describe those technologies that enable the spatial impression to be created using the basic means of

[12]Or sometimes even deliberately false, in order to artificially exaggerate a particular technology (usually one's own) in the media response.

stereoscopy without the need for 3D glasses.[13] The following definition is taken from the Handbook of Autostereoscopy (Handbuch der Autostereoskopie) [16].

▶ Autostereoscopy refers to the entirety of all procedures in which a 3D spatial illusion is achieved by binocular display without the use of 3D glasses.

The term "automultiscopy" occurs sporadically in the English-speaking world[14] and basically means the same thing. Here, the word creation is intended to refer in particular to the simultaneous use of several (multiple) perspectives. As a result, this term is an unfavorable conflation of the automatic process of 3D perception (auto + scopy) and the technology used (multi).

▶ For all procedures as defined above, the use of the term "autostereoscopy" is recommended.

A description of different technologies for autostereoscopy can be found in Sect. 4.7.

Light Field

Mainly in the USA and more recently, the term "light field" is used for a spatial distribution. Basically, this type of description uses the physical model of a field in which various measurable physical quantities[15] are arranged in a spatially defined manner. In 1936, the Russian physicist Gershun described the photometric distribution of light as a measurable intensity. Using the English translation "The Light Field" [17], the idea was reactivated with the beginnings of computer graphics and received current attention at the latest with Lytro's light field camera. Special emphasis is now placed on the property that the perception of the light field is dependent on the viewer position and the viewing direction. A light field viewed with both eyes is therefore three-dimensional. The term light field is often mentioned in connection with autostereoscopy and especially integral photography.

▶ A light field describes the distribution of light intensities in a room in which measurements are taken from all points in all directions.

In practical applications, the light field is often recorded by means of multiple cameras or a microlens array or reproduced via a lenticular. A light field data set can therefore be recorded and reconstructed in two dimensions. In this sense, light field technology is

[13]Hence, the term "3D without 3D glasses" or "glasses free 3D".

[14]The German equivalent "Automultiskopie" is not used at all.

[15]In the electric field the field strength, in the light field the light intensity.

similar to holography, but holography uses the wave properties of light—light field uses the pure intensity effects of radiation.

Following the term "holography", one could also speak with some justification of "lumigraphy"[16] for a process using the light field. In fact, in 1996, a team around Gortler proposed a calculation of perspectives based on the theory of the light field, which was entitled "The Lumigraph" [18].

The light field is discussed again in detail in Sect. 4.8.

Holography

When Gabor tried to improve the resolution of an electron microscope[17] on an Easter day in 1947, he started from the basic idea of separation between recording and reproduction in imaging. In the first step, a photographic image is taken of the entire information of a scene, which is then fully reconstructed in a second step [20].[18] If suitable light is used in recording and reproduction, the image impression is then even three-dimensional. Gabor initially proposed the name "holoscope[19]"for this, which describes the facts quite compactly. However, since he already referred to the images as holograms[20] at this time, the name "holography" is clearly more appropriate for the process.[21]

▶ Holography refers to the totality of all processes in which a three-dimensional image is created using the wave properties of light.

Holography will be discussed again later.

[16] Luminis (lat.) = light.

[17] In the first publication in 1948, he proposed the name "electron interference microscope" for his novel method [19].

[18] Gabor reports on this in his lecture on the occasion of the award of the Nobel Prize.

[19] "Moreover three-dimensional objects may be recorded in one photograph, hence the suggested name "holoscope" which means "entire" or "whole" vision." (from [21]).

[20] "holograms" (from [21]).

[21] "-gram" the recorded, "-graph" the process of recording.

Virtual Reality

Reality[22] refers to the totality of what is real. Things in reality actually exist. Virtuality, on the other hand, denotes the existence of significant properties of reality that are comparable to the real but not real. Things in virtuality only seem to exist. Virtual reality is therefore the alleged existence of reality, an illusion of reality.

▶ Virtual reality describes the sensory impression of reality through artificial stimulation of perception.

In Plato's Allegory of the Cave,[23] Socrates, in order to illustrate the perception of reality, tells his pupil Glaucon a fictitious story of cave-dwellers who, "bound from childhood by head and thighs," only ever see the shadows of deliberate deceptions and never the things themselves. Socrates now supposes that these prisoners cannot be able to distinguish reality from fiction, since the true shape of things is wholly unknown to them (from [22], p. 362):

> In no way, then, can these take anything for the true but the shadows of those works of art?

This form of creating an apparent reality presupposes a massive encroachment on personal rights and can hardly be achieved by legal means.

To Ovid we owe the story of Pygmalion and his exceedingly realistic ivory sculpture, with which Pygmalion falls in love and which then comes to life by the grace of Venus [23]. This story inspired numerous poets again and again in the centuries to come, and finally also seduced Goethe to write his youthful story Pygmalion.

Goethe's companion Herder also used this fictitious event in a philosophical treatise. There he divided things into their real form and sensual perception and then distinguished between the senses feeling and face for the sensation of body and form in sculpture and painting [24].

A technical solution for the pretence of false facts without the actual captivation of the audience and with the renunciation of divine assistance would be the use of deception glasses.

In 1930, in his short story "Pygmalion's Spectacles," [25] Weinbaum has the strange Professor Ludwig describe a peculiar device by which a dream can become reality (from [26], p. 174):

---

[22] Lat. realitas.

[23] In the seventh book of the Politeia ("The State" ["Der Staat"]).

> I photograph the story in a liquid with light-sensitive chromates. . .I add taste chemically and sound electrically. And when the story is recorded, then I put the solution in my spectacles—my movie projector. I electrolyze the solution, the story, sight, sound, smell, taste all!

These fictional glasses for the perception of complete virtual reality were a literary prelude to the technical feasibility of such fictions.

In his "Summa technologiae" [3], published in 1964, the Polish writer Lem attempted to predict the unforeseeable future. Lem devoted the entire sixth chapter to phantomology,[24] which deals, among other things, with the technical problem of phantomatics[25] (from [3], p. 321):

> Is it possible to create an artificial reality that is completely similar to the natural one, but cannot be distinguished from it in any way?

Lem also describes in his book a pair of glasses as an attachment for the eyes[26] ([3], p. 325) and thus anticipates the later development of virtual reality glasses. The term artificial reality became known in its English form through Krueger's book "Artificial Reality" [27] and has been used as "Virtual Reality" since the 90s of the twentieth century.[27]

Literature

1. Bereczky A. Wandel der Fernsehtechnik in 50 Jahren ZDF. Bewegt seit Generationen 50 Jahre ZDF. ZDF-Pressestelle; 2013. pp. 39–41.
2. Bundesministerium für Verkehr, Bau und Stadtentwicklung. Verordnung über die Zulassung von Personen zum Straßenverkehr. Fahrerlaubnis-Verordnung Dec 13, 2010.
3. Lem S. Summa technologiae. 7. Aufl. Frankfurt a. M.: Suhrkamp; 2016.
4. Encyclopædia Britannica. Definition of Holograph by Merriam-Webster. Merriam-Webster 2017. https://www.merriam-webster.com/dictionary/holograph. Accessed on 30. Dec. 2017.
5. Ministry of the Attorney Ontario, CAGeneral. Holograph will. Glossary of terms. 2015. https://www.attorneygeneral.jus.gov.on.ca/english/glossary/?id=470#results. Accessed on 30. Dec. 2017.
6. Pokorny J. Indogermanisches Etymologisches Wörterbuch. Francke: Bern; 1959.
7. Walde A. Lateinisches Etymologisches Wörterbuch. 2. Aufl. Heidelberg: Carl Winters Universitätsbuchhandlung; 1910.
8. Köbler G. Altgriechisches Herkunftswörterbuch. Online; 2007.
9. Scheimpflug T. Über Orientierung von Ballonaufnahmen. In: Dolezal E, Editor. Internationales Archiv für Photogrammetrie, Organ der Internationalen Gesellschaft für Photogrammetrie. Wien: K. u. K. Hof-Buchdruckerei Carl Fromme; 1911. pp. 34–54.

[24] Lem refers to the scientific discipline of investigating artificial reality as "phantomology".

[25] "Phantomatics" refers to the technique used to create artificial reality.

[26] The "opposite eye".

[27] Also by Krueger himself, e.g. in his later book Artificial Reality II [28].

10. Ventress JA. Description of a Stereometer. In: Brewster D, editor. The Edinburgh Journal of Science. Edinburgh: John Thomson; 1827. p. 143–5.
11. Bibliographisches Institut, Editor. Meyers Konversations-Lexikon. 4. Aufl. Leipzig; 1885. http://www.retrobibliothek.de/retrobib/stoebern.html?werkid=100149
12. Seliger P. Die stereoskopische Meßmethode in der Praxis. Berlin: Julius Springer; 1911.
13. Nobel Media AB. The Nobel Prize in Physics 1964. Nobelprice.org. 2017. https://www.nobelprize.org/nobel_prizes/physics/laureates/1964/. Accessed on 31. Dec. 2017.
14. Gould G. Page from Gould's Lab Notebook (Internet). 1957. https://photos.aip.org/history-programs/niels-bohr-library/photos/gould-gordon-h1.
15. Wheatstone C. Contributions to the physiology of vision. Part the First. On Some Remarkable, and Hitherto Unobserved, Phenomena of Binocular Vision. Philos Trans R Soc Lond. 1838;128: 371–94.
16. Grasnick A. 3D ohne 3D-Brille: Handbuch der Autostereoskopie. Springer; 2016. http://www.springer.com/de/book/9783642305092#aboutBook.
17. Gershun A. The light field. J Math Phys. 1939;18:51–151.
18. Gortler SJ, Grzeszczuk R, Szeliski R, Cohen MF. The Lumigraph. ACM Press; 1996. pp. 43–54.
19. Gabor D. A new microscopic principle. Nature. 1948;161:777–8.
20. Lundqvist S, editor. Nobel lectures, Physics 1971–1980. Singapore: World Scientific; 1992.
21. Gabor D. Brief an Dr. J. B. le Poole (Delft) in response to le Pooles letter of 21. Jan. 1948, in which he proposes the name "Holoscope". 1948.
22. Platon. Der Staat. Schleiermacher F, Editor. Berlin: G. Reimer; 1828.
23. Ovid. Verwandlungen nach Publius Ovidius Naso. Berlin: Friedrich Vieweg der Ältere; 1798.
24. v. Herder JG. Plastik. – Einige Wahrnehmungen über Form und Gestalt aus Pygmalions bildendem Traume. Riga: Johann Friedrich Hartknoch; 1778.
25. Weinbaum SG. Pygmalions spectacles. Wonder Stories. 1935;7:28–39.
26. Weinbaum SG. Pygmalions Brille. Die besten Stories von Stanley G Weinbaum. Moewig Verlag; 1980. pp. 171–196.
27. Krueger MW. Artificial reality. Reading: Addison-Wesley; 1983.
28. Krueger MW. Artificial reality II. Reading: Addison-Wesley; 1991.

Discovery of Perspective

2

The Natural Perception of Perspective and Its Reproduction

Summary

After the eye had developed in evolutionary sub-steps over millions of years, living creatures were able to see their surroundings in true light. The visual perception was an evolutionary advantage in the eternal selection process, the "Survival of the Fittest". The anatomical uniqueness, the combination of vision, upright gait and thus freely available hands, coupled with the cognitive abilities of the brain, had enabled humans to fabricate inventive creations. At some point, humans had begun to reproduce what they saw. The initial murals and sculptures were superseded in ancient times by works of art that had an amazing realism. Craft art went hand in hand with a thirst for knowledge. Is what you see what is real? Is the world real or virtual and recognizable at all? The first reflections on seeing still showed a strong analogy to the sense of touch, whereby seeing was understood as touching things with the eyes. However, the image in the eye occurs without touch. The lens of the eye images objects onto the retina of the eye. Every image is always the projection from a certain viewing position. By changing the position, the perceived perspective also changes. The knowledge of the perspective in turn allows its construction—even without real templates and thus the realistic visualization of virtual objects.

2.1 Sense of Light

About the Development and Structure of the Eyes

The perception of the environment is closely linked to the development of the senses and especially to the development of vision. But the development of such a complex organ as the human eye seems to be a masterpiece of evolution. Even the founder of the theory of

A. Grasnick, *Basics of Virtual Reality*,
https://doi.org/10.1007/978-3-662-64201-6_2

evolution had some difficulty in explaining to his critics the origin of the eyes by natural selection. In "On the Origin of Species "([1], p. 212) Darwin writes:

> To suppose that the eye, with all its inimitable contrivances for adjusting the focus to different distances, for admitting different amounts of light, and for the correction of spherical and chromatic aberration, could have been formed by natural selection, seems, I freely confess, absurd in the highest possible degree.

In the end, however, he concludes "...if we know of a long series of gradations in complexity, each good for its possessor, then, under changing conditions of life, there is no logical impossibility in the acquirement of any conceivable degree of perfection through natural selection". ([1], p. 231–232).

Animal eyes and thus the function of vision can be traced in numerous fossils after the Cambrian explosion 540 million years ago.

The Cambrian Explosion

The Earth's Early Period (Precambrian) begins with the solidification of the Earth's crust and the development of an atmosphere. The obsolete term "abioticum" labels what the term is intended to define: A period of time without visible, surviving traces of life. Consequently, this is followed by the time of the appearance of the first fossils—the Palaeozoic era, which begins with the Cambrian period 541 million years ago.

The terms Paleozoic[1] and Cambrian[2] were coined by the English geologist Sedwick in the mid-nineteenth century, who associated the term with the Paleozoic in particular with the appearance of animal fossils. This way of looking at things can no longer be fully agreed with the present state of knowledge. Thus, not only different species of bacteria or unicellular organisms existed in the early Earth period, but even multicellular soft-bodied animals, which are known today under the name Ediacara fauna.[3] A three-dimensional fossil of such organisms was found in southern China in 2005 [5], attesting to the diversity of species already present at that time. Even though various fossils from this period have been preserved, it seems that the precursors of today's living organisms only became established during the Palaeozoic era.

(continued)

[1] The term "Cambridge Palaeozoic Fossils" is used by Sedgwick for classification (e.g. in [2], p. 139), from Greek palaios = ancient and zōikos = animal.

[2] "...now named by Professor Sedgwick the 'Upper Cambrian'" (from [3], p. 60). Sedgwick chose the name after the Latin "Cambria" for the Welsh "Cymru", the name of Wales.

[3] Named after the site where the Precambrian fossils were found, the Ediacara Hills in Australia, discovered in 1947 by Australian mining engineer Sprigg [4].

The Cambrian marks the beginning of the age of aquatic life. From this time on, numerous fossils are now preserved, which now also show hard-shelled parts. Within only ten million years, the most diverse animals with limbs, skeletons and even eyes developed virtually out of nowhere. In 1991, Gould speaks of the "Cambrian explosion" [6]. This "big bang of evolution"[4] was the exuberant prelude to evolution. Or as the "New Scientist" put it a few years ago [8]:

> Life was single-celled and boring for billions of years, then BOOM!

The simplest living beings after the Cambrian explosion had only light-sensitive cells, which are also called eyespots. Since primeval times the unicellular eye- or flagellate animals "Euglena" live on earth and still populate almost every pond. The protozoologist Ehrenberg made it his business to document these microscopic "infusion animals[5]" on his travels through Africa and Asia [10]. In his book "Die Infusionsthierchen als vollkommene Organismen" (The infusion animalcules as perfect organisms) there are numerous illustrations of the smallest creatures—of course, there is also an eye animalcule ("Augentierchen") among them. In Fig. 2.1 the red-eye point can be seen very nicely.

This red spot of the green eye animal "Euglena Viridis" is the so-called stigma, an accumulation of reddish pigments. These color pigments prevent the incidence of light on the photoreceptor from certain directions, so that Euglena can orientate itself according to the light (phototaxis, see also [12], p. 311). Since these amazing creatures are also capable of photosynthesis, swimming according to the light is essential for their survival.

The photoreceptor,[6] known as the paraflagellar body, generates a signal molecule when irradiated with blue light, which then signals the incidence of light from a specific direction. Since the photoreceptor is an ordered structure, it already performs the task of a flat sensor. An eye that is formed only by a two-dimensional sensor and a shadowing of pigment cells can be called a flat or plate eye. Such primitive eyes can still be found in starfish or jellyfish. Certain jellyfish are particularly interesting because they have a certain assortment of different eyes. A specimen worth mentioning is certainly the box jellyfish,[7] which has a total of 24 eyes of different types in groups of six (at 4 positions, each offset by 90°, see

---

[4]"Evolution's Big Bang," for example, was the December 4, 1995 cover story of Time magazine [7].

[5]The Damen Conversations Lexikon of 1835 explains the name as follows: "Infusion animalcules, the smallest living creatures visible to us, whose knowledge we have only come to through the invention of the compound microscope. They are perhaps the most wonderful and finest thing that nature has produced. They are developed from any substance which is subject to putrefaction, by infusion of water (hence infusion animalcules)" [9].

[6]These blue light receptors were first discovered in 2002 in the ophthalmic animal *Euglena gracilis* [13]. The somewhat unwieldy name "photoactivated adenylate cyclase" (PAC) refers to the enzymes that make up the ordered lamellar crystalline structure of the light receptors.

[7]Cubozoa or Cubomedusae.

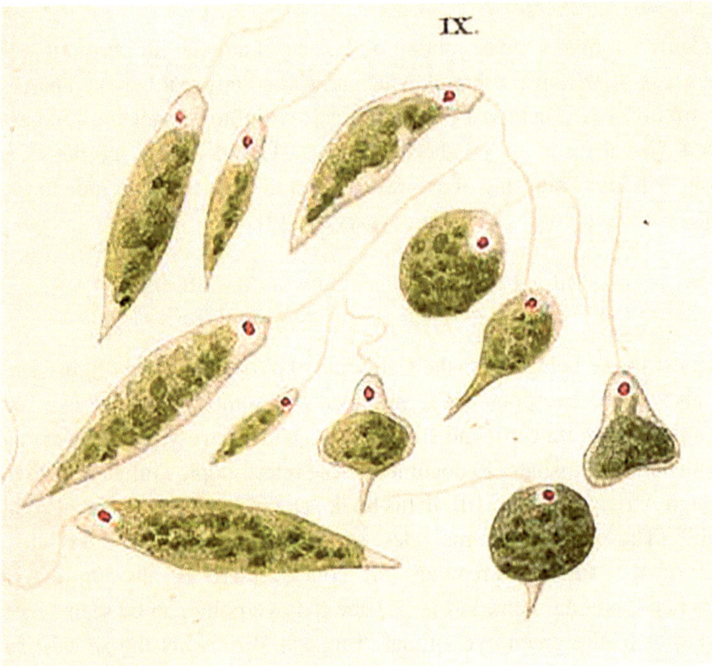

Fig. 2.1 Augentierchen (Euglena viridis), Christian Gottfried Ehrenberg (1838) in [11], T. VII, Fig. IX

also Fig. 2.4) [14]. Such a jellyfish has indeed also "slit eyes". In principle, this slit eye (see Fig. 2.2) is only a pit eye in an asymmetrical form. Here, the pit only takes over the function of direction-selective shading, so that the jellyfish with slit and pit eye perceives the direction of light incidence.

The proximal[8] and distal[9] lens eyes have a much more complex structure and already have imaging lenses and a retina as receiver [14]. Thus, optical imaging and downstream analysis are already possible. However, box jellyfish do not have a central brain, which makes it difficult to use the image data. The diversity of the eyes, the multitude and their arrangement nevertheless allow navigation through the simplest linking of the neuronal circuits, the nerve tracts [16].

The Jena evolutionary scientist Haeckel was also a connoisseur and admirer of jellyfish[10] and raved about them in his travel notes (from [17], p. 61, news from Lanzerote, Puerto de Arrecife, January 27, 1867):

[8]Located towards the center of the body.

[9]Lying away from the center of the body.

[10]As is manifested, among other things, in the designation of his residence in Jena's Berggasse as "Villa Medusa".

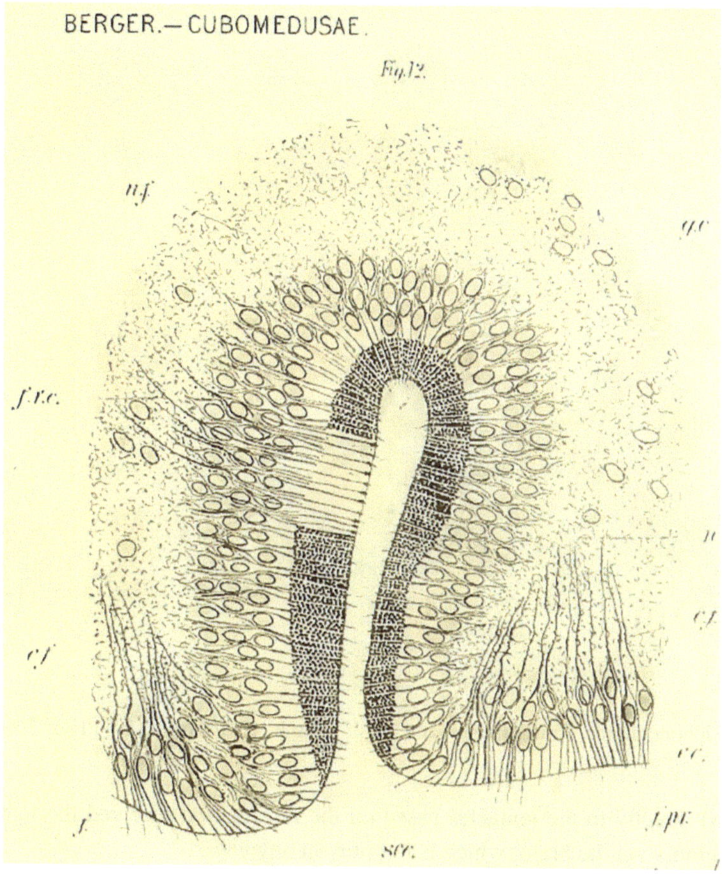

Fig. 2.2 Slit eye "Simple Eye" of the box jellyfish, Edward William Berger, 1900 (from [15], Plate II)

> All the other animal forms of the sea here, however, are surpassed in beauty and daintiness, as in high, scientific interest, by the magnificent siphonophores,[11] which I myself have chosen as the special object of my investigation.

These Siphonophores had been so wonderfully illustrated for Haeckel by the Jena lithographer Giltsch that Haeckel's enthusiasm for these animals is quite understandable (Fig. 2.3).

Giltsch also drew the eyes of the box jellyfish, showing the arrangement of the eye groups in the heart-shaped niches above and central to the tentacles (Fig. 2.4). In the accompanying description, Haeckel notes that the four eyes are connected to each other and to the tentacles by a strong nerve ring. Thus, the desired direction of movement can be

[11] "Siphonophorae" named after the organs for feeding and digestion known as siphons or suction tubes (see also [18], p. 21).

Fig. 2.3 Siphonophorae, Ernst Haeckel and Adolf Giltsch (1899), from [19] Plates 7 and 17

transmitted directly to the tentacles based on the information received through the eyes, without detours via the brain, which is not present anyway.

Surprisingly, fossils of even these gelatinous medusae of prehistoric times have been preserved from the time of the Cambrian explosion, which is only the case under extremely rare circumstances. Such circumstances occur when a creature not usually capable of fossil preservation suddenly finds itself in a deposit made of ice or peat, for example, and can thus be conservated for long periods of time. These "conservate deposits" represent the few sites for fossil jellyfish today [20].

In contrast to the jellyfish, which consist almost entirely of water, fossils of hard-shelled trilobites are much more common. The trilobites were extremely widespread in the Cambrian with innumerable species and are sometimes also eponymous for the "Age of Trilobites" ([21], p. 116). These arthropods had an exoskeleton, to which we owe the present distribution as fossilized fossils.

Fig. 2.4 Cubomedusae, Ernst Haeckel and Adolf Giltsch (1899), from [19] Plate 78

Background Information

The name Trilobite is derived from the "Systema minerale" of the preacher Woltersdorff, published in 1748. Woltersdorff attempted a classification of minerals based on the "Systema naturae" of Mr. Linné [22]. Woltersdorff did not fully agree with Linné's classification of minerals. Thus, Linné used the name "Entomolithus

(continued)

paradoxus[12]" ([23], p. 162) for this fossil petrifaction, which only has the meaning "paradoxical fossilized insect".

Woltersdorf, on the other hand, places under the heading "Petrified Bodies" the third order "The Petrified Shells of Snails and Shells". There we find the "shell stone" and as a subspecies the beetle shell stone "Conchites trilobus". The description "Round or elongated in 3 elevations divided off, looks a beetle somewhat similar" ([24], p. 42) is appropriate to the external form of a trilobite. Walch, a scientist from Jena, was both an expert in geology and a master of language. After an intensive analysis of all the names for this petrefact,[13] he arrives at a name for the stony creature that is still valid today, based on its external shape: "We want to give it the most universal name and call it a **trilobite**." ([25], p. 172).

From the numerous fossils of the trilobites it is known that they already (partly) had eyes. A team led by the palaeontologist Clarkson has intensively studied the trilobites [26], focusing in particular on the compound eyes.[14]

A trilobite whose eyes Clarkson describes was found by Portlock [27] as early as 1843 (Fig. 2.5).

Schizochroal or Holochroal?

In 1888 Clarke divided the eyes of the trilobites into two large groups according to their anatomical structure (from [28], p. 254):

> The character of the visual area in the trilobites is twofold;
>
> (a) it may be covered by a smooth continuous epithelial film[15] or cornea, through which the lenses of the ommatidia[16] are visible by translucence; and
>
> (b) the cornea may be transected by the protrusion[17] of the sclera[18] and limited to the surfaces of the ommatidia.
>
> The first group may be designated by the term holochroal[19]; the second group by the term schizochroal[20]

(continued)

[12]Entomolite from Greek entomon = insect and lithos = stone.

[13]Rare for petrification, from Latin petra = stone and factum = made.

[14]A compound eye is composed of a large number of individual globules (ommatidia).

[15]Cover tissue.

[16]Single globules of the compound eye.

[17]From Latin protrusio = to protrude, to shift outward.

[18]Sclera of the eye (eye tissue), from Greek skleros = brittle, hard.

[19]From Greek holos = whole + chroa = skin, color.

[20]From Greek schizein = to split + chroa = skin, color.

The group of "abathochroal"[21] eyes added by Jell in 1975 [29] was discussed by Zhang and Clarkson in 1990 and represents what is now the third group of trilobite eyes.

The lenses are maximally densely set in a hexagonal arrangement, which results in better light utilization (Fig. 2.7). As Kepler already suspected, the densest sphere packing of a single layer is a hexagonal[22] arrangement [30].

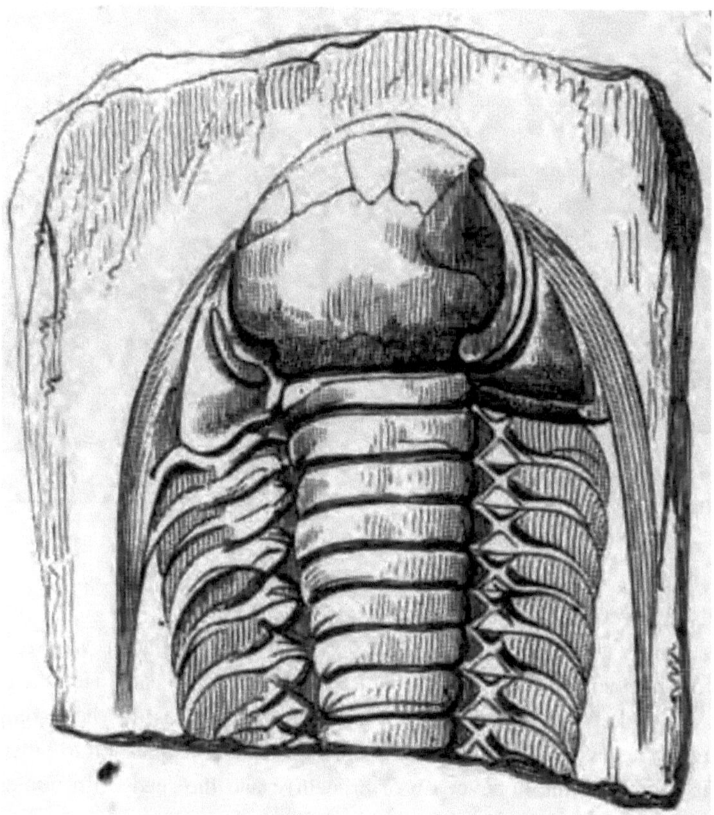

Fig. 2.5 Trilobite, Portlock and Du Noyer (1843), from [27] PL 1

[21] Abatochroal, from Greek abathos = flat, flattened + chroa = skin, color.

Jell: "Since it cannot be classified within either of the well established eye types, I propose the name abathochroal for this new type of eye."

[22] Six-sided.

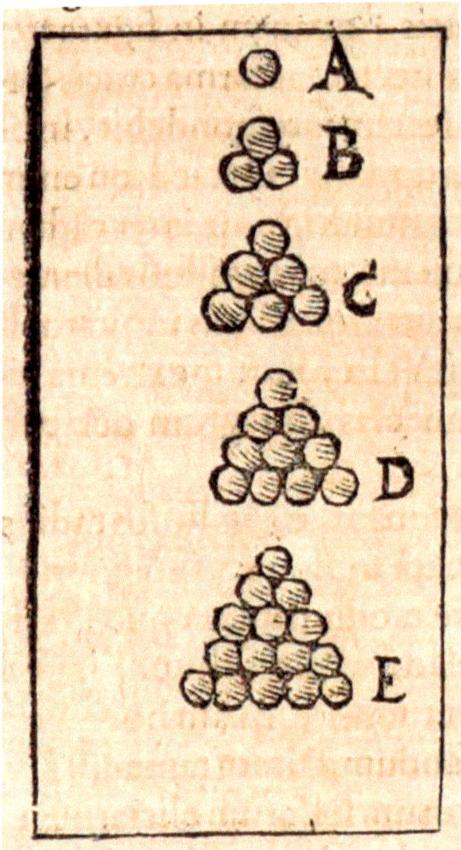

Fig. 2.6 Densest sphere packing, Johannes Kepler (1611), from [30], p. 753

Kepler's Conjecture

Kepler tried in an essay to explain to his friend and benefactor Wackenfels that certain arrangements always arise in nature for different objects.[23] He asked himself why a snowflake is always hexagonal and whether perhaps this hexagonal figure arose from a necessity of the material. Kepler extended the question to other known materials, choosing the honeycomb of a beehive and the seeds of a pomegranate. Simplifying the objects into spheres, the packing is densest when the spheres are arranged hexagonally in a layer. Viewed in a section, the spheres are arranged in pyramids (see Fig. 2.6).

(continued)

[23] In the work entitled "Strena seu de nive sexangula" (On the Snowflake, or the Six-Angled Crystal) as a New Year's gift.

This problem, now known as the "Keplerian conjecture", is generally considered to be solved in the Keplerian sense and yet remained unproven for centuries. Lagrange had reduced the problem to a two-dimensional plane in 1773 and proved that the hexagonal packing of circles is indeed optimal in terms of packing density [31]. In higher dimensions, the problem is rather unproven. In 1831, Gauss had pointed out in a review of a theory of Seeber that Seeber had proved in principle the densest packing by a defined hexagonal lattice arrangement [32]. Things become difficult, of course, when the lattice becomes non-uniform and the sphere diameters different. Toth proposed the use of computer analysis instead of classical proof as a proposed solution. Hsiang's geometric-analytic proof [33] was first celebrated, then criticized, and finally rejected by the scientific community. Since 1998, a team led by Hales has now been trying a computer proof [34], which was published only in 2005 (7 years after submission) and despite still existing doubts [35]. In 2017, Hales has now published another formal proof which, by means of a computer-aided evaluation over many years, confirms the computer proof of that time [36].

All this may serve as an illustration of the difficulty of this proof, but equally well of the power of Kepler's cognition. As it seems, Kepler had discovered by his brief observation a fact whose inherent theory is, on the one hand, absolutely fundamental and simple, but, on the other hand, is extremely difficult to prove because of the complexity that quickly arises when simple arrangements are deviated from. Nature evidently likes to arrange objects in a dense packing, having the highest degree of filling for the same volume. This not only increases the efficiency in function but also reduces the effort in creating the structure—all structures to be built are optimally small.

This incredible genetic intelligence of nature has been evident in the structure of living beings for millions of years. The facets of the trilobite's eyes are also laid out in an optimal arrangement (see Fig. 2.7).

Although trilobites have been extinct since prehistoric times, their complex eye structure has not. All insects living today have compound eyes. An example of a bee's eye is shown as a close-up in Fig. 2.8. Bees can, for example, perceive colors and shapes with their eyes, as the bee researcher Frisch demonstrated in extensive experiments [37]. The spatial resolution of the compound eyes and thus the quality of the sense of form depends on the number of facets. Their number can vary in insects, from just one hundred facets in ants to 30,000 in dragonflies [38]. In flying insects, additional compound eyes arranged in a triangle probably serve as an organ of balance or horizon detector and to control rapid flight movements [39]. The compound eyes of insects are also repeatedly a template for technical solutions. If a microlens grid is placed in front of an image sensor, an artificial, light-

Fig. 2.7 LDUCZ-J422 Dalmanites caudatus close-up, University College London, Tannis Davidson (2016), courtesy of the Grant Museum of Zoology, University College London

Fig. 2.8 Close-up of a bee eye, Frank Fox (2012), courtesy of Frank Fox (mikro-foto.de)

sensitive superposition eye is obtained [40]. A fine example of biologically-inspired computer vision.[24]

[24]s a. [41, 42].

Apposition or Superposition

In 1891, in his work "Die Physiologie der facettirten Augen von Krebsen und Insekten" (The Physiology of the Faceted Eyes of Crustaceans and Insects), Exner had divided insect eyes into two groups that are still valid today. In doing so, Exner started from the retinal image of the respective type and subdivided into apposition[25] and superposition[26] (from [43], p. 17):

> In my experience the compound eyes may be divided into three types according to their optical effect; all design an erect retinal image, but in different ways. Two of these types act dioptrically,[27] one mainly catoptrically.[28] The retinal images of the first two types, of which we shall first speak, I shall distinguish according to their mode of formation as the apposition image and the superposition image.

The basic design of the apposition eye is simple. Each individual facet has a separate lens and a rod-shaped viewing element, which means that each individual eye captures only a single image point. Thus, an image is formed only by a multiplicity of pixels. The image resolution of an apposition eye is low. An ant's eye would have an image resolution of only 10 × 10 pixels.[29] Thus, according to current knowledge, an ant can only detect objects, i.e. determine that there is something there [44]. The ant is far away from object recognition or even identification. A dragonfly with an image resolution of 170 × 170 pixels[30] sees better.

In Fig. 2.9, the ant's view is shown on the left, and that of the dragonfly on the right.

The superposition eye is constructed somewhat differently. Exner: "In the eye now to be discussed, the light masses belonging to the individual facet members fall over each other in the plane of the retina to a large extent. I therefore call this type of retinal image the superposition image."

For illustration, Exner extracted the lens structure of the compound eye of a firefly,[31] illuminated it, and recorded the resulting images. The images correspond to the AA′ (Fig. 2), CC′ (Fig. 3), DD′ (Fig. 4), and EE′ (Fig. 5) setting planes. The

(continued)

[25] From lat. Apponere = to add.

[26] From lat. Super = over + positio = position.

[27] Refractive.

[28] Reflective.

Exner refers here mainly to the reflection of light in the crystal cone within the individual ommatidia.

[29] At 100 ommatidia and a virtual square image.

[30] At approximately 30,000 ommatidia and a virtual square image.

[31] Lampyris splendidula.

optical function and especially the focal point near the DD′ (Fig. 4) plane can be easily seen (see Fig. 2.10).

Superposition eyes are significantly more light sensitive than apposition eyes and are therefore particularly common in nocturnal insects (e.g. fireflies and moths).

Fig. 2.9 Echinacea seen with compound eyes, adaptation Grasnick (2018), original Sappl (2007)

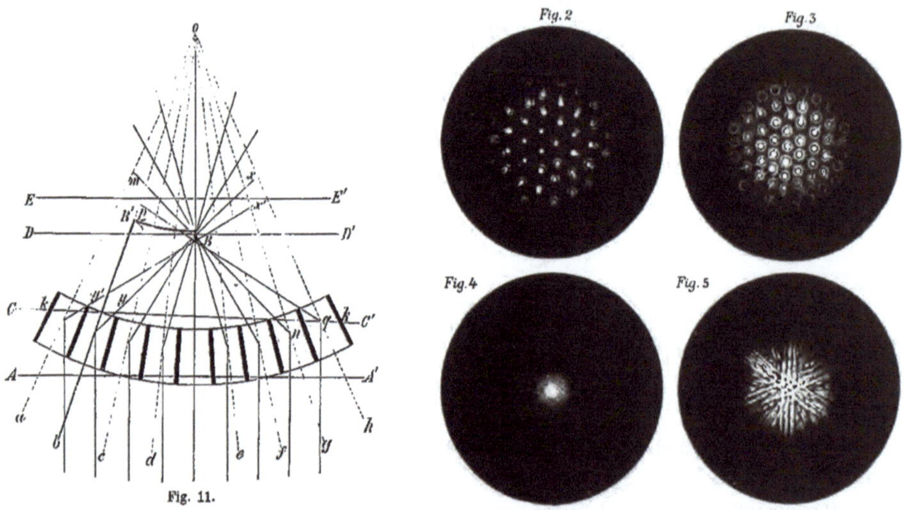

Fig. 2.10 Superposition eye with retinal images, Sigmund Exner (1891), from [43], p. 40 Fig. 11 and Plate 1 Figs. 2–5

Now that it has become clear that even the earliest living creatures had more or less complex eyes, it is certainly interesting to turn once again to Darwin's doubts about the origin of the eye through selection. How is it possible that such a complicated structure could have arisen through mutation and selection?

If one assumes that only a tiny change takes place in each generation, a distinct change can nevertheless take place over a large number of generations even with the smallest variations. Nilsson and Pelger had calculated in 1994 that even with the smallest deviation of the parameters of the light receptors in thickness, width or receptor density, a pit eye and finally a lens eye can develop from a plate eye over many generations [45]. However, even in the Swedish researchers' model, this still takes 364,000 generations. However, since the Cambrian period more than 500 million years ago, enough time has passed for even the most complicated eye shapes to form.

Any mutation must be genetically heritable to reliably change the blueprint of the new eye. A change in an individual without a reshaping of the genes does not contribute to the intergenerational trait change. The group that wants to pass on a mutation must of course survive, reproduce and never fall below a certain population size.

In his book "Sophie's World", the writer Gaarder has Alberto summarize Darwin's theory of evolution as follows: "Constant variations among individuals of one and the same species and high birth rates form the raw material or material for the development of life on earth. Natural selection in the struggle for existence is the mechanism or driving force behind this evolution. Natural selection ensures that the fittest or best adapted always survive." (from [46], p. 491).

Our eyes of today are the product of evolution and amazingly almost unchanged since primeval times. Goethe formulates this in the poem "Metamorphosis of the Animals" (Metamorphose der Thiere) (from [47], p. 89) thus[32]:

> All limbs are formed according to the eternal laws..,
> And the rarest form secretly preserves the archetype.

The physician Galen was not only an authority in his time, but due to his extensive publications[33] later also influential and powerful. Since in Galen's lifetime the dissection of humans contradicted the common rite of the dead[34] and was therefore taboo, Galen used the numerous opportunities to treat the most severe injuries during his work as a gladiator physician [48] to deepen his knowledge of the inner workings of the human being.

[32] "Alle Glieder bilden sich aus nach ew'gen Gesetzen / Und die seltenste Form bewahrt im geheimen das Urbild."

[33] Galen wrote hundreds of writings, some of which have been repeatedly reproduced or translated and have thus been preserved to this day.

[34] The burial of a deceased was an elaborate, centuries-old ritual and the sacred duty of the bereaved. The possible dismemberment and non-burial of a deceased was completely unthinkable and certainly drew the wrath of the gods.

However, even the treatment of a severe injury is in no way comparable to the dissection of a human being in terms of knowledge gained. Therefore, Galen's anatomical findings were mostly based on the dissection of animals and were basically the projection of animal anatomy onto humans. Nevertheless, Galen's anatomical studies were the basis for medical teaching and practice for more than a millennium.

Of particular interest in Galen's dissections are those of the eye. Gradually, Galen's dissection of the eye reveals the individual components known to us. A few examples give Galen's descriptions, with common descriptions added in brackets (from [49], pp. 26–28):

- the liquid resembling liquid glass (corpus vitreum)
- the net-like blanket (retina)
- the circle from which someone looking at the eye sees the image (pupil)
- the hailstone-like liquid (lens)

Galen lists the various tissues and fluids of the eye, creating a general model of internal fluids and bounding membranes.

This model became adopted by the Arab scholars of the early Middle Ages. The Christian physician Hunain was the head of the "House of Wisdom"[35] in Baghdad and one of the most important translators of Galen in his time. He wrote a treatise on ophthalmology, which was translated into Latin by Constantine the African as part of his "Omnia opera "[50].

A drawing from Meyerhof's translation [51] illustrates the schematic character of Hunain's eye model (Fig. 2.11). The designations still closely follow Galen. In the text the terms "ice-like fluid" for the lens or "net-like tunica" for the retina can be found.

This basic character remained with other Arab scholars as well.

From the "Book of Optics"[36] written at the end of the first millennium A.D. by the Arab scholar Alhazen, Risner has taken over a drawing of Alhazen in the Latin translation "Opticae thesaurus" [52]. There an eye is shown in cross-section, which is also composed of different skins (tunica) and fluids (aqueus). The tunica cornea[37] and tunica uvea[38] enclose the "humor aqueus"[39] and form the anterior chamber of the eye. The eyeball is completely covered by the sclera[40] and filled with another fluid (humor vitreus[41]). At the posterior (in the picture lower) end of the eyeball the "Tunica reti similis" forms the retina

[35] Arab. Bayt al-Hikma, a center for scientific translations of antiquity into Arabic, academy and library.

[36] Arabic "Kitab al-Manazir", Latin "De Aspectibus" or "Perspectiva".

[37] Transparent front part of the eye.

[38] Vascular middle layer of the eye.

[39] Also humor aquosus (aqueous humor), transparent water-like fluid.

[40] The white of the eye

[41] Vitreous body, glass-like gel.

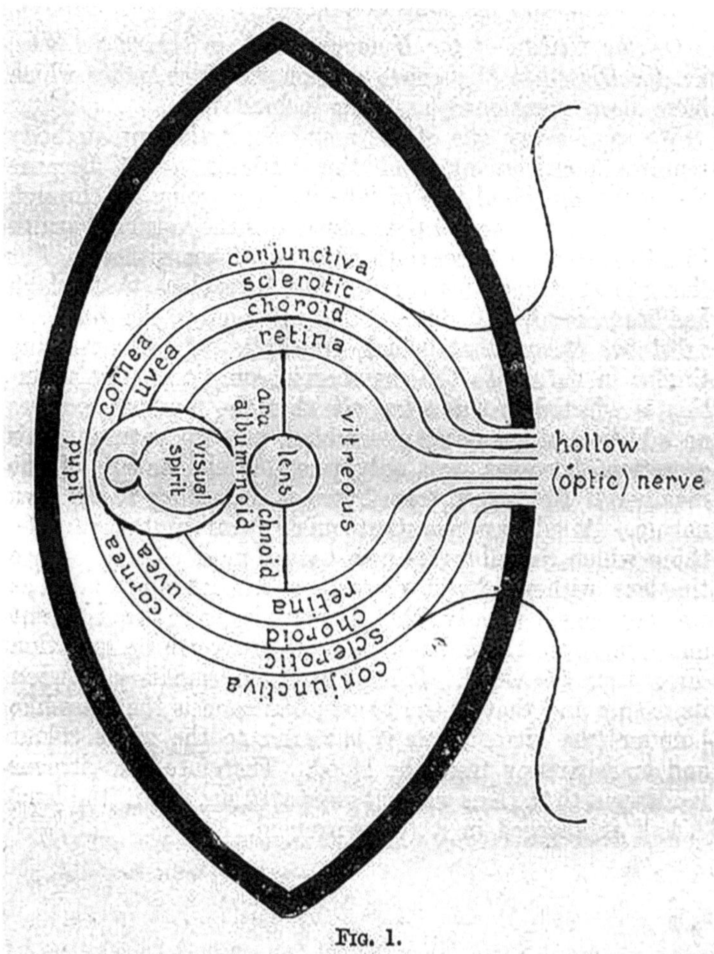

Fig. 2.11 The nature of the eye and its structure, Max Meyerhof (1928) after Ḥunain Ibn-Isḥāq (before 877) from [51], p. 5

of the eye. The lens is represented in the middle of the eye by the "humor crystallinus" (Fig. 2.12).

The translation of Alhazen's work by Risner [52] also contains the optical observations of Witelo,[42] who already sketches the structure of the eye in his thirteenth century manuscript "Opticae Thesaurus" [53]. This drawing is reproduced in a printed edition from 1535 (Fig. 2.13).

[42]Risner also calls Witelo "Thuringopoloni [52]" in the title of Alhazen's "Opticae thesaurus", which goes back to Witelo's own designation as "filius thuringorum et polonorum" (son of the Thuringians and the Poles) [53].

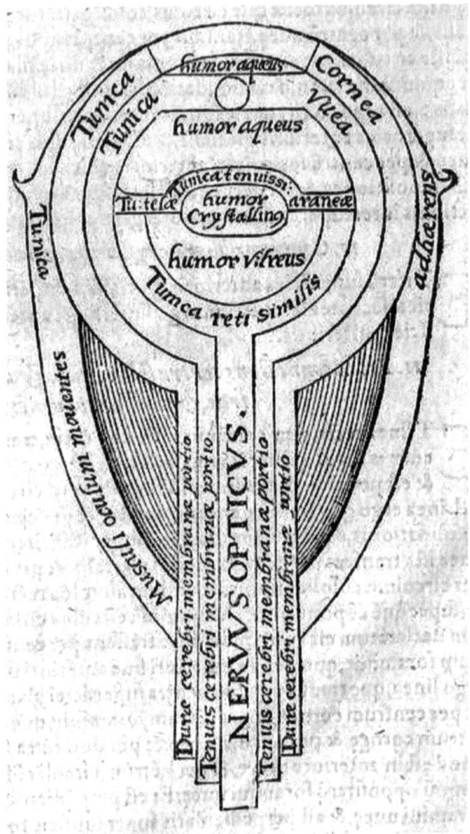

Fig. 2.12 "In totius oculi…", after Alhazen, Friedrich Risner (1572), from [52], p. 6

From the observations of Hunain, Alhazen and Witelo it becomes clear that a precise knowledge of the inner structure of the human eye can only be achieved by means of anatomy.

Galen's writings based on animal anatomy were still medical teaching material in the sixteenth century. However, the deviation of Galen's anatomy from the actual constitution of the human body became obvious with the increasing spread of anatomy and was finally made public by Vesalius in his monumental work "Seven Books on the Structure of the Human Body" (De humani corporis fabrica libri septem) [55]. In this book, the structure of the eye as an instrument of vision is also illustrated (Fig. 2.14). Vesalius is considered the founder of modern anatomy. With Vesalius, anatomy gained public recognition and "anatomical theatres" became a matter of course at medical schools.

The Swiss physician Platter also profited from this. From 1559 onwards, Platter carried out dissections, some of them in public [56], and thus acquired a great deal of knowledge about the human body, which he subsequently published in analogy to Vesalius in the richly illustrated anatomy book "On the Structure of the Human Body" (De corporis

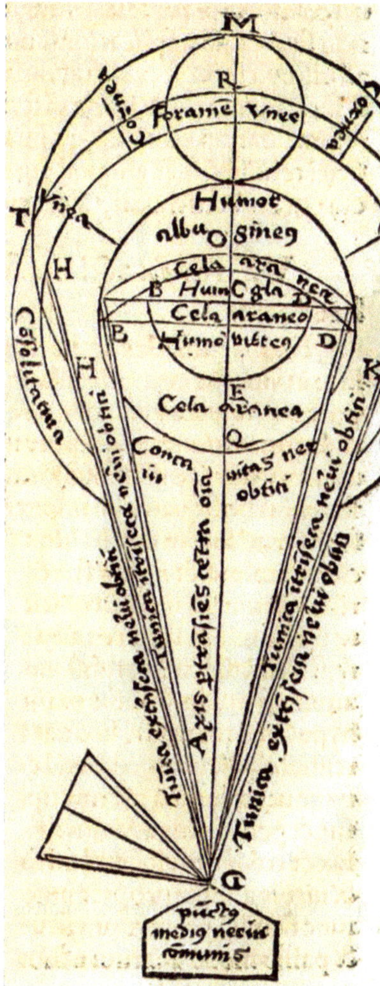

Fig. 2.13 Oculus, ex tribus humoribus, Georg Tannstetter and Peter Apian 1535 (from [54], Liber Tertius p. 55) after Witelo ([53], 56v) before 1280

humani structura) [57]. Of the numerous illustrations, of particular interest here are the illustrations of the eye, which not only show the anatomical structure of the eye in great detail, but also provide an anatomical cross-section of the eye (Fig. 1 in Fig. 2.15).

Platter had worked as an ophthalmologist on the side and had also dealt with the treatment of cataracts [58]. In his practical handbook for physicians [59], the chapter "de Visus Laesione" is dedicated to the description and treatment of eye diseases.

In 1619, Scheiner, an astronomer from Ingolstadt, described the eye in his work "Oculus" [60], not only with regard to its structure, but especially with regard to its function. One drawing shows the complete structure of an eye in section (see Fig. 2.16), another the optical image on the retina of the eye (Fig. 2.17).

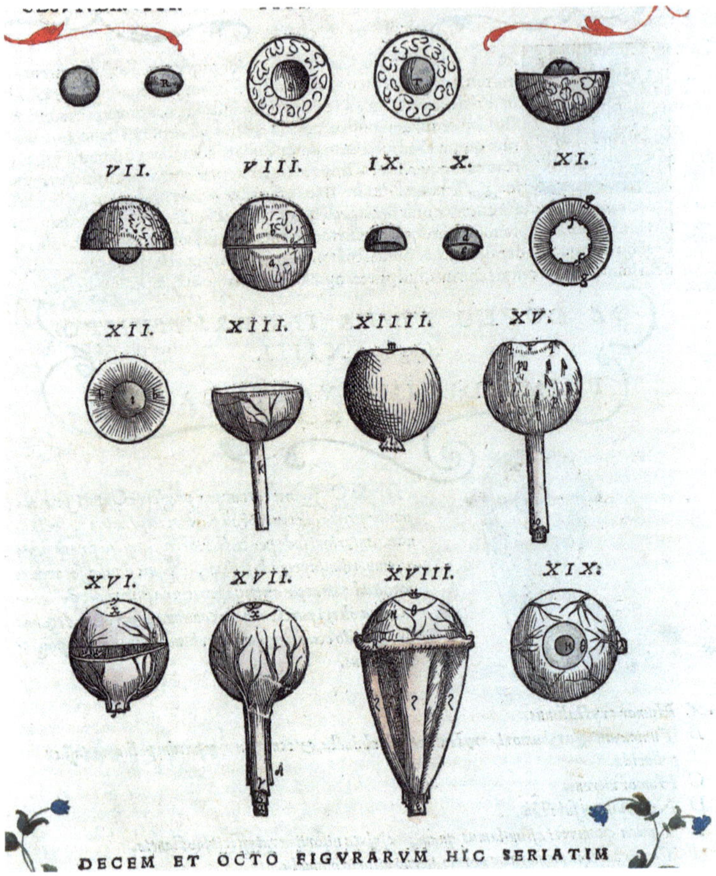

Fig. 2.14 De Oculo Visus Instrumento, Andreas Vesalius (1543) from [55], p. 644

In the "Rosa ursina" [61] Scheiner had presented his findings and equipment for the observation of sunspots. Remarkable is his knowledge of optical imaging as well as the precise reproduction of the findings. Scheiner compares the mode of action of an eye with that of a camera obscura. This is not remarkable in itself, but Scheiner has realistically depicted the cross-section of the eye and even combined it with various attachment lenses (Fig. 2.18).

In the illustration, the image position is compared in each case on the basis of natural vision (Natura) and the artificial representation (Presentatio). By choosing suitable lenses, Scheiner succeeds in creating an upright image (Visio recta) or its inversion (Visio versa) as required. Fig. 4 of Fig. 2.18 shows what is known to us today as the Galilean telescope: a converging lens as objective with a diverging lens as eyepiece for producing an upright image.

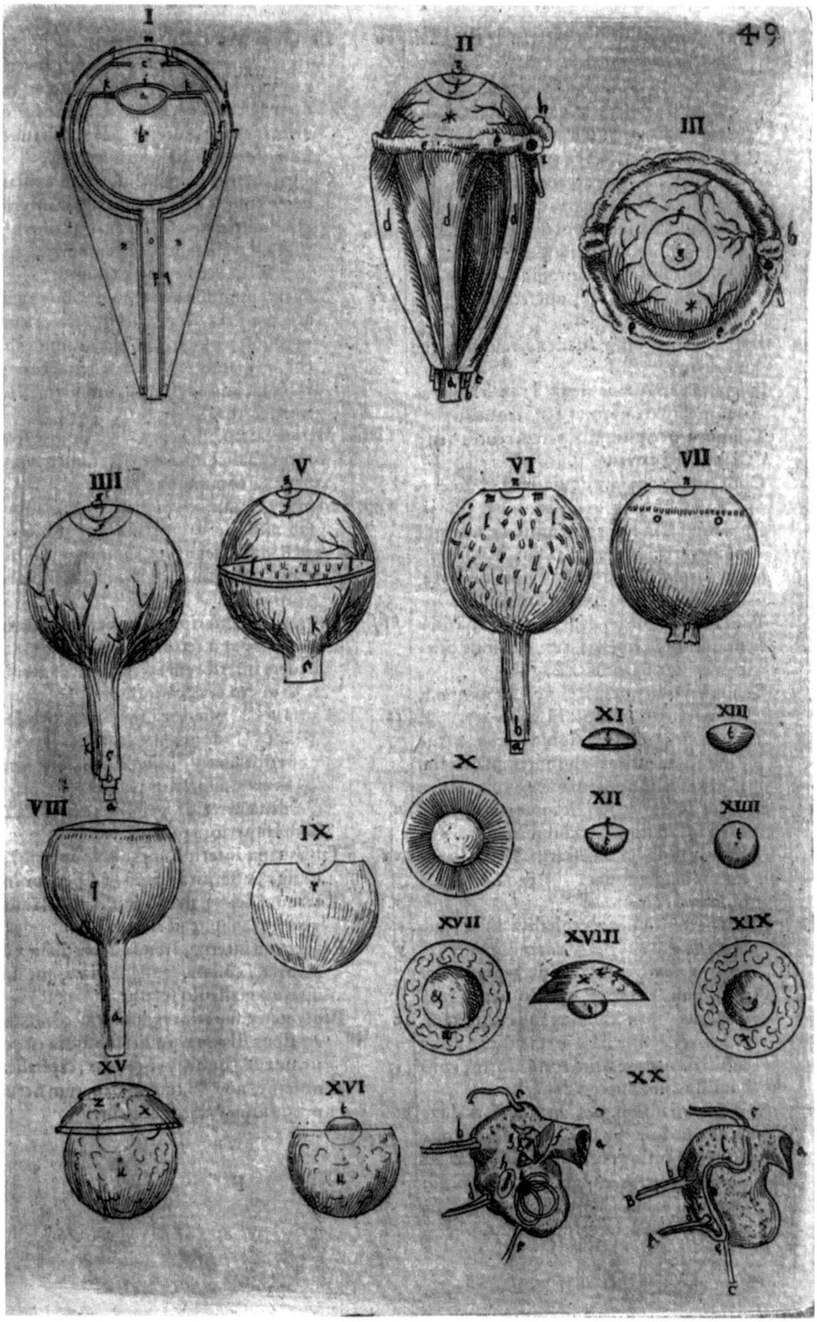

Fig. 2.15 Tabula XLIX Oculus, Felix Platter (1583), from [57], p. 49

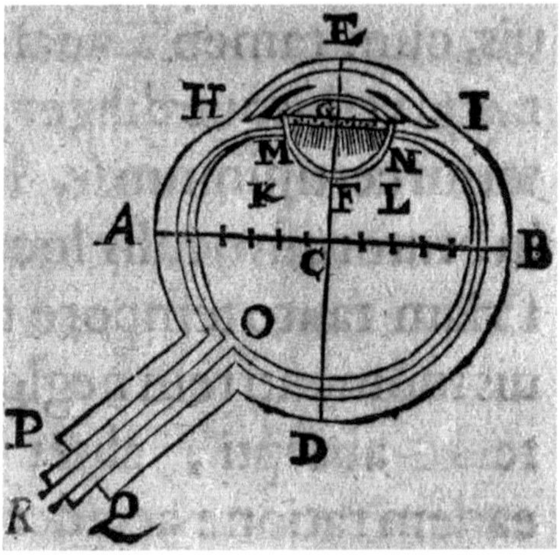

Fig. 2.16 Oculi per axem et nervum opticum, Christoph Scheiner (1619), from [60], Liber Primus, p. 17

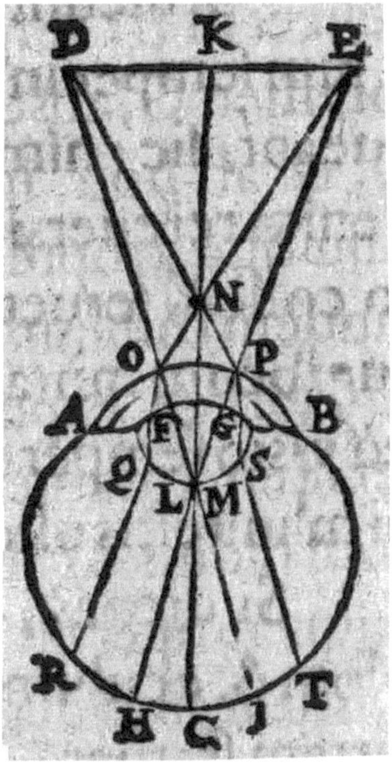

Fig. 2.17 Vera difficultatis caussa redditur (The true cause of image reversal.), Christoph Scheiner (1619), from [60], Liber Tertius, p. 112

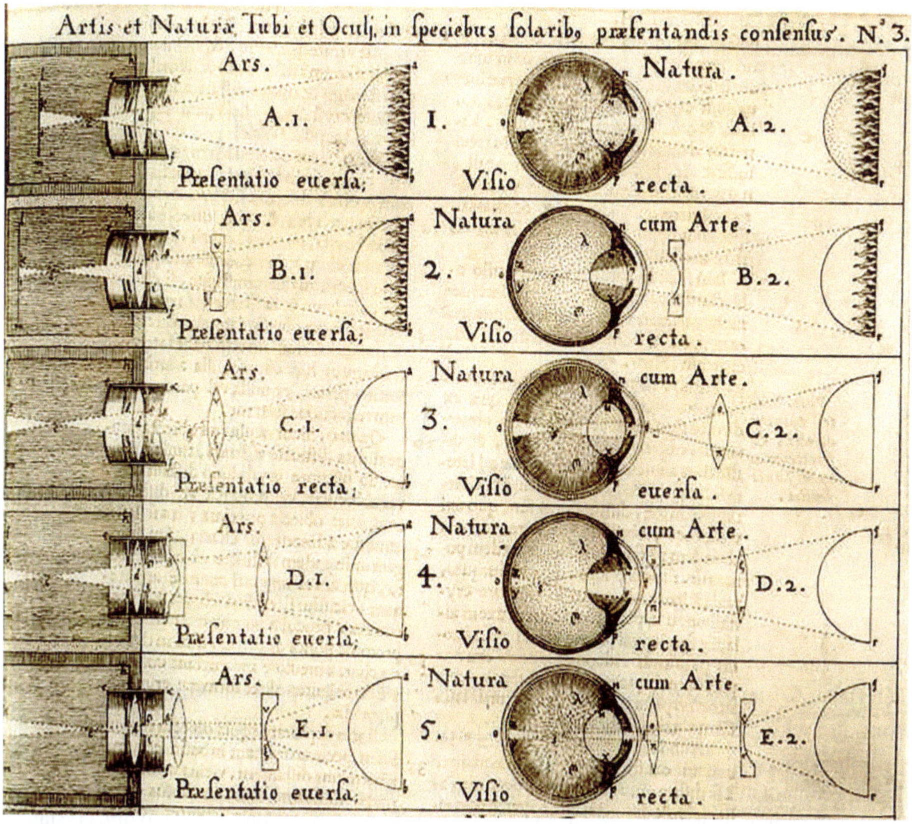

Fig. 2.18 Tubi et Oculi, Christoph Scheiner, 1630 (from [61], p. 107)

Huygens explains in the "Dioptrica" [62] the structure of the eye and the optical relations at a reduced eye. This has only one refractive surface, which is represented in the illustration by the upper small semicircle. The constriction AC represents the pupil, and the larger lower semicircle forms the eyeball with the retina. With this simple construct Huygens can make refraction and imaging in the eye comprehensible. If the large semicircle is three times as large as the small one and the inside of the construction is filled with water, Huygens eye works quite reasonably (see Fig. 2.19).

Huygens contrasted the structure of his reduced eye with the cross-section of a more natural eye (Fig. 2.20), in which the various components involved in imaging can now be seen clearly and in function.

Scheiner's and Huygens' representations were characterized by great clarity and simplification in their depiction and certainly fostered interest in the structure and function of the human eye. In the seventeenth and eighteenth centuries not only the study of anatomy became 'on vogue' but also the possession of elegant anatomical models made of valuable

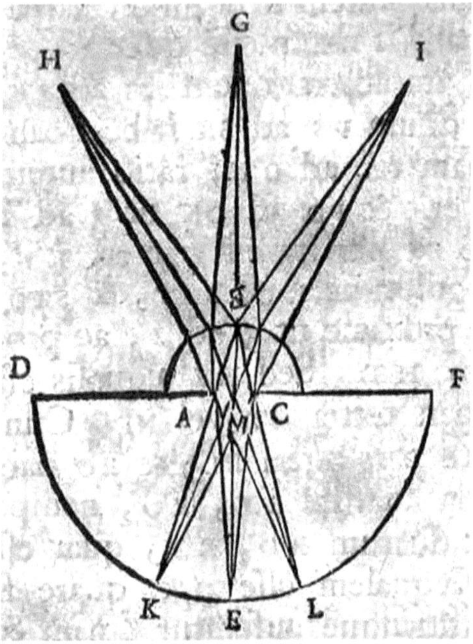

Fig. 2.19 Huygens reduced eye, Christiaan Huygens (1703) from [62], p. 112

materials. A fine example of this is the artificial eye of the optician Muth, who gives his work an awkward but nevertheless smoothly descriptive title [63][43]:

> Dissection of the human eye, as also an exact representation of all the parts of the same, where everything is accurately preformed, and manufactured by art, consisting of horn, ivory, crystal, and glass. By which the whole anatomy of the eye can be dissected and precisely represented with the musculis, tunicis & humoribus in a completely pleasant manner.

Muth's construction set of the eye dates the earliest edition of his notebook to 1709, only a few years after Huygens' bequeathed manuscripts were published as 'Opuscula Posthuma'.

However, a reasonable calculation of the optical imaging must also take into account the shapes and refractive indices of the various layers and liquids.

Therefore, in 1845 Listing proposed an optical definition now based on three individual media bounded by spherical surfaces (from [64], pp. 56–57).[44]

[43] "Zergliederung des menschlichen Auges, wie auch eine genaue Darstellung aller desselbigen Theile woselbsten alles accurat vorgebildet, und durch Kunst verfertiget, bestehend aus Horn, Helffenbein, Crystall und Glas. Wodurch die gantze Anatomie des Auges auf das allergenaueste mit denen Musculis, tunicis & humoribus auf gantz angenehme Weise zergliedert und dargestellet werden kann."

[44] "Der menschliche Sehapparat kann ... einem System von drei verschieden stark brechenden Mitteln verglichen werden, ... Das erste Mittel ... wird gebildet von der wäßrigen Feuchtigkeit

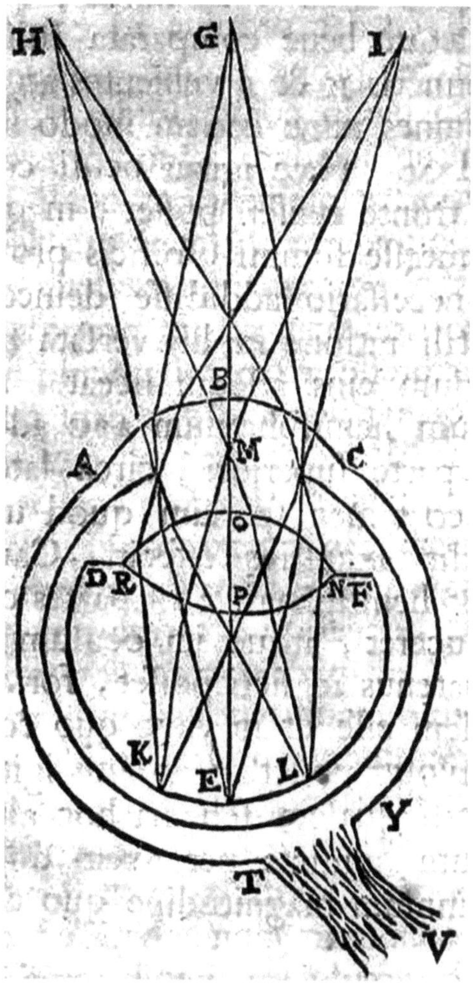

Fig. 2.20 Huygens realistic eye, Christiaan Huygens (1703) from [62], p. 114

> The human visual apparatus may be likened ... to a system of three refractive agents of different strengths, ...
>
> The first means ... is formed by the aqueous moisture together with the cornea, the second ... by the crystalline lens together with its capsule, and the third ... by the vitreous moisture.

In this simplified system of the "schematic or so-called 'average' eye", Listing gives details of radii of curvature, distances and refractive indices, which are used for illustration in this paper.

---

nebst der Hornhaut, das zweite ... von der Krystalllinse sammt ihrer Kapsel, und das dritte ... von der Glasfeuchtigkeit."

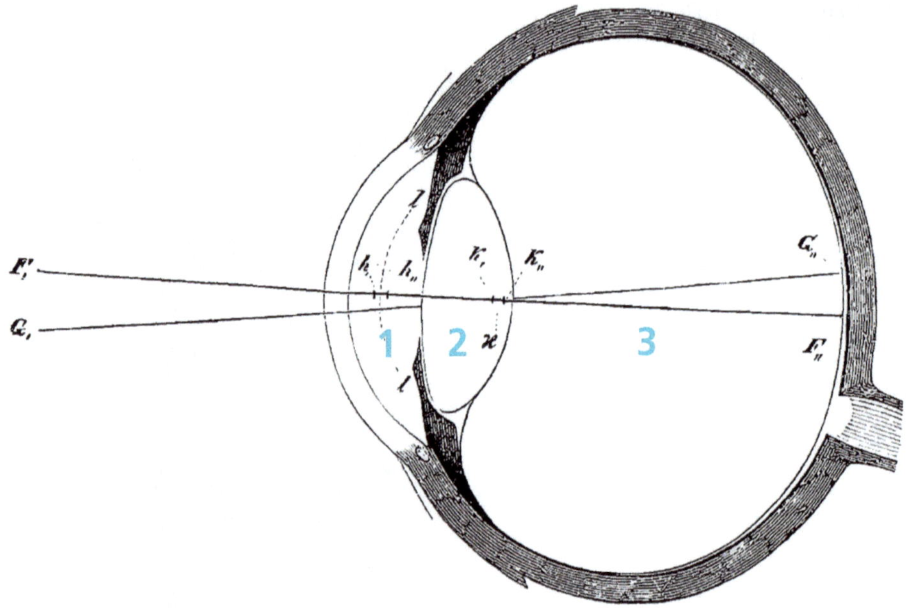

Fig. 2.21 Schematic and reduced eye after Listing, Hermann v. Helmholtz (1867), from [67], p. 77, Fig. 44

Helmholtz, a physiologist and physicist, had studied Listings' preparatory work and adopted Listings' schematic eye.[45] Helmholtz examined Listing's statements and confirmed the fundamental correctness of the scheme. The term "schematic and reduced eye"used by Listing and Helmholtz has survived the times and is still used in current textbooks.[46]

Helmholtz's basic structure was composed of three different optical elements according to Listings simplified system. The elements are (1) the aqueous humor (anterior chamber of the eye), (2) the crystalline lens and (3) the vitreous body (see Fig. 2.21).

The physiologist Gullstrand was not only an editor of Helmholtz' "Handbuch der Physiologischen Optik", but in this work he himself contributed to the "Dioptric of the Eye". Here Gullstrand presents, among other things, the optical values of the schematic eye (see [67], pp. 300–303) and establishes with these values the "Gullstrand eye". Gullstrand's values are still used today as a reference for calculating eyeglass lenses[47] and for

[45] Helmholtz's "Handbook of Physiological Optics" (Handbuch der physiologischen Optik) [65] was the standard work in ophthalmic optics for decades.

[46] e.g. in the current textbook "Physiology of Man" ("Physiologie des Menschen") [66] (31st edition).

[47] e.g. in the optician's textbook "Eye—Spectacles—Refraction" ("Auge—Brille—Refraktion") [68].

determining the refractive power of the cornea of the eye.[48] Although there are now a large number of eye models,[49] the Gullstrand eye is certainly the best known because of its simplicity.

From the Dissection of an Eye

Herophilos is considered by tradition to be the most famous physician of Hellenism. It is believed that Ptolemy I allowed him and the Greek anatomist Erasistratos to systematically dissect[50] convicted criminals alive in Alexandria. The authoritative source for this is Celsus, who, however, was born more than two centuries after Herophilos and Erasistratos died. Celsus believed that it was absolutely necessary to open living bodies and praised the two physicians for their drive for knowledge and action (from [75], p. 24):

> Most excellently, therefore, would Herophilos and Erasistratus have acted by opening alive criminals whom you received from kings in prisons, and so, while even breathing still continued, viewed those parts which nature had previously withdrawn from the eye...[51]

The Byzantine physician Aëtios also mentioned a special work of Herophilos on the eye in his extensive 16-volume treatise on all medicine [78]. This work, like the numerous dissections of living and dead human beings, found its way into the writings of later physicians, even though the works themselves have not been preserved.

The scholars of the early Middle Ages owed much of their knowledge of the anatomical structure of the eye to the aforementioned Greco-Roman physician Galen. He resumed the (allegedly) Hellenistic tradition of vivisections, but had to make do with monkeys, goats or pigs. He not only dissected these poor animals alive down to the last detail,[52] but also dissected the individual organs and even the eyes.

(continued)

[48] e.g. for calculating the necessary ablation in eye laser surgery [69].

[49] e.g. Lotmar [70] (eye model with aspheres), Kooijman [71] (wide-angle eye), Liou and Brennan [72] (anatomically correct eye) or Navarro [73] (normally sighted aging eye). A nice overview on different optical eye models is provided by Atchison and Thibos [74].

[50] Vivisections.

[51] In the Latin original "...longeque optime fecisse Herophilum et Erasistratum, qui nocentes homines, a regibus ex carcere acceptos, vivos inciderint, considerarintque, etiamnum spiritu remanente, ea., quae natura ante clausisset, ..." from [76] (pp. 4–5), the text of the 1465 manuscript (see also [77], p. 3) deviates somewhat from this.

[52] The dissections were performed on the living animal to observe the reactions and functions.

Galen described each step in such detail and vividly that his explanations remained the standard of anatomy for many centuries.

There is no doubt that da Vinci, who died shortly after the birth of Vesalius,[53] also dissected some[54] of his contemporaries to refine his knowledge of human anatomy. In his anatomical drawings of manuscripts A, B and C [80] there are amazing insights into the inner workings of the human being. When dissecting the eye, however, a problem arose when cutting through the vitreous body,[55] even for da Vinci. The vitreous body contains a fluid[56] that naturally leaks out when it is cut open. Da Vinci provides a "practical" solution to this problem as well ([79], p. 153):

> In the anatomy of the eye, in order to see the inside well, without spilling its watery juice, you should put the whole eye in egg-white, and bring it to the boil, and let it set; and then cut the egg and eye crosswise, so that no part of the middle piece may flow out.

Helmholtz later compared the eye to a camera obscura, as did da Vinci, and laid hands on some "freshly cut out" eyes for proof (from [65], p. 64):

> The eye behaves against the light essentially like a camera obscura. . . . one can make it visible on freshly excised eyes if one carefully removes the posterior middle part . . . and now turns the cornea of a thus prepared eye against bright objects.

Finally, the complete structure of the eye is shown here in cross-section, as it is familiar to us today. In this eye model (Fig. 2.22), all optical media and interfaces involved in imaging are now shown.

2.2 Paleo VR

From Virtual Reality in the Stone Age

The upright gait and the development of manual skills of prehistoric man and the increasing knowledge of the environment have already led in the Stone Age to the first attempts to reproduce as realistically as possible their own living environment. Our current image of this epoch is essentially shaped by the few sculptural figures that have survived the millennia.

---

[53] and consequently could have no knowledge of modern anatomy.

[54] ". . . more than thirty men and women of all ages" [79], p. 357.

[55] The vitreous body (corpus vitreum) defines the shape of the eye.

[56] The vitreous body is completely filled with a watery fluid (humor vitreus), which actually consists mainly of water.

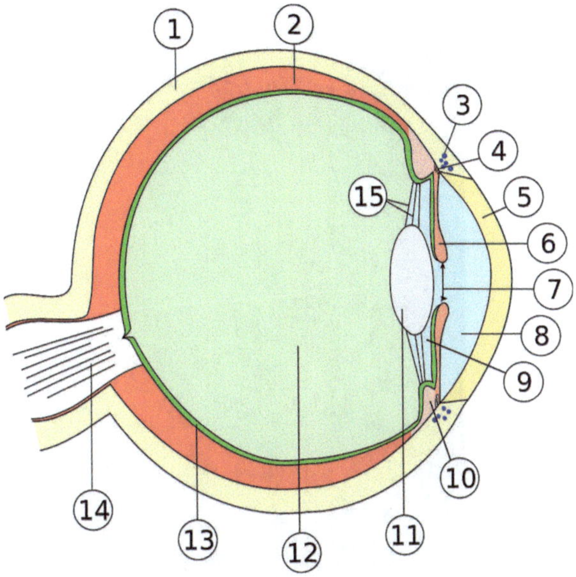

Fig. 2.22 Anatomy of the eye (1. Sclera, 2. Chorioidea, 3. Sinus venosus sclerae/Plexus venosus sclerae, 4. Radix iridis, 5. Cornea, 6. Iris, 7. Pupilla, 8. Camera anterior bulbi, 9. Camera posterior bulbi, 10. Corpus ciliare, 11. Lens, 12. Corpus vitreum, 13. Retina, 14. Nervus opticus, 15. Fibrae zonulares.), Joël Gubler "Talos" (2008), Wikimedia Commons

The striving for a realistic reproduction of the immediate surroundings has been a tangible drive of artistic activity at least since the Palaeolithic Age. The "Venus of the Hohle Fels" discovered in a cave in the Achtal near Schelklingen unmistakably shows the characteristics of a woman. The Tübingen professor Conard, who had discovered the figure [81], recently placed this small sculpture together with other mammoth ivory carvings in his book "Als der Mensch die Kunst erfand" (When man invented art) [82] in a direct context with the beginnings of art. In the meantime (since 2017), the cave and Ice Age art in the Swabian Jura has been recognized as a Unesco World Heritage Site [83]. The original figure can be admired in Blaubeuren in the prehistoric museum there [84], but is also available as a 3D model with other 3D scans of various Ice Age figures and the respective find caves (for the scanning technique, see [85]) [86].

To illustrate the scan quality, the 3D models of Venus and the hall of the "Hohle Fels" cave were rendered and combined into one overall image (see Fig. 2.23).

That this figure, made about 40,000 years ago, is not a strange isolated case, is proved among other things by the discovery of the "Venus of Willendorf" whose higher quality is obvious, possibly can be traced back to the younger date (about 25,000 years ago[57]) [89].

3D scans or 3D models exist for this figure, so that a comparison of the real artwork with a virtual replica is possible (Fig. 2.24). The 3D model here is optimized for 3D printing and thus somewhat simpler.

[57] The Natural History Museum in Vienna, where the Venus is kept, gives an older dating [87]. The basis of this newer age determination was here the re-calibration of the radiocarbon dating for the Willendorf loess profile [88], in which the figure was found.

Fig. 2.23 Venus in the Hohle Fels, collage of 3D models [86], Grasnick (2018)

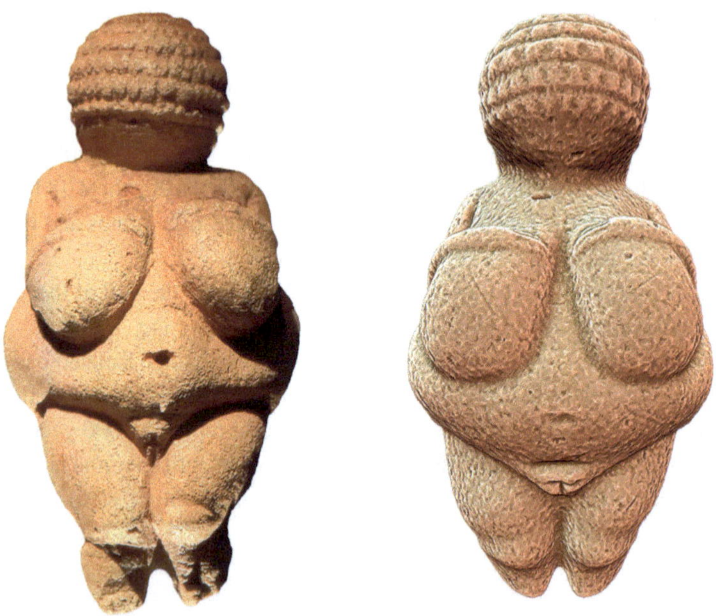

Fig. 2.24 Venus of Willendorf: left photograph Matthias Kabel (2007) Wikimedia Commons, right 3D model "Scan the World" (2015) [90]

Table 2.1 Art styles of cave painting according to Leroi-Gourhan

	Art	Start		End	
Style I	Primitive period, incised drawings	ca. 30,000 B.C.	Aurignacian	ca. 23,000 B.C.	Gravettian
Style II	Phase of stylized representation in painting and sculpture (Venus)	ca. 23,000 B.C.	Late Gravettian	ca. 17,000 B.C.	Solutrean
Style III	Perspective representation of scenes	ca. 17,000 B.C.	Late Solutrean	ca. 13,000 B.C.	Magdalenian
Style IV	Realistic representation of shape and movement	ca. 13,000–11,000 B. C.	Middle Magdalenian	ca. 10,000–8000 B. C.	Late Magdalenian

The depiction of entire scenes on cave walls already represents a projection of the three-dimensional environment onto a two-dimensional surface and testifies to a high level of abstraction. These wall paintings are found not only in Europe, but also in Africa and even in America.

The silhouettes of various hands in the Stone Age caves of El Castillo[58] or La Garma[59] are not in themselves an artistic masterpiece. And yet these images, known as hand negatives, represent such an exact copy of the hand that even after 37,000 years it is still possible to determine whether the hand depicted belonged to a man or a woman. Snow determines a surprising result on a basis of 32 handprints: 75% of the hand negatives are female [91]. In his analysis "New views on old hands" [92] Pettitt also comes to the conclusion that mostly female hands were depicted.

Of course, the art was not limited to the depiction of body parts. Hand negatives are found in other caves at Chauvet, but also much more elaborate drawings.

Today, such cave paintings are classified in different styles. A common classification is into four art styles according to Leroi-Gourhan [93]. The subdivision is naturally not very clear-cut, the chronology is rough, but nevertheless allows a certain dating of found artefacts on the basis of the style (see Table 2.1).

Jean Clottes summarizes the artistic achievements of early man in the article "Art in the Morning Light of Mankind" (Kunst im Morgenlicht der Menschheit) thus [94]:

[58] El Castillo Cave near Puente Viesgo in Cantabria, Spain, discovered in 1903.

[59] La Garma cave complex near Omoño, Cantabria, Spain, discovered in 1995.

In fact, however, people of the late Paleolithic period were apparently already able to draw in perspective, knew various painting techniques, and were able to reproduce animal behavior in a lifelike manner.

2.3 Historical Theories of Vision

From the Real Object to Its Perception

The eye is indisputably an extraordinarily amazing evolutionary development and enables humans to perceive. Even if the human eye is an actually "unattainable device" in the Darwinian sense compared to the simplest primal eyes, the quality of the imaging is nevertheless limited. A natural eye does not have precise dimensions. The external shape of the cornea, vitreous body and lens does not follow a perfect mathematical curve, the refractive media are inhomogeneous and the position of the optical surfaces in relation to each other is not ideal for perfect image formation. The color and image defects caused by this imperfection are further worsened by scattering and internal reflection in the eye. There are areas on the retina where there are no nerve cells and thus no image is recorded at all. Nevertheless, one sees a brilliant and high-resolution image. This shows that vision is more than the result of refractions on different media.

In 1871, Hermann von Helmholtz described the numerous imaging defects and limitations of the eye in an essay entitled "The Optical Apparatus of the Eye" and concluded ([95], p. 21):

> Now it is not too much to say that to an optician who wished to sell me an instrument which had the latter faults, I should think myself perfectly justified in using the harshest expressions about the negligence of his work, and in returning his instrument to him with protest.
>
> With regard to my eyes, of course, I will not do the latter, but on the contrary, I will be glad to be able to keep them with their faults as long as possible. But the fact that they are irreplaceable to me in spite of their faults obviously does not diminish the magnitude of these faults, if we place ourselves for once on the admittedly unilateral but justified position of the optician.

It is already evident from this simple consideration that the process of seeing cannot be reduced to optical imaging. Here the transition from imaging to perception takes place, which in the visual field is also Helmholtz's "sight perception" ([95], p. 98):

> The correspondence between sight perceptions and the external world is therefore based entirely, or at least in the main, on the same ground on which all our knowledge of the real world is based, namely, on experience and the continual testing of its correctness by means of experiment, as we do with every movement of our body.

So what is vision really?

The history of epistemology usually begins with the Seven Sages of Greece. Plato, in Socrates' dialogue with Protagoras, describes these sages as a group of "old men"[60] who were perfectly instructed in the art of laconizing.[61] But already Plato describes the instruction of Solon, one of the Seven Sages, by a "very aged" Egyptian priest, who points out to Solon the recent history of the Greeks in comparison with the Egyptians (from [97], p. 13):

> O Solon, Solon, you Hellenes always remain children; there is no such thing as an old Hellene.... You are all young in spirit; for you preserve in it neither a view that is already old because of ancient tradition, nor any knowledge that has grown gray through time.

In the following, the aged priest points to the ancient records in the Egyptian temples and finally tells Solon about the island of Atlantis. Although the priest's last statement makes him hardly credible as a witness, he is right in his opening statement.

The "Papyrus Ebers"[62] from the time of the Pharaoh Amosis is one of the forty-two hermetic books,[63] which are said to be inspired by the Egyptian god Thot (Greek Hermes). Ebers himself felt justified in calling the annotated scroll "The Hermetic Book on the Medicines of the Ancient Egyptians" ("Das hermetische Buch über die Arzeneimittel der alten Ägypter") [99]. The paper also contains "The Book of the Eyes" ("Das Buch von den Augen"), in which numerous remedies[64] for the relief of eye ailments are described. Some of these remedies are quite suitable for the relief of complaints and were still used in the nineteenth century, as Hirschberg explains in the "History of Ophthalmology"

[60]"...Thales of Miletos, Pittakos of Mytilene, Bias of Priene, our Solon, Cleobulos of Lindos, Myson of Chenae, and as the seventh was enumerated among these the Lacedaimonian Chilon." (from [96], p. 288)

[61]By "laconicizing" or "laconic short speech" Plato describes the unerring, linguistic brevity of the Lacedaimonians (Spartans) in the formulation of short memorable (wise) sayings.

[62]Named after the Egyptologist Ebers, who acquired this ancient Egyptian papyrus and published it annotated as the "Hermetic Book on the Medicines of the Ancient Egyptians" ("Hermetischs Buch über die Arzneimittel der alten Ägypter") [98, 99].

[63]According to Clemens of Alexandria, also Titus Flavius Clemens, Clemens reports in his collection of texts "Stromata" 42 writings, of which 36 books contain the entire Egyptian philosophy, but the remaining six deal with medical content: anatomy, diseases, surgical tools, medicines, and women's diseases ([100], p. 758).

[64]Almost always in conjunction with eye makeup and often with an incantation.

("Geschichte der Augenheilkunde")[65] [101]. The use of crocodile dung[66] in eye ointment recommended in the Book of the Eyes, on the other hand, fortunately did not catch on.[67]

The eye was an important symbol in ancient Egypt. Of special importance were the eyes of the sun-god Re (or also Ra). Re is the child of the "Mehet-weret[68]", the all-encompassing and star-bearing ocean of heaven, which has risen from the primordial waters. The "Mehet-weret" as mistress of heaven and mother of the sun was conceived as the celestial nourishing mother and quite naturally represented as a cow. The celestial primordial cow gives birth daily to her son Re as the sun and carries him between her horns. The Egyptian divine legend of the Heavenly Cow has come down to us from the tombs of the pharaohs Tutankhamun, Sethos I, Ramses II and Ramses III and is known as the "Book of the Heavenly Cow" ("Buch der Himmelkuh") [105]. The eyes of the sun god have a special significance in mythology. Humans were created from the tears of the sun god, and his gaze can destroy them. In the Book of the Heavenly Cow is found the lament of Re and a proposal for the punishment of mankind ([105], p. 38).

> You oldest god, from whom I arose, and you gods and ancestors—behold, men, sprung from my eye have devised plots against me!
>
> Let thine eye wander, that it may expose to thee them that conspire as evildoers. There is no eye superior to it to smite them for thee—may it descend as Hathor!

Re thus hands over the function of the sun's eye to the goddess Hathor so that she can punish the people. The sun's eye in the person of Hathor thus kills the rebels of the desert on behalf of Re.

It can be seen here that the eye in this myth takes on an active function and can in principle act actively at a distance. This observation forms the basis for the early assumptions about the remote effect of vision and, like water, is given a role in philosophy. Already the sky-cow Mehet-weret rose from the waters of the primeval ocean to give birth to the sun-god Re. The water as the beginning and the end of everything is a common motif in the history of creation and is found as the Flood on the clay tablets of the Gilgamesh epic [106] and in the first Book of Moses, the "Genesis" [107].

---

[65] e.g. the eyelid ointment made from black spit shine (Antimonium crudum).

[66] However, crocodile dung was for a long time a recognized ingredient in medicines, e.g. described by Pliny the Elder in the XXVIII. Book of the "Historia naturalis" ([102], p. 35–36). The disgusting climax of the treatment with excrements is certainly the "Healing Dirt Pharmacy" ("Heilsame Dreck-Apotheke") [103] of the physician Paullini with numerous, allegedly healing applications of excrement and urine.

[67] In ancient medical practice, other animal ingredients were also used. Galen used, among other things, dried vipers in the panacea he described, "Theriacus andromachi senioris" (theriac after Andromachus, Nero's personal physician), with opium as its main ingredient, which explains its generally soothing effect (recipe from the Edinburgh Dispensatory [104], pp. 554–555).

[68] Literally, "great flood."

Aristotle reports that Thales, another member of the "Seven Sages", also considered water to be the substance from which everything is formed (from [108], p. 30):

> He probably came to this thought because he saw that the nourishment of everything is moist, and that even that which is warm arises from it and is nourished by it (and that from which everything becomes is the beginning); also he saw that the seed of all things is of a moist nature, and that water is the beginning of that which is moist; so he came to this assumption.

In Greek mythology, the waters of the river Styx formed the boundary between the world of the living and the realm of the dead, Hades. Even the gods swore in the name of the water, as Homer tells us in the 14th canto of the Iliad ([109], p. 49):

> Now therefore, conjure it by Styx's woeful waters,
> Stirring with one hand the food sprouting earth,
> And with the other, the shimmering sea; that all of them now
> Witness the lower gods gathered around Cronus.

Anaximander, a disciple of Thales, invents the primordial substance of the cosmos; the Apeiron ([110], p. 13):

> The beginning of things is the infinite [Apeiron[69]]. But whence is their birth, thither also goeth their death according to necessity. For they pay punishment and penance to one another for their wickedness according to the order of time.

The imperishable substance of the apeiron is thus indeterminable and yet responsible for life and death. Nietzsche sees in the idea of the "indeterminate as primordial substance" also a critical assessment of existence and consequently considers Anaximander to be the first philosophical pessimist ([111], p. 275).

From the Apeiron Via the Ether to the Quantum Field

A mysterious substance like the apeiron of Anaximander has the advantage of always being able to complete that part of a theory which would otherwise remain inexplicable. At first water served the Egyptians, and later the Greeks also the air, as the original elements from which everything arises. For the following philosophers, however, the conventional myths and God-given models were no longer sufficient to describe the everyday. The explanation of the processes of perception seemed to call for a mediating substance that would transmit an effect over great distances. Such indeterminate substances as aids to physical models also found renowned advocates

(continued)

[69] Apeiron, from Greek peirata (limit) and the negation "a", means the opposite of limit - the unlimited, indeterminate or infinite.

in science. In his work "On the Heavens",[70] Aristotle described, in addition to the four elements of fire, water, earth and air, a fifth substance (Latin quintessence), an immaterial animate substance, which he, like his predecessors, called the ether. Thus, for example, one finds in Aristotle's "Four Books on the Heavens and Two Books on Coming into Being and Passing Away": "...undeveloped and imperishable, it is a divine thing, which was not unjustly called ether by the earlier ones..." (from [112], p. 11/12).

The ether, after the Greek word "aither"for the upper sky, was the seat of the gods in ancient mythology. Plato already used the name as a designation for the "...brightest part of the air..." (from [97], p. 64).

Huygens, too, needed a medium in which the light waves could propagate in order to explain the propagation of light. Huygens used the etheric term "l'Ether"in several places of his "Traité de la lumière" [113] and thereby makes use of Descartes' idea of the fine materiality "matière subtile" (from [114], in the translation simply the "fine matter" [115]) with which, according to Descartes, any space is filled. According to Huygens' disguise, it is "...the ether which mediates the light..." (from [116], p. 32).

Light waves can now propagate in the fine matter of the ether just as well as waves in water. Newton also used the ether model, but developed increasing misgivings about the nature of this mysterious substance and finally wrote in the third book of his Optics "...for I do not know what this Æther is..." (from [117], p. 326). Einstein's theory of relativity is closely linked to his abandonment of the ether theory of that time. While on vacation from his Bern patent office duties, Einstein delivered a remarkable lecture [118] "On the Development of Our Views Concerning the Nature and Constitution of Radiation" at the Salzburg Naturalists' Meeting in 1909. In it Einstein begins with explanations of wave-particle dualism, the "...fusion of undulation—and emission theory of light..." and the remark "Today, however, we must regard the ether hypothesis as point of view that has been overcome." ([119], p. 817). Subsequently, Einstein nevertheless devoted some contributions to the ether[71] and finally insisted on the necessity of a "continuum endowed with physical properties", since his general theory of relativity excluded a direct action at a distance and he

(continued)

[70]Greek Peri uranu, Latin De caelo.

[71]e.g. "Ether and Relativity" ("Äther und Relativitätstheorie") [120] or "On the Ether" ("Über den Äther") [121].

concluded "...every theory of near action, however, presupposes continuous fields, thus also the existence of an 'ether'." (from [121], p. 14). This certainly partly justifies Einstein's rejection of an inexplicable action at a distance and the doubts about the indeterminacies of quantum theory.[72] Einstein spent much energy proving the incompleteness of quantum theory to the physicist Bohr, asking in an essay,[73] "Can the quantum mechanical description of the physical world be complete?" [123]. Quantum mechanics is a difficult terrain to understand even for insiders. Thus, Born, who later won the Nobel Prize, wrote in a letter to Einstein: "Surely the quanta are a hopeless mess!". Born had later obviously made his peace with these quanta, since he was awarded the Nobel Prize in Physics expressly for his fundamental research in quantum mechanics [124].

The distribution of physical properties in space is usually described mathematically as a field. Since Planck[74] and Einstein,[75] all elementary particles can be described as quantum objects, which can mean the smallest amount of energy of the respective field as well as the interaction. All matter known to us consists of quantum objects like photons, electrons and other elementary particles, whose interactions in space can therefore be described as a quantum field.

Anaximenes, taking up the doctrine of Anaximander, holds that air is not only the breath of life, but also the primordial element from which the other elements arise by dilution or concentration. Two fragments of his doctrine mark his world-view (from [110], p. 21):

> As our soul is air, and thereby holds us together, so the whole order of the world embraces breath and air.
>
> That which contracts and condenses matter is the cold, while that which is thin and flaccid is the warm.

Anaximenes tries to put his observations on a physical basis. According to tradition, Anaximenes is said to have been an attentive eye-witness of extraordinary natural

[72] In particular, Heisenberg's uncertainty principle (two complementary properties cannot be determined exactly at the same time), Bohr's complementarity (two different observations of a process exclude each other) and Born's rule (a particle is not in a single place, but in many places at the same time).

[73] Bohr describes some of these disputes in his "Discussions with Einstein" ("Diskussionen mit Einstein") [122] and also mentions Einstein's figurative language using the example of the "ghost fields".

[74] The natural constant which Planck calls "h" in 1900 (from [125], p. 240) is now called Planck's constant.

[75] Einstein describes the energy of a light quantum (the photon) as the product of Planck's constant and the frequency of light (Planck–Einstein relation).

spectacles (such as milky seas effect or moon rainbows) and he is credited with the knowledge of the reflection of sunlight by the moon (see [126], p. 26).

The desire for the greatest possible gain in knowledge seems to have been especially manifest in the Greek Pythagoras. Heraclitus wrote on Pythagoras' passion for collecting ([110], p. 80):

> Pythagoras, son of Mnesarchus, of all men was the most devoted to research, and having worked through these writings, he made of them wisdom of his own: polytheism, rabulism.

This is not to be understood in an entirely positive way, for rabulism initially means only the art of being right, and is to be understood as pettifoggery. Heraclitus, moreover, does not automatically translate being a polyglot with being clever, but he means: "Being a polyglot does not teach one to have understanding. Otherwise Hesiod would have learned it, and Pythagoras, further also Xenophanes and Hecataios." ([110], p. 68).

Pythagoras is subsequently credited with various profound insights, of which the Pythagorean theorem[76] is certainly one of the best known. Laertius has Alexander[77] tell him about Pythagoras. His view of perception is interesting ([129], pp. 106–107):

> He said that perception in general, and the vision in particular, is a kind of particularly warm breath. Therefore, it is also said of it that it passes through the air and through the water. For the warm experiences a counter-pressure of the cold. For if the breath of the eyes were cold, it would remain separate from the like air; but so contrary air is found in the eyes, which he [Phytagoras] calls the gates of the sun.

Here we find the idea of seeing as a contradiction of heat and cold, which resulted from the four-element worldview of that time. From the apeiron of Anaximander arise the four elements water, air, earth and fire. According to the understanding of the Pythagoreans, the eye emits a fiery breath, which experiences a tangible counter-pressure when the breath hits a cool object. The element of feeling places sight on the level of touch. As in touching an object with the hands, an object is touched with the fire of the eye and thus a visual impression is conveyed even over long distances. This analogy seems explicable as long as light emanates from the eye.

---

[76] "In a right triangle, the sum of the areas of the squares over the cathetes is equal to the area of the square over the hypotenuse". However, this basic relationship of the triangle calculation was already known to the Babylonians, as the clay tablet "Plimpton 322" (from the nineteenth–sixteenth century BC, i.e. at least 1000 years before Pythagoras) proves [127].

[77] Alexander Polyhistor, Greek scholar, "polyglot and a prolific writer" ([128], p. 21).

Tetractys—the Divine Fourfoldness

Four was an extremely important number to the Pythagoreans. In addition to the four elements from which the ancient Greeks constructed the entire world, these elements could be described with only four properties, the primary qualities of cold, warm, dry and moist. The group of numbers from 1 to 4, which add up to 10, the Pythagoreans called the "tetractys" (fourness). These four numbers were, according to the understanding of the time, the key to the mathematical description of the world.

The Pythagoreans took an oath not to reveal the teachings of the school of Pythagoras, and Pythagoras is praised as the one who discovered the tetractys:

> Surely not! By Him who gave the tetractys to our race... ([130] from p. 243)

The symbolic representation of the tetractys was usually done by arranging the numbers in triangular form as individual symbols in rows one above the other. Such a triangular number is depicted in the painting "The School of Athens" by the Italian Renaissance painter Raphael. A huge fresco on one wall of the "Stanza della Segnatura" ("Room of the Segnatura"; a hall in the Vatican) depicts a fictional as well as illustrious assembly of ancient philosophers, including Plato, Aristotle, Socrates, Ptolemy, and Diogenes (Fig. 2.25). At one place Pythagoras sits writing, while a student hands him a slate. At the bottom of the slate is the tetractys of Roman ones (I) with the sum 10 (X, see Fig. 2.26) written underneath.

The eye exudes as a light that encompasses the objects and thus creates a visual impression. According to the theory of Diogenes, the image develops in the pupil with the spirit (the air) of the eye from the interaction of spirit and picture the visual sensation.[78] Diogenes also believed that eye color had a significant effect on visual performance ([132], p. 94).

According to Theophrastus, Anaxagoras held the principle that something can be perceived only if it differs in composition from that of the particular sense organ (e. g., eye or ear), the similar being insensible to the similar. Vision is effected by a reflection of the object in the pupil, but this occurs only if the eye and the object were dissimilar in color ([133], p. 350).[79] Anaxagoras, as an inveterate materialist, taught the view that the sun was only "a lump of matter of nothing but fire" ([134], p. 105), and for this was accused of sacrilege and thrown into prison ([135], p. 300).

[78] According to Theophrastus as edited by Philippson: "At oculos cernere, imagine in pupilla facta, qua cum aëre interno mixta sensum fieri putat." (from [131], pp. 117–119)

[79] According to Theophrastus: "Cerni enim repraesentatione pupillae: repraesentari vero non in con coloribus, sed in diversis colore. At multis die, aliis nocte esse discolora (acutius vero cernere)" (from [131], p. 109).

Fig. 2.25 Stanza della Segnatura in the Vatican for Pope Julius II, wall fresco: The School of Athens, Raphael (1509) from The Yorck Project (2002)

Is Seeing Nothing But Hot Air?

It may seem surprising at first sight that the pre-Socratics and Pythagoreans should have noticed little of the influence of light on the process of vision, but should have spoken again and again of fire. But the essence of light in the philosophers is characterized by heat or fire in analogy to the sun and air in the function of mind and soul. This conception is by no means absurd even from the modern point of view. In ancient times, fire and light were inseparable; the production of light without fire was completely unthinkable. However, infrared heat radiation, like visible light, is part of the electromagnetic spectrum and follows the same laws. The mind of the brain is certainly still necessary today to generate the visual sensation from the optical image of the eye.

Empedocles finally recognized the nature of light. In one of the preserved fragments Empedocles explains the nature of light using the example of a lantern (from [110], pp. 196–197):

Fig. 2.26 Pythagoras with tetractys, detail from Fig. 2.25

> As when one intends a walk in the winter night, and for this purpose prepares a light and kindles the burning fire's brilliance, from all sides shielding lanterns from the wind; they indeed divide the blowing 'winds' blowing, but the light penetrated outwards, because it was so much finer, and shone to the firmament with unweakened rays: thus hid the primeval fire ... behind the round pupil enclosed itself in skins and thin garments pierced with divinely furnished pores going straight through. These kept out the depth of the water that flowed all around, but they let out the fire because it was so much finer.

Even if it is not immediately obvious: Empedocles compares the lantern with the eye in his lantern allegory. In the lantern is the fire of light, which is shielded from the wind by the lantern, but nevertheless spreads out in rays unattenuated and infinitely far. In the eye, so Empedocles assumes, is enclosed from birth just such a fire, which finds its way out through naturally (divinely) arranged pores, while the liquid of the eye remains reliably shielded by skins inside.

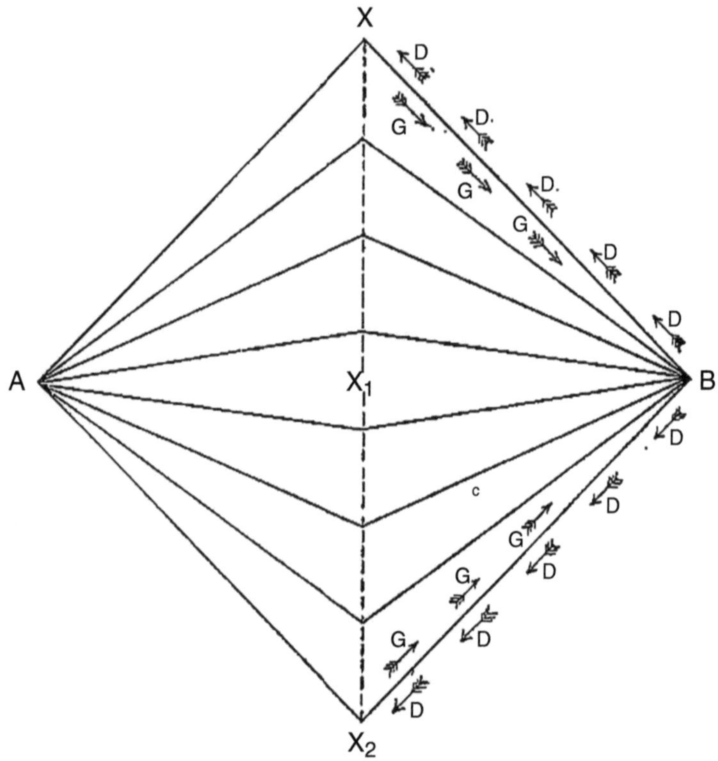

Fig. 2.27 Representation of the act of seeing according to Empedocles-Plato, Magnus (1901) (from [132], p. 106)

Empedocles, moreover, was of the opinion that all bodies constantly emit effluvia, which in turn penetrate through the pores into our eyes. From the mixture of the rays of sight and the rays of the body the impression of sight arises for Empedocles ([136], p. 362).

This sounds strange, but becomes comprehensible if one makes a geometrical analysis concerning the process of seeing according to Plato and Empedocles. Such a consideration had been made by Magnus in his "Ophthalmology of the Ancients" ("Augenheilkunde der Alten") [132]. The consideration assumes that both from the eye and from the object to be seen, divergent bundles of rays emanate. This corresponds to Empedocles' theory. These now come into contact with each other in an imaginary plane at the points of intersection. When the "sensing rays" of the eye collide with the "luminous rays" of the object, the sensing rays are trembled at the point of collision. The vibration now propagates in the direction of the eye and leads there to the perception of the light emanating from the object.

Magnus provides the following picture (Fig. 2.27):

The image resembles a ray path through a lens. If we replace the line X-X_1-X_2 with a converging lens, we would actually have the image of a point A (the luminous object) on a point B (here the eye).

In his "Megas Diakosmos" (Great World Order), Leucippus rejected the previous idea of the heavenly bodies as living beings. Furthermore, he declares that all things are made up of countless particles—the atoms ([137], p. 19). He is not only regarded as the founder of atomism but also as a forerunner of materialism with his theorem about the spirit (from ([137], p. 19):

> No thing comes into being without a cause, but everything comes into being for a definite reason and under the pressure of necessity.

Background Information
Leucippus' leitmotif of determinism is still surprisingly relevant after more than 2500 years. Nobel laureate 't Hooft takes a very similar view in December 2018. In an interview, he was asked by the reporter of the journal "Spektrum der Wissenschaft" whether he believed in free will ([138], p. 23). He answered quite in the spirit of Leucippus:

> I believe that all the laws of nature are completely deterministic. Nothing happens without a reason.

The atomistic idea of Leucippus was refined by his pupil Democritus in his "Mikros Diakosmos" (Small World Order)[80] to a theory of sight. Basically Democritus distinguishes between the actual object and its perception (in [137], p. 60).

> We perceive in reality nothing veritable, but only what changes according to the particular condition of our body and the influences flowing towards or against it.

This statement is the primacy of virtual reality: what we see is never the object itself, but always only what has somehow been presented to our senses. This form of insight Democritus calls the inauthentic cognition, to which he counts vision, hearing, smell, taste, and feeling [137], p. 61). The knowledge of Democritus's doctrine of sense perception comes mainly from the transmission of Theophrastus's "De sensu et sensibilibus" (in [139], pp. 207–213). All things continuously emit atoms, which as "eidola" ("little pictures") reach the senses and can be perceived in this way.

In his dissertation Karl Marx had dealt with the differences in the philosophy of Democritus and Epicurus, and inevitably came to speak of Eidola (from [140], p. 297):

[80] Diogenes Laertios reports in his ninth book that Democritus wrote his "Small World Order" 730 years after the fall of Troy ([129], p. 154).

> The Εἴδωλα [Eidola] are the forms of the bodies of nature, which, as it were, skin themselves from them as surfaces and carry them into appearance. These forms of things constantly stream from them and penetrate the senses, and by this very means, make the objects appear.

"Eidola," then, are little images emitted by things and finally striking the senses of the observer. Lucretius, a philosopher in the tradition of Epicurus, is with his work "de rerum natura" [141] one of the main sources for Epicurus' philosophy of nature. Epicurus' theory of images is there described in numerous verses, how images detach themselves from things and float invisibly up to the eye, there to produce the effect of sight. An example of Epicurus's (or Lucretius's) extraordinary poetry will be given here from Knebel[81]'s translation of "Of the Nature of Things" ("Von der Natur der Dinge") ([143] Fourth Book, p. 120):

> These images, therefore, of which I speak, spread everywhere, and throw themselves to every side: but since we are able to see with the eye alone, it is only from whence the eye turns that it is hit by the form and color of things. Images also cause us to see the distance between things and to be able to distinguish between them. For as soon as the image is detached, it drives and pushes away all the air between the eye and it; this then penetrates to the eye, caresses it, as it were, around the pupil first, and then slips into it.

Although this approach seems strange at first, it contains insights that are still valid today. The doctrine "pictures throw themselves to any side" can be reconciled with the current definition "light spreads to all sides". The observer sees the shape, color and distance of things, i.e. also the arrangement of objects in space, by means of the pictures. The perception arises when the (invisible) image detaches itself from the object and slips through the pupil into the eye. This may well be considered a poetic paraphrase for an optical image.

The Breslau ophthalmologist Magnus points out a special peculiarity of the Epicurian theory of vision: Through the possible fusion of images whirling around in the air, whimsical beings (e.g. centaurs) could also be perceived (from [132], p. 104).

Socrates is said to have mastered the art of questioning the apparently firmly established truths and to make the discussants and the audience doubt. No works of his own have survived, so that knowledge of his philosophy and person is based mainly on the writings of his pupils Plato and Xenophon.[82]

---

[81] Known as Goethe's "original friend" (according to a letter "An meinen alten Weimarischen Urfreund, Major von Knebel" of Nov. 30, 1817 ([142], p. 197).

[82] Xenophon handed down some dialogues of Socrates (e.g. [144]), in his "Oikonomikos" Xenophon even lets Socrates philosophize about economics [145].

The Apology of SOCRATES

Socrates is fondly quoted as saying, "I know that I know nothing," which comes from his defense speech during the prosecution in 399 BC.

Socrates, as an old man, was accused by his adversaries of corrupting the youth and not believing in the gods of the state. The trial took place before about 500 judges drawn by lot (the Heliasts). The plaintiffs' request was for the death penalty. Plato reports in the "Apology of Socrates" an anecdote of Socrates about wisdom during his defense.

Socrates' friend Chairephon had asked the oracle in Delphi whether anyone was wiser than Socrates. The Pythia, the prophesying priestess of the oracle at Delphi, declared that no one was wiser than Socrates himself.

Socrates, contrary to the oracle's saying, was in doubt about his own wisdom and immediately set out to investigate the truth. He turned to a statesman who had a reputation for wisdom among many people. Socrates got the impression that this man seemed most wise to himself, but was by no means so. To this Socrates ([146], p. 30):

> I am, however, superior to this man in wisdom; for it seems that neither of us knows anything right and proper, but he, notwithstanding his ignorance, imagines that he knows something, while I, conscious of my ignorance, do not imagine that I know anything either. It seems, then, that I am a little wiser than he, namely, for this: what I do not know, I do not imagine I know.

At the end of the trial, Socrates was sentenced to death, drank the cup of hemlock,[83] and died.

In the "Phaedo" ("Phaidon") Plato lets Socrates philosophize in a dialogue about sense perception and rational knowledge. Socrates talks about how knowledge can be gained and that what is righteous, beautiful and good, i.e. the true essence of the object of observation, cannot be grasped with the senses ([147], p. 41):

> Will not he of us come nearest to the true knowledge of any thing who makes it his business most rigorously and acutely to inquire by pure thought into the very essence of the object under consideration?

According to Plato, vision is based on three decisive foundations: the eyes, color, and light (from [148], pp. 353–354).

> Even if there is vision in the eyes, and he who has them tries to use them, and even if there is color for them, you know that unless a third being is added, which is specifically there for this

[83] A poison cocktail popular in ancient times for execution.

> purpose by nature, the vision will see nothing, and the colors will remain invisible.—What then is this that thou meanest? he asked.—What you, I said, call the light.

If the light does not shine, then the eyes are "stupid" and almost blind. Only the sun gives "the visible the ability to be seen". Light and vision are used in this dialogue as an allegory for truth and knowledge. Plato distinguishes between the visible and the imaginable, thus separating the recognizable from the imaginable, things from images and shadows. Insufficient sensory impressions lead to cognition only with a certain probability; the belief in the correctness of the senses can sometimes even lead to a misinterpretation of the actual situation, mistaking the mirages of shadows and mirrors for images of real things. This would make it a non-real perception—or already a simple virtual reality. The combination of the senses with reason and logic, with mathematics and geometry, increases the accuracy of knowledge. But only the knowledge of things, the connections, problems or contradictions, causes and effects and their dialectical analysis can lead to the insight of reason. Reason as an epistemological faculty leads Plato to his ideas and conclusions through the use of thesis, antithesis, rhetorical discourse, and subsumption.

And so, at the end of this discourse in the sixth book of the "Politeia" ([148], p. 360), Plato has his brother Glaucon answer Socrates "reasonably":

> I understand . . . and concede it, and arrange it as you say.

Plato certainly had an influence on his student Aristotle. Equally undisputed, however, is Aristotle's decisive contribution to the emergence of the systematic sciences. Cognition and perception theory are also of central importance in Aristotle's work. In his work "On the Soul",[84] Aristotle described the senses of sight, hearing, taste, smell and feeling[85] as tools of cognition. There, the senses are only sensitive to certain objects and the objects thus perceived are consequently possible to sense. The process of seeing is apparently described very simply: "What can be seen is visible to us" ([149], p. 126). But this statement must be understood with what has been said before: Only those things which we can perceive with our senses (precisely the sensible things) become visible to us. It is not the things themselves that are visible, but the color of the things, which only becomes visible through light. But what is the light?

It is, first of all, that by means of which one sees the colors and, moreover, something transparent (like water, glass or ice). Aristotle described seeing in such a way as to come very close to present-day cognition (from [149], p. 129):

> But the efficacy of the transparent is light. This is also revealed by observation. . . . For color moves the transparent, (what is between the eyes and the colored object,) e.g. the air. From this, as a continuum, the organ of sense (the eye) is moved (afficirt.)

[84] Greek "Peri psyches", Latin "De anima".

[85] "the colors, the sound, the tastable, the olfactory, and the tactile" ([149], p. 125).

Aristotle clearly distinguishes himself from the earlier doctrines of Plato, Empedocles and Democritus. Light is "neither a fire nor any other body, nor the effluence of a body".

According to Aristotle, the function of seeing is therefore based on the effect of the object on the eye by means of light. The quintessence of Aristotle is the sentence "Visio non fit extramittendo",[86] which the Swiss medical historian Koelbing even considered an "Aristotelian law" ([151], p. 7).

Aristotle's doctrine was authoritative for millennia. At least as noteworthy as Aristotle was unquestionably Euclid. His mathematical textbook "The Elements" ([152]) was, next to the Bible, the most popular book until the nineteenth century (from [153], p. 18). At first sight, the great mathematician and geometrician nevertheless seems to have been a supporter of the emission theory. In the postulates to Euclid's "Optica" [154] there are clear indications of this.[87] Euclid presupposed that the straight lines emanating from the eye can traverse large spaces. Further, the lines proceed from the eye in a cone shape, whereby this cone has its apex (the origin of the rays) in the eye and sees things with the other side (the base plane of the cone).[88] Euclid's straight rays of vision allow a geometrical view of seeing and allow the explanation of perceived sizes and directions in Euclidean space.

Euclidean Space

Until the nineteenth century it was generally assumed that Euclid's mathematical and geometrical descriptions were completely sufficient to explain the three-dimensional physical space known to us. The addition of "Euclidean" became necessary only after Euclid's explanations were no longer sufficient to fully describe all of geometry.

Like many generations of mathematicians, Lobachevsky had struggled with Euclid's axiom of parallels. In it, Euclid had explained that a point lying outside a straight line can only draw a single parallel to the straight line through this very point (in this plane). In "The Mysterious Flame of Queen Loana" ("La Misteriosa Fiamma della Regina Loana") Umberto Eco lets his protagonist give a more understandable explanation ([159], p. 10):

(continued)

[86] "Seeing does not happen by emission" (from the index of a Latin edition of Aristotle from 1530 [150]).

[87] Of course, in the following two millennia after the compilation of the "Optica", copies of the original Greek text were repeatedly made and translations into Latin (e.g. [155]) and Arabic (e.g. [156]) were undertaken. According to Lindberg ([157], p. 210), the most common variant is a direct translation from the Greek, beginning with the incipit "Ponatur ab oculo eductas rectas lineas..." (or a variant thereof, e.g. [158]).

[88] Lat.: "1. Ponatur ergo, radios ab oculo emissos, in rectam lineam ferri, aliquoque interuallo inter se distare. 2. item, figuram a radiis comprehensam, esse conum, qui verticem habeat in oculo, basim vero in extremis rerum visarum." (from [154], p. 4)

> The desperate loneliness of parallel lines that never meet...

These desperate parallels seduced Lobachevsky into studying them. In February 1826, he submitted an essay to the Kazan College in which he claimed to have "brought the dispute to a head".[89] Unlike all others before him, however, he found a solution. First of all, Lobachevsky recognized the unprovability of the axiom of parallels and proposed a solution independent of the axiom, which Gauss had already called "non-Euclidean" a few years earlier in his letters to his friend Bolyai.

The mathematician Wolfgang Bolyai implored his son Johann not to concern himself with the axiom of parallels (from [161], p. 56):

> Thou must not try the parallels on that path; I know that path to its end—I too have traversed that bottomless night, every light, every joy of my life have been extinguished in it. I beseech thee by God! leave the doctrine of the parallels in peace ...

The mathematician Riemann used the Gaussian idea of curved surfaces [162] and extended it to a higher number of dimensions [163]. A higher number of dimensions resists the human space of experience and is therefore actually unimaginable. But if one understands these higher-dimensional spaces not as esoteric peculiarities, but as mere analytical fictions and aids to simplify calculations, things can be handled again. In the lecture of Carl Cranz "Gemeinverständliches über die sogenannte vierte Dimension" ("Common sense about the so-called fourth dimension") [164], which is worth reading, these higher dimensions are called "mental aids to visualization" or "physical auxiliary ideas". For Einstein and Minkowski, who proposed an elegant description of the theory of relativity with the Minkowski space named after him [165], finally a higher-dimensional space with time as fourth dimension was a necessity for the description of space-time and thus formed a basis of the theory of relativity.

Aristarchus advocated the hypothesis of a heliocentric view of the world, as Archimedes of Syracuse tells us: "Then he sets the fixed or adherent stars together with the sun as immovable; but the earth movable around the sun..." (from [166], p. 2). However, not Aristarchus, but the name Ptolemy is closely connected with the favored geocentric view of the world, which shaped the understanding of the universe for almost one and a half millennia.

---

[89]The document was entitled "Exposition succincte des principes de la géométrie avec une démonstration rigoureuse du théorème des parallèles" ([160], p. 371) and is found in the treatise "Ueber die Anfangsgründe der Geometrie" ([160], pp. 1–66).

But Ptolemy had also worked in the fields of mathematics and optics. His book of optics was saved from oblivion by Eugenius of Palermo,[90] who translated the available manuscripts into Latin as "Liber de Optica" [168].[91]

Ptolemy's theory of vision, like that of his predecessors, was also fundamentally based on the idea that there must be physical contact between the eye and the object. Plato united the two seemingly conflicting theories of the active emission of fire by the eye (emission theory) and the atomistic perception of light or images (immission theory) with the fusion of vision and light. This Platonic collision of two streams of matter, the double emanation of matter known as "efflux" also existed in Ptolemy's understanding. The emanation, the outflow of light matter from the eye, produces a pneumatic propagation of visual flow in the air, which strikes objects and is noticed by the eye. However, visual flow alone is not sufficient to see the objects; a certain brightness density[92] at the point of vision is also required to actually see the objects. Like Aristotle, Ptolemy believes that the most important function in this regard is color, in order to see the object.

The division into ray of sight and ray of light used in Ptolemaic theory sometimes leads to difficulties in the subsequent explanations of reflection and refraction. Nevertheless, his book is an early example of the scientific approach of supporting a theory by experiment.

Air as a transmitter of sight was very common at the time of Ptolemy. Also Cicero had written in the second book "On the Nature of the Gods" ("De natura deorum") about the importance of air for sustaining life and as a medium of the senses: "...and the air itself sees with us, hears with us...". (from [171], p. 600). The physical effect of seeing is described by Stoics through the pneuma.[93]

Stoics

The Stoics were followers of the philosophical doctrine of the Stoa (founded by Zeno of Kition around 300 BC), according to which the world and nature are based on a universal principle that can be known by means of physics, logic and ethics. The Stoic strives for wisdom with serenity and peace of mind; the ideal of the Stoic state of mind is "apeitheia", the absence of affect. The adjective "stoic" still denotes the qualities imperturbable; equanimous and serene. Famous Stoics were Seneca and the Roman emperor Marc Aurel.

---

[90] According to Smith, Eugenius' text is based on a "badly battered" Latin translation of the then available Arabic translation of the lost original Greek text (from [167], p. 7 ff.).

[91] Although the text has been repeatedly annotated (e.g. by Jean-Baptiste Delambre in 1812 [169]), surprisingly the very first printed version of the Latin text (in the translation of Eugenius of Palermo) was published only in 1885 by Gilberto Govi [170].

[92] "lucida spissa" (from [170], p. 8).

[93] Greek pneuma, air or fire-like substance as the basic principle of nature and life.

The concept of sense perception, in Greek "aisthesis", is explained in the very first sentence of Aristotle's Metaphysics: "All men by nature desire knowledge; a sign of which is their love of sense perceptions, which, even apart from benefit, they love for their own sake; especially that of the seeing." (from [108], p. 9).

For the Stoics, the pneuma is a cause of perception. The pneuma produces sense perception in the eye by interacting with the air, thus embracing an object as it were with polyp arms. The visual pneuma interacts with the air, illuminating and stretching[94] it in a wedge shape, the base of the wedge palpating the visual object as it were with a stick.[95]

For the physician Galen, the pneuma was the fundamental principle of life.[96] In physiological terms, the pneuma functions as the cause of all vegetative life activity, such as nutrition, metabolism, growth or reproduction, here as "pneuma physicon" with its seat in the liver. Galen hoped that one day it would be possible to extract from the air that part of the pneuma which was absorbed into the blood during respiration. This actually exists: it is oxygen. The proof did not come until the end of the eighteenth century with the discovery of the importance of oxygen in combustion and respiration by Lavoisier and Laplace [174]. Another form of pneuma, the "pneuma psychikon" with its seat in the brain, is responsible for the mental processes of thought, sensation, or will, and thus serves as the soul or spirit. Finally, a third form was the cause of pulse or heartbeat. The "Pneuma zotikon" with its seat in the heart controlled the so-called "sphygmic functions", which were caused by the pulse (Greek Sphygmos).

Galen, in his work on the theories of Hippocrates and Plato, "De placitis Hippocratis et Platonis," [175] discussed the theories of vision and recognized two principal alternatives:

> A body that is seen does one of two things: either it sends something from itself to us and thereby gives an indication of its peculiar character, or if it does not itself send something, it waits for some sensory power to come to it from us.[97]

The first possibility seemed extremely unlikely to him, since he could not explain how the images of large mountains are dramatically reduced in size to fit through the pupil, or how the images of these mountains can reach several observers at the same time. Thus, if the mountain does not reach the observer, the observer must reach the mountain. Galen concluded that, consequently, the sensory power of the sighted must somehow reach the mountain.

According to Galen, the pneuma comes into contact with the air from the eye and changes it in a certain way, so that the pneuma can now pass from the object into the eye all the way to the optical nerves. Thus, the ambient air, together with the illuminating light of

[94] Pneumatic tension, Greek tonos, which "propagates itself timelessly in spherical waves ..." (from [136], p. 382); an imagery already prefigure Huygens' elementary waves.

[95] see also Ingenkamp "Zur stoischen Lehre vom Sehen" [172].

[96] see also Haeckel "Die Welträthsel." (The Riddle oft he Universe, [173], pp. 48 and 49)

[97] According to the English translation by DeLacy (from [175], p. 453).

the sun, becomes the image-transporting medium that stands between the nerves of the brain and the object and thus "touches" it only in a figurative sense. With this, Galen distances himself from the image of the palpating stick of the Stoics, which he also makes clear

> The Stoics, then, must not say that we see by means of the surrounding air as with a walking-stick.[98]

Alhazen described in the first millennium AD. In his essay "Treatise on Light" ("Risalah fi al-Dawa") [176],[99] Alhazen described the propagation of light on lines along which light propagates. Of interest here is first of all the statement: "The manner of the progress of light ... happens ... in such a way that light progresses from any point to all sides in the direction of straight lines until it reaches where it ends." (from [177], p. 349). In this the universal form of the propagation of light is explained: light propagates in a straight line and in all directions. However, this insight, which may seem immediately clear and obvious, forms the basis for a model of imaging that is still useful today (from [177], p. 340):

> The light that progresses on straight lines in the transparent body is called a "ray". The straight lines are imaginary (mathematical) lines that cannot be felt. They themselves, together with the light that advances on them, form the rays. The rays are an image of the essence to move forward on straight lines.

The "image of the essence" is a model, the "move forward on straight lines" a ray. In the translation by Baarmann, who had another manuscript at his disposal [178], the model character of the rays of light becomes even somewhat clearer (from [179], p. 214):

> But the straight lines on which light advances are imaginary lines, not perceptible to the senses. Now the imaginary lines together with the light advancing on them, these two together are what is called a ray.

The rays of light of Alhazen are clearly different from the rays of sight or vision of Euclid or Aristotle. Their basic view was that the rays of vision emanate from the eye thanks to a fiery power of sight and touch the object with the light of fire, as it were. The Stoics also came into contact with the object with the tense pneuma or Galen with the somehow feeling pneuma. Alhazen now radically separated himself from this tactile view and altogether from the theory of the ray of sight and recognized "... that seeing takes place by means of an image which propagates itself from the seen to the eye..." and "... meets in the centre of the eye from the side of the light" (from [177], p. 341).

---

[98] According to the English translation by DeLacy (from [175], p. 475)

[99] From the German translation "Abhandlung über das Licht" by Wiedemann [177].

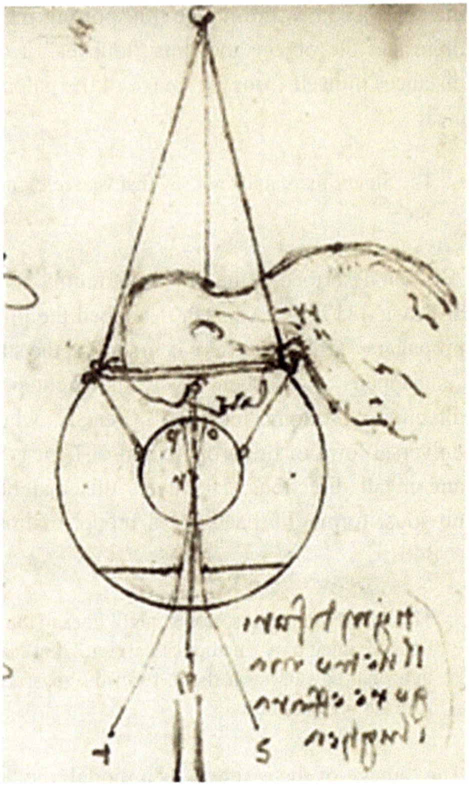

Fig. 2.28 Eye apparatus, Leonardo da Vinci (1508) from manuscript D, fol. 3 verso, Paris, Institut de France

This is a groundbreaking insight. Light exists independently of the observer and makes its way from the object to the eye through the transparent medium of air, water or glass. The media change the path of the light through refraction and reflection and lead to an optical image.

An optical image was also considered by da Vinci to be essential for seeing. Not only have numerous cross-sections through the human eye or images through rays been preserved by his hand,[100] but even the drawing of an apparatus with which human vision can be imitated Fig. 2.28.

However, in da Vinci's opinion, the process of imaging was completely different from our current understanding. Similar to the notion of "eidola", in da Vinci's mind objects were constantly sending out images of themselves that converge in a point-like fashion at every point around the object. Space is completely filled with point-like images of the object. Thus, from the eye, there is always a possibility of drawing a visual pyramid to the object (with the pupil at the apex of the pyramid). Da Vinci himself describes it this way (from [181], p. 43):

[100] s. e. g. Fehrenbach "Der oszillierende Blick. Sfumato und die Optik des späten Leonardo" [180].

> As the stone, thrown into the water, makes itself the centre, and causes various circles, so the sound produced in the air spreads itself in circles; so every body, placed in the luminous atmosphere, spreads itself in circles, and fills the surrounding parts with innumerable of its images, and appears as a whole entirely and completely in every part.

Looking at the results of da Vinci's work, the simplicity and clarity of these point images, which move on straight lines in space, seem to have quite obviously satisfied the perspective requirements for design. Presumably, however, this theory clashed with the increasing knowledge of optics through books by Bacon, Pecham, or Witelo, [180] and not least with the practical anatomy of the eye by da Vinci himself. It is not clear whether da Vinci duly appreciated the optical effect of lenses, but it is clear that—especially in his optical sketches (e.g. in manuscript D [182])—he dealt with it intensively. The lack of optical effect[101] (viewed with today's knowledge) in one of his most famous paintings, the "Salvator Mundi",[102] had raised doubts about da Vinci's authorship. However, it is by no means certain that Leonardo da Vinci always transferred all his scientific knowledge to his artistic work [184]. A partially distorted representation in an image of the Saviour would hardly have been appropriate at that time.

In any case, the anatomist Platter realized a little later that optical imaging in vision takes place via the lens of the eye, but primary vision[103] functions via the retinal nerve.[104]

In his great work "Ad Vitellionem Paralipomena: Astronomiae Pars Optica Traditur"[105] [185], Kepler reproduces Platter's anatomical drawings of the eye (see Fig. 2.15), but also criticizes the physician somewhat for his lack of mathematical knowledge (see [185], p. 208). Kepler finally says, somewhat conciliatory, that Platter was no further from true knowledge than the profession of medicine is from mathematics.[106]

Kepler developed a theory of imaging by means of lenses from the action of ground glasses and transparent crystals in "Dioptrics" [187] published in 1611. Kepler recognized the operation of the lens of the eye as the imaging part of the eye and the image reversed by the lens on the retina. For Kepler, the visual impression is created by imaging the visible things on the retina filled with mental matter (from [188], p. 28):

---

[101] e.g. Walter Isaacson in a biography of Leonardo da Vinci: "But Leonardo failed to paint the distortion that would occur when looking through a solid clear orb at objects that are not touching the orb." (from [183], p. 333)

[102] Lat. "Saviour of the World "

[103] "pars primaria visionis" ([57], p. 187).

[104] "nervus retiformis" from Latin rete = net + forma = shape ([59], p. 218).

[105] Supplements to Witelo, in which the optical part of astronomy is continued.

[106] In the original: "Confer iam modum verum visionis à me propositum cum illo PLATERI, videbis, clarissimum virum non longiùs à vero abesse, quàm quantum professione Medicum, qui Mathematica non dedita opera. Tractat, abesse consentaneum est." (from [186], p. 187)

> Seeing is a feeling activity of the retina, which is excited and filled with the spirit of seeing; or, to see is to feel the excitation of the retina, so far as it is excited. The retina is painted by the colored rays of the visible world.

Kepler's theory of vision is essentially the one that shaped the understanding of the physical process of vision that endures to this day.

Descartes had as clear an understanding of the process of vision as Kepler. In his richly illustrated account of optics and vision [114], the process of imaging in the eye is very well illustrated. Here, special attention should be drawn to the inverse mapping of the objects V, X and Y to their retinal positions T, S and R (Fig. 2.29).

In his capacity as a philosopher, Descartes questions the reality of sense impressions and doubts the purely physical correctness of the senses (from [189], p. 99):

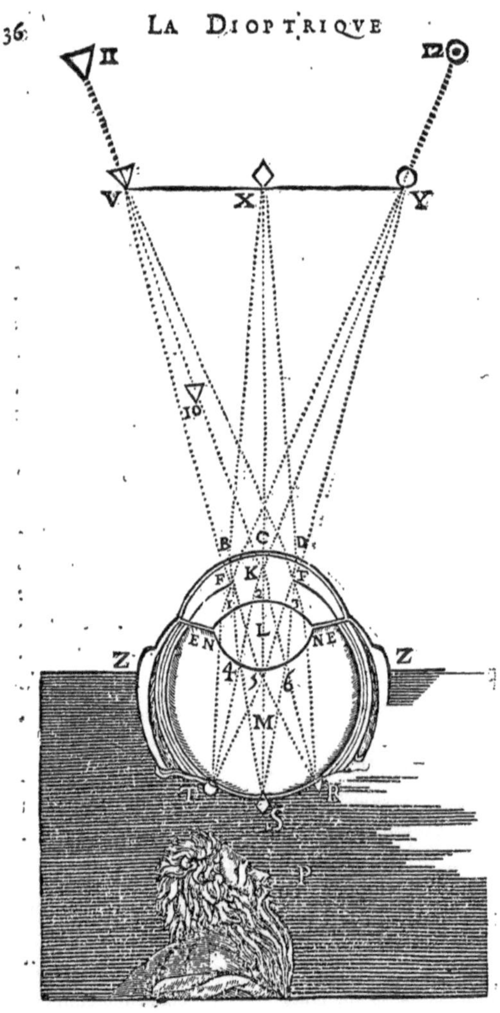

Fig. 2.29 Reverse image in the eye, René Descartes (1657), from [114] "La Dioptrique", p. 36

> Everything else, too, that I assumed about the objects of the senses, nature seemed to have taught me; ... Later, however, various experiences gradually weakened all the confidence I placed in the senses.

Doubts about everything form the basis of Descartes' epistemology. The reality of the factual, the so-called formal reality, becomes the representational reality of the imagination through perception and thought ([189] from p. 52). In principle, his recognition of the possibility of imperfect, erroneous perception also already contains the possibility of deception, of creating a virtual reality.

Descartes' idea that body and mind, or even heart and soul, are two independent substances that can only lead to cognition together (Cartesian dualism) has as its core the quite mechanical idea of a logically explainable structure of things and their function. Only reason, the ability to think and speak, makes man out of the machine.

Newton broke away from the philosophical-hypothetical consideration of natural phenomena and verified his assertions by experiment (in the first book of Optics [190], p. 5):

> It is not my intention in this book to explain the properties of light by hypotheses, but only to state them and confirm them by calculation and experiment.

Newton thus placed optical imaging on a firm physical foundation. Nevertheless, the insight of many philosophers before him remains correct, that seeing is not limited to the pure image, but only leads to a perception with the mind and through thinking. And therefore Descartes deserves the last sentence of this section (from [189], p. 45).

> I think, therefore I am

2.4 Ancient Perspectives

The Use of Perspective in Antiquity

Generally known are the numerous works of art from Egyptian and Assyrian antiquity or from Greek and Roman antiquity, of which the bust of Nefertiti will serve here as an example of the high craftsmanship of this epoch. Since its discovery, the statue has been in the Egyptian Museum on Museum Island in Berlin (Fig. 2.30).

The bust of Nefertiti was found in 1912 in the ruins of the workshop of Thutmosis and probably served him as a model for the production of further statues.

Fig. 2.30 Bust of Nefertiti in the Neues Museum, Egyptian Museum and Papyrus Collection of the National Museums in Berlin—Prussian Cultural Heritage, photographer Philip Pikart (2009), Wikimedia Commons

Background Information

What is astonishing is not only the overwhelming realism of the figure, but also its colorfulness. As a result, the Swiss writer Stierlin, in his book "Le buste de Néfertiti: une imposture de l'égyptologie?"[107] [191] even a fraud. The head of the excavation at that time, Borchardt, had the figure made by the Berlin sculptor Marcks only to test antique pigments on it. Borchardt himself describes in his book "Portraits of Queen Nefertiti" ("Porträts der Königin Nofret-Ete") anyhow the finding of Nefertiti quite detailed (from [192], pp. 30–31):

> When on December 6, 1912, soon after the lunch break, I had been hastily summoned to house P47.2 by a note from Prof. Ranke, who was in charge of the supervision, I found

(continued)

[107] "The Bust of Nefertiti: A Fraud of Egyptology?"

> in room 19 east of the "wide" hall . . . already the fragments of a life-size colored bust of Amenophis IV that had just come to light. . . . Then . . . before us at first only a flesh-colored neck with painted red bands became bare. "Life-size colorful bust of the queen" was announced and written down, the hoe was put aside, and work continued gently with the hands. . . .Then the colorful bust was first lifted out and we had in our hands the most life-size Egyptian work of art.

If Stierlin's thesis were correct, then not only Ranke would have to have been involved in the fraud, but also the other members of the excavation team. This possibility is of course not excluded, but it decreases with the growing number of secret bearers.

Regardless of whether this sculpture is a real treasure or just a common forgery, there are countless other works of art from ancient times, which in all probability are not all forgeries and in their entirety may be regarded as proof of the high skills and the need for maximum fidelity to reality. The bust of Nefertiti was measured in 1992 and 2006 by means of computer tomography [193] and in 2008 again with a 3D strip light scanner, which enabled a significantly higher measurement accuracy. The 3D data set together with high-resolution images of the Rathgen Research Laboratory [194] served the plaster moulding workshop of the Berlin State Museums as a model for a particularly realistic replica [195].[108]

The high quality of such sculptures from Egyptian, Greek or Roman prehistory allowed even today a vivid idea of the appearance of dignitaries of that time. This type of reproduction is naturally limited by the cost of production. The reproduction of large groups of figures is therefore rare, although not entirely unknown. The first Chinese emperor Qin not only had the Great Wall built, but also a huge mausoleum for his planned afterlife. His tomb had a complete army of thousands of terracotta warriors, chariots, horses and officials, just everything that seemed important to him for a decent afterlife.

About the same time as Emperor Qin in China, Archimedes lived in Syracuse, Greece, who is considered a great mathematician and whose physical calculations and inventions are partly still in use today. To Archimedes the astronomer Fabricius also attributes a treatise on perspective, which existed in Arabic (see also [196], p. XI), but has not been found to this day. It seems more probable that Archimedes was skilled in the art of mirrors[109]; at least later commentators refer to such a treatise ([197], p. 45).

[108] The hand-painted replica GF 539 of the Bust of Nefertiti in the highest quality can be purchased from the Berlin State Museums for 8900 EUR (as of 2019).

[109] Catoptrics.

The Burning Mirrors of Archimedes

For two millennia the story has been told that Archimedes set fire to the fleet of an attacking enemy with burning mirrors. Thus, Galen (in the translation of Wilde, from [198], p. 26) writes: "So, however, as I believe, Archimedes is also said to have burned the fleet of the enemy by means of ignition machines." Although this anecdote may seem plausible and true at first glance, its truthfulness has been repeatedly questioned and its feasibility verified. As early as 1747, Buffon had made a burning mirror of 168 plane mirrors with which he could set fire to wood and melt metals (from [198], p. 30). The experiment was repeated in 2010 in the television series "MythBusters". The American President Obama had asked the Mythbusters to debunk the myth. In the test, the Mythbusters could not ignite a ship even with 500 mirrors and had to tell the president that the myth was destroyed [199].

Busted!

To Agatharchus, in the biography of Pericles, Plutarch has Zeuxis say that he himself boasted of his slowness in painting, and that as a result his works were more beautiful and more long-term in design ([200], p. 118). This leads to two findings. The painter Agatharchus could paint faster than Zeuxis (which obviously annoyed the latter, but is rather unimportant today) and Agatharchus lived at the time of Pericles and Zeuxis. This is of greater interest because Agatharchus later reappears in Vitruvius' books, where he is associated with perspective painting. In the ten books on architecture ("De architectura libri decem") Vitruvius writes (from [201], p. 325):

> Agatharchus at Athens, when Aeschylus was giving way to tragedy, first produced the scena (stage wall), and left a treatise on the same. Inspired by this, Democritus and Anaxagoras wrote on the same subject, namely, how the lines must be made to correspond to the optical effect and the divergence of the radii from a certain assumed point in a manner similar to nature, so that by means of deception the representations on the stage sets (decorations) receive the appearance of reality.

The Greek term for stage painting is derived from the representation of the scene by graphic means: "Skenographia". This term has already been handed down in Aristotle's "Poetics", in which Aristotle also writes about the technical structure of tragedy. Aristotle first mentions Aeschylus, who introduced a second actor and gave the leading role to the dialogue. Then the innovations of Sophocles are discussed (from [202], p. 19)[110]:

---

[110] In Butcher's English translation the word Skenographia does not occur, but in the Greek text it does ([202], p. 18, at the end of the 13th paragraph).

> Sophocles increased the number of actors to three and added scene painting.

Thus, it now seems established that skenographia came into the world with the great Greek tragedies, and through them an early form of perspective painting became known to a wide public. The works of Aeschylus and Sophocles were first performed in the famous Dionysus Theatre on the southern slope of the Acropolis in Athens. These facilities seated between 14,000 and 17,000 spectators per performance (see [203], p. 44), which is an incredibly large number even from today's point of view. A modern theatre seats only a few thousand spectators. The Metropolitan Opera in New York has 3800 seats, the Deutsche Oper in Berlin even only 1860 seats. Since the total number of inhabitants in Athens in the fifth century was probably not more than 30,000 citizens (see [204], p. 230), one can assume that a large part of the population had attended the performances and thus also came to enjoy the skenographia.

But what is behind the name? Let's take a closer look at the Dionysus Theatre. For the performance of the plays, a venue as realistic as possible was needed—the playhouse, in Greek "skene". In order to represent the various environments (e.g. the rocky mountains in "Prometheus" or the cave in "Philoctetes") the front of the playhouse had to be adapted to the theme. This required skene painters such as Zeuxis or Agatharchus, who added appropriately realistic-looking landscapes to the corresponding backdrops in order to transport the audience into the illusion of the respective act. Whether these representations were already of a perspective nature is disputed. Presumably, however, there will have been an attempt to depict a realistic environment on the play background (see also [203], pp. 211–212). Dörpfeld and Reisch have reconstructed what the skene of the Dionysus Theatre may have looked like (see Fig. 2.31).

The painters of this period understood admirably the use of light, shade, and color, and painted pictures which, as Cicero informs us, produced in the spectator pleasant sensations like those produced by natural vision (from [205], p. 327):

> Likewise, by means of the eyes, we obtain almost innumerable pleasurable sensations; and although the objects which touch our eyes are very dissimilar, yet it is always the same sense which affords us the pleasure.... The art and skill of painting is only one thing; but dissimilar as Zeuxis, Aglaophon, and Apelles are to each other: yet none of them can be accused of any defect in the art.

No paintings by any of the artists have survived, and yet a picture of Apelles has been handed down to our time. Lucian, in his treatise "The Defamation," told of the mischief which proceeded from calumny, and illustrates this mischief by a painting which Apelles had made of Ephesus ([206], pp. 1441–1459). Apelles had been slandered to King Ptolemy IV by a rival, the painter Antiphilos, that he had been part of the treason of Theodotos at Tyre.[111]

[111] Polybios knows the story ([207] fifth book, ninth chap.): Theodotos had surrendered the cities Tyros and Ptolemais as well as a fleet of 40 ships to Antiochos without a fight in the fourth Syrian

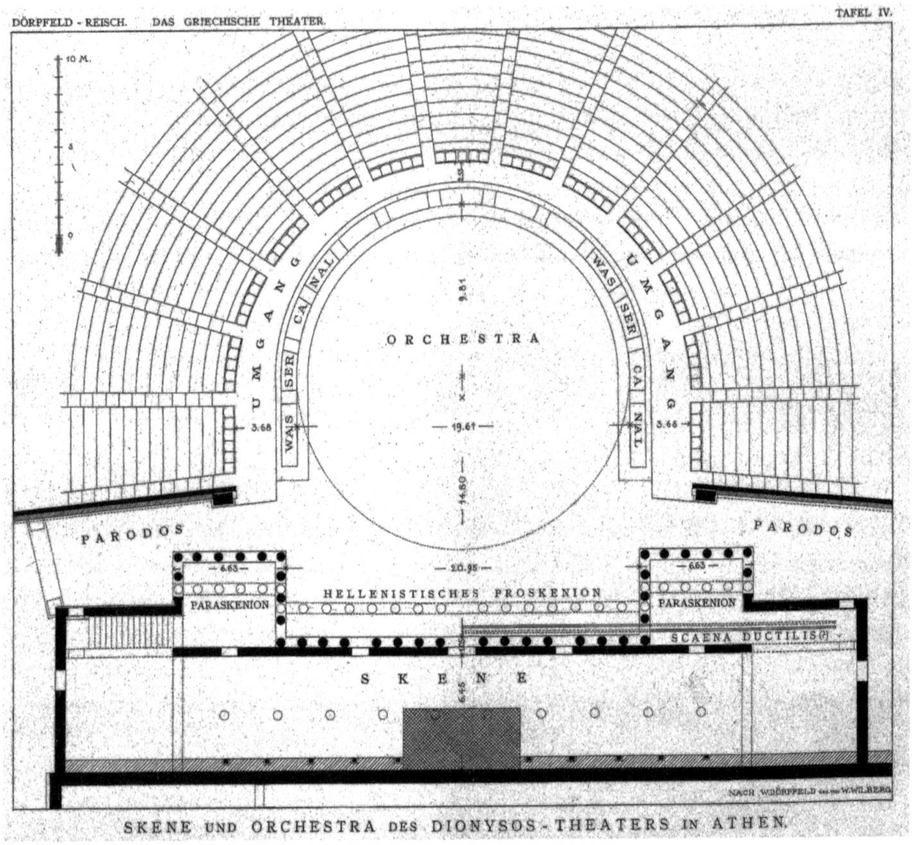

Fig. 2.31 Skene and orchestra of the Dionysus Theatre in Athens, Wilhelm Dörpfeld and Emil Reisch (1896) from [203] Plate IV

Ptolemy was foamed with rage, but was finally convinced by one of the prisoners that Apelles was innocent. Ptolemy was ashamed of his anger, gave Apelles 100 talents[112] and made Antiphilos his serf. Apelles processed this event in a painting, which Lucian describes as follows:

> On the right side sits a man with long ears that lack little to pass for Midas ears: his hand is outstretched towards the defiance coming at him from afar. Beside him stand two female figures, whom I take to be ignorance and distrust. From the left side, the defamation approaches him in the form of an extraordinarily charming, but heated and agitated girl,

war, in which the Seleucid Antiochos III. and the Egyptian Ptolemaios IV. were opposed. In return, Antiochos appointed him commander-in-chief of the Seleucid army.

[112] According to Klein ([208], p. 88) a talent had a weight of about 26 kg. That would be for 100 silver talents after today's value a pain money of more than one million euro.

Fig. 2.32 The Calumny of Apelles, Sandro Botticelli (c. 1495), from The Yorck Project (2002) 10,000 Masterpieces of Painting (DVD-ROM)

> whose features and gestures reveal anger and rage: in her left hand, she holds a burning torch; with her right, she drags a young man by the hair, who holds up his hands to heaven and calls the gods to witness.
>
> In front of her walks a pale, ugly man with a sharp gaze, who looks as if a long illness had emaciated him, and whom everyone will recognize for envy. Behind him walk two female figures, who seem to be the delinquents, and to dress and adorn them: these are, as the interpreter of the painting told me, the guile and the deceit. At the very back follows a mourning figure in black torn garb, namely, turning backward weeping, and casting bashful glances at the approaching truth."

According to Lucian's description, Botticelli had painted "The Calumny of Apelles" two millennia later. Lucian's description is here literally unset (Fig. 2.32).

The coloring of the surrounding architecture is remarkable. The reliefs here are monochrome and seem to be based on an ancient, monochromatic[113] style. Zeuxis[114] is actually

[113]Pliny writes about this (in [209], p. 116) "...in the beginning the shadow of a man was drawn around with lines, the second step was done by applying only one color each time, this process, after it had been more developed, was called monochromatics, and this very branch of art still exists to this day."

[114]See, e. g., Pliny ([209], p. 133) "He also made pieces of one color, and that in white."

considered to be an avid user of this style, which is why the art historian Zöllner suspects that Botticelli may have fallen for an erroneous translation of Pliny ([210], p. 103).

The Competition of the Painters

In the thirty-fifth book of his Natural History Pliny reports a contest between the painters Zeuxis and Parrhasius (from [209], p. 133):

> The latter [Parrhasius] is said to have got into a contest with Zeuxis; for Zeuxis had painted grapes so accurately that the birds would have flown to the stage, but Parrhasius had put up a linen curtain so naturally painted that Zeuxis, made proud by the verdict of the birds, at last insisted on removing the curtain and showing the painting; when he now saw his error, he is said, from sincere shame, to have thought himself defeated, and to have said that he had deceived birds, but that Parrhasius had deceived him, the artist.

Here we find again the idea of artificial reality: the birds consider Zeuxis' grapes to be real, the latter could not distinguish Parrhasius' painting from reality.

The convincing reality of painting was perhaps also achieved by the means of geometry. At least Pamphilos, a teacher of Apelles, is quoted by Pliny with his view on the amalgamation of art and science (from [209], p. 136):

> He [Pamphilus] was a Macedonian by birth, but the first painter of extensive scientific education, especially in arithmetic and geometry, without which, in his opinion, one could never get far in art. . . .

Even without geometrical knowledge of projection, the first works of art were created in this period by means of optical imaging. Pliny reports of the potter Dibutades, whose daughter traced the shadow of her lover on a wall (from [209], p. 158, see also Fig. 2.33):

> This art of forming likenesses out of earth and clay was invented in Corinth by the potter Dibutades of Sicyon with the help of his daughter; for the latter was in love with a young man and, when he went abroad, painted the silhouette of his face with lines on a wall in the light, and the father covered the outline with clay, thus obtaining an imprint. . . .

Sandrart adds to this historical love story the legend of Gyges ([211], p. 11), who traced his own outline, cast by the fire on the wall, with charcoal.[115]

Although the ideas of Gyges or the Butades family are commonly considered the beginning of the art of drawing, it is actually a simple form of shadow projection. This involves the outline of an object being projected onto a screen by means of a powerful light

[115] But the story may simply go back to a faulty reading of Pliny, as Schürer points out. Gyges is listed there as the inventor of the ball game, not of the art of painting (s. [212], p. 203).

Fig. 2.33 Dibutades' daughter draws her lover, Joachim von Sandrart (1774) from [211] plate A, bottom

source. This shadow painting ("skiagraphia") was a simple redrawing of the shadow on a surface. "The shadow produces painting by means of light." ([213], p. 135).

Overall, one can now note that the theatrical perspective was increasingly viewed through the eyes of projection. In this context, perspective is mainly found as decorative painting for the decoration of representative villas. Many of these works of art can only be admired today because they were buried for centuries by the eruption of Vesuvius in 79. Thereby, among others, the places Herculaneum and Pompeii were buried by a thick layer of ash and thus preserved in the state they were in at that time. In Pompeii in particular, various decorative wall paintings have been preserved. On the basis of these finds Mau was able to divide the painting into four styles [214]. First of all, there is the incrustation style (masonry style), in which large stone blocks or valuable materials were imitated by painting or plastic stucco work of the wall. An example of this is the House of the Faun in Pompeii (see Fig. 2.34).

The second style was altogether a bit more elaborate. Here, an attempt was made to build up a mock architecture and simulate views into virtual landscapes. Consequently, this style is called architectural style. The subsequent ornamental style divided the wall into flat fields bordered by different stripes and decorated with ornaments, including decorative architecture, such as columns or cornices. The ornamentation of this period gives the third style its name: ornamental style. The fourth style is the last Pompeian style, which came to

Fig. 2.34 Incrustation style Casa del Fauno, August Mau (1882), from [214] Plate I

an abrupt end with the fall of Pompeii. In this style, the represented architectures are animated by figures and actions and decorated with numerous ornaments and garlands.

A wonderful example of this period is the painting of the tablinum[116] from the Casa di Apollo at Pompeii. In the tablinum is told the story of the contest between Apollo and the satyr Marsyas, in which Marsyas played on Athena's flute and Apollo on his lyre ([215], pp. 85–86). Apollo, who added song to his playing, was declared the victor. Apollo skinned the defeated Marsyas alive.

The unpleasant outcome of this competition will not be discussed further here. The painting serves here merely as an illustration of the actual use of perspective in the Roman Empire. The painting is shown as a superimposition of two images. First, a photographic image was used showing the mural in its present state. This photograph was superimposed with a drawing by Mastracchio made during the excavations at Pompeii [216]. Mastracchio's[117] drawing, in turn, served Blazeby at Kings College in London in the "Skenographia Project"[118] [218] as the basis for an analysis of the vanishing points contained in the image, with which the entire image was finally superimposed (see Fig. 2.35).

The superimposition of the mural with the perspective lines reveals the simultaneous use of several vanishing points lying on a vertical axis of symmetry (magenta in the picture).

[116]In a Roman house, a representative room adjoining the atrium and usually facing the entrance.

[117]Michele Mastracchio was a draughtsman during the excavations in Pompeii and made many drawings of the wall paintings, which represent the state of the walls shortly after the excavation.

[118]"The Skenographia Project: Investigating Roman Wall Paintings through Digital Visualisation" [217].

Fig. 2.35 Architectural style at Casa di Apollo, superimposition of photograph and analysis, Michele Mastracchio (1847), Martin Blazeby (2007), Mentnafunangann (2014), Grasnick (2019)

The white lines of the upper vanishing point refer mainly to the outer parts of the picture, which thus appear to have been taken from a greater height. The light blue lines form the authoritative central perspective of the important part of the picture in the middle. Furthermore, the red lines show an additional use of a parallel perspective. At a cursory glance, such a multi-perspective painting seems incorrect with today's knowledge; however, one should not underestimate the size of the painting. Since the painting took up the entire wall, a viewer presumably did not have a constant opportunity to grasp the entire picture, but only parts of it and thus a single perspective. Another explanation of the existence of multiple vanishing points on a vertical axis is Panofsky's herringbone thesis [219]. By using a projection circle instead of a projection surface and its subsequent "unrolling" on a flat wall, one would obtain an image with several projection centers. However, Panofsky already refers to the fact that the preserved perspective pictures "... are probably without exception not strictly constructed at all" ([219], p. 106).

Finally, the quality of Roman painting at the beginning of the Christian era will be demonstrated by an example from Herculaneum.[119] The drawing shows an excellent perspective representation of a (presumably unreal) candelabra (Fig. 2.36).

Skenographia was an important element of theatrical culture in Greek times and later became a part of Roman architecture. A replica initiated by the Bavarian King Ludwig I with numerous wall paintings is the Pompejanun in Aschaffenburg [221].

[119]Roman city on the Gulf of Naples, buried in 79 during the eruption of Mount Vesuvius.

Fig. 2.36 Herculaneum. Ornaments in colors, Candelaber, Wilhelm Zahn (1842) from [220], III. Heft. Plate 25

To Damianos we owe a description of skenographia from the point of view of late antiquity. Damianos divides optics into proper optics, catoptrics, and skenographia, and finally asks "What is skenographia?" The answer is given by Damianos himself (in [222] from p. 29):

> The skenographic part of optics asks how the views of buildings are to be drawn.... Because the task of the architect is to make his work pleasing to the eye in its proportions, and as far as possible to find remedies for the deceptions of the eye, and to strive not for the real, but for what appears to the eye to be the equality and pleasingness of the proportions.

Damianos explains that sometimes a circle should be drawn as an ellipse or a square as a stretched quadrilateral in order to be pleasing to the eye. This does not mean anything else than that the two-dimensional drawing of a three-dimensional building should correspond to how it would be seen in reality. This is the basic description of perspective.

Whether the construction of a real central perspective was already possible with this information may be doubted. In any case, it is certain that the projection of a spherical surface onto a plane surface was known at that time. Ptolemy describes this very fact in his "Planisphaerium" [223]. In the Latin edition of the Planisphaerium of 1558, the translator and editor Commandino refers to the central projection (see p. 6 of Commandino's commentary in the Planisphaerium [224]). This projection of the spherical surface onto a plane had importance in cartography as a stereographic projection.

The skenographies of Agatharchus and his contemporaries may represent the beginning of scientific perspective [225]. The writings of Euclid or Ptolemy provided geometric guidance for making correct drawings. The basic knowledge for the creation of perspective drawings seems to have been present until late antiquity, as preserved wall paintings from late Roman times and not least the books of Vitruvius prove.

With the transition from late antiquity to the early Middle Ages began a process of forgetting knowledge, which was cemented for a long time by the significant book losses of late antiquity. However, literate monks reproduced and translated the works of Vitruvius,[120] Euclid or Ptolemy and thus preserved knowledge from complete annihilation.[121] The transpositions of the perspective ideas of the ancient Greeks and Romans succeeded the monks perhaps also therefore still to some extent, as for example the wall paintings from the church of St. George in Oberzell on the island of Reichenau showed [228].

2.5 Projections from the Middle Ages

The Renaissance of Perspective

The term "Renaissance" is a loanword from the French and means "rebirth". And indeed, from the fourteenth century onwards, the rediscovery of antiquity in art and science began in Italy. Vasari used this term[122] in his biographies of famous artists "Le Vite",[123] to denote the progress and rebirth of painting, sculpture and architecture.

The form of the image had changed after the disintegration of the Roman Empire and the increasing iconization of the visual arts in the wake of the Christianization of Europe. The realistic, perspective representation had disappeared from the images and was replaced by surface painting. This is particularly evident in icon painting, where perspective not only

[120] Reginbert (librarian of the monastery of Reichenau), in a list concerning the books written during the tenure of Abbot Ruadhelm (838–842), also mentions Vitruvius: "Item Liber Vitruvii magistri de Architectura comprehensa X. libris." [226]

[121] The Bavarian Academy of Sciences and Humanities (Bayerische Akademie der Wissenschaften) has set up a project entitled "Medieval Library Catalogues of Germany and Switzerland" ("Mittelalterliche Bibliothekskataloge Deutschlands und der Schweiz") in which the writings available at that time are being catalogued as a whole [227].

[122] Ital. "rinascita".

[123] In the Ital. original from 1550 "Le vite de piu eccellenti architetti, pittori, et scultori italiani, da Cimabue insino a' tempi nostri" [229], the current German complete edition comprises 45 volumes [230].

Fig. 2.37 Troitsa (Trinity icon), Andrei Rublev (1411) now in the Tretyakov Gallery in Moscow, from The Yorck Project (2002)

simply disappears, but is sometimes even replaced by an inverted perspective.[124] In this case, the vanishing point in front of the image is on the viewer's side, which was intended to create a symbolic overemphasis of the venerable characters by means of the unrealistically emphasized "false" representation (see Fig. 2.37).

This form of representation was widespread in the Middle Ages as Byzantine painting.[125] As with philosophy and the sciences as a whole, real optical representation became the focus of interest for the scholars of the early Middle Ages. The writings of Vitruvius,

[124] see also Pawel Florenski's "The Reversed Perspective" ("Die umgekehrte Perspektive") [231].

[125] A practical guide to 12th/thirteenth century painting in the style of the icon painter Manuel Panselinos was written by the Greek monk Dionysius of Phourna on Mount Athos [232].

Euclid, and Ptolemy had made their way from antiquity to Renaissance Italy. In addition to the extant Greek and Roman works of antiquity, Arabic writings were now being translated into Latin. Around this time Alhazen's "Books of Optics" was translated from the Arabic "Kitab al-Manazir"[126] and it seems that this translation had a great influence on the scholarly opticians of the time.

Peckham's "Perspectiva communis" [234], Witelo's "Peri Optices"[127] [54] and also Bacon's "Perspectiva" [235] indicate knowledge of Alhazenz optics. In all these, subsequently influential works, perspective is geometrically analyzed in addition to optics and the process of seeing.

The Renaissance began for Vasari with Cimabue, Giotto's discoverer and teacher. Giotto, in turn, was the painter who increased realism in his paintings through the use of perspective elements. In this way, the Renaissance is linked to the rebirth of perspective.

It is not known whether Giotto, who according to legend came from a humble background, knew the ancient writings. What is undisputed is that Giotto's work rose above his predecessors and heralded a new quality in painting.

The Legend of Giotto's Childhood

The sculptor and writer Ghiberti contributed to the creation of legends about Giotto's origin with a narrative in the second book of his commentaries ([236], p. 35). There Ghiberti tells the story of Giotto's discovery (in German translation after Schwarz from [237], p. 14):

> The art of painting began its rise in Etruria: In a village near Florence called Vespignano, a boy was born full of wondrous talent. Once he drew a sheep after nature. The painter Cimabue came along the road leading to Bologna and found the child sitting on the ground drawing the sheep on a stone slab. He was astonished that he could do so well at such a tender age. When he saw that he had received his gift from the hands of nature itself, he asked him his name. He answered: 'I am called Giotto. My father's name is Bondone, and he lives in the house next door.' So Cimabue went with Giotto to the father and was well received. He requested the boy from him. And as the father was very poor, he gave him to Cimabue. The latter took him with him, and so Giotto became Cimabue's apprentice. The latter still adhered to the old Greek manner, and in it enjoyed great renown in Etruria. Giotto, however, made himself great in the art of painting.

The story points to an important aspect of Renaissance painting. The preoccupation with nature and its depiction as realistically as possible became significant again.

(continued)

[126] According to Lindberg at the end of the twelfth century/beginning of the thirteenth century (from [157], p. 209). For the sometimes expressed assumption that Gerhard of Cremona (1114–1187) was the translator (see e.g. [233], p. 35), the final proof is missing ([157], p. 210).

[127] Preserved as a manuscript from this period, kept in the Biblioteca Apostolica Vaticana [53].

A second point is the social perspective. In the Middle Ages, the possibilities for social advancement were rather slim, and the chances of each individual were shaped above all by his birth status. Giotto had made it in his life only through talent from a simple country child to a recognized, wealthy citizen who kept company with Dante[128] and was praised by Boccaccio.[129]

Giotto thus finally overcame the limitations of Byzantine painting and re-established a lifelike image with vivid colors and realistic perspective (Fig. 2.38).

Giotto's high fidelity to the depicted persons and architecture was the result of careful observation. In the paintings of the younger Lorenzettis, knowledge of the Roman Republic was also incorporated, as can be seen in the example of the fresco cycle "On Good and Bad Government[130]". From a drawing point of view, the representations of the buildings are particularly interesting. The city is depicted in perspective, although the buildings are obviously not constructed by vanishing point, but presumably painted as closely as possible to reality. In principle, this kind of painting, which appears to be perspective, but somehow does not seem to be constructed correctly, is quite common in fourteenth century painting.[131]

The transition from the late Gothic style of painting[132] to the early Renaissance was influenced, among other things, by the realistic nature painting of Pisanello or Fabriano.

---

[128]Dante mentions Giotto in the 11th canto of the Divine Comedy (Purgatory): "In painting Cimabue thought he held the field/but now it's Giotto has the cry/so that the other's fame is dimmed." (from [238], p. 183)

[129]In the Decameron (Sixth Day. Fifth Story.) Boccaccio reproduces a conversation between Giotto and Forese, and introduces the anecdote with a hymn of praise to Giotto. "... whilst the other, whose name was Giotto, had so excellent a genius that there was nothing of all which Nature, mother and mover of all things, presenteth unto us by the ceaseless revolution of the heavens, but he with pencil and pen and brush depicted it and that so closely that not like, nay, but rather the thing itself it seemed, insomuch that men'svisual sense is found to have been oftentimes deceived in things of his fashion, taking that for real which wasbut depictured." (from [239], p. 217)

[130]The fresco was commissioned by the republican government of Siena for the Palazzo Pubblico (it can still be viewed there [240]) and demonstrates the city fathers' republican worldview of Siena. The pictorial contents suggest knowledge of Roman history (at least among the patrons) (see "The fresco cycle by Ambrogio Lorenzetti on good and bad government" [241]).

[131]e.g. by Taddeo Gaddi in the fresco "Presentazione della Vergine al tempio e Sposalizio della Vergine" of the Cappella Baroncelli or by Maso di Banco (d. 1348) in the fresco "Miracolo del santo che chiude la fauci al drago e resuscita due maghi uccisi dall'alito del mostro" of the Cappella Bardi di Vernio (both pictures can be found in the Basilica di Santa Croce in Florence).

[132]The focus was neither on the realistic nor the perspective-correct depiction, but rather on the religious perspective of the motif as a meaningful element.

Fig. 2.38 The expulsion of the merchants from the temple, fresco in the Arena Chapel in Padu, Giotto di Bondone (1304–1306), Zeno.org

The Codex Vallardi[133] contains various studies illustrating Pisanello's attempt to depict plants, animals and people as faithfully as possible to nature. Fabriano's altarpiece "The Adoration of the Magi" seems almost like a transposition of Pisanello's sketches into a real painting. Animals and people are depicted exquisitely and partly in natural movement[134] (Fig. 2.39).

Dvořák called the Italian artists Masaccio, Brunelleschi and Donatello "patres of the Renaissance" (in [246], p. 45). Donatello, who as a young sculptor assisted Ghiberti in his workshop, had catapulted the style of Roman antiquity into the early fifteenth century. Donatello's sculptures appear unusually realistic; the depiction of hair or the drapery of

[133] according to Zoege v. Mateuffel (s. [242], p. 53, note 1) the codex was sold by Vallardi in 1856 erroneously as a compilation of da Vinci's (s. also Vallardi's catalog of 1855 [243]) to the Louvre and exhibited there in 1996 [244].

[134] The golden frame, which seems very realistic in the picture, is so: it is not painted, but carved, and belongs to the altar of the Strozzi family, now in the Uffizi in Florence [245].

Fig. 2.39 The Adoration of the Magi, Gentile da Fabriano (1423), Gli Uffizi Florence, from The Yorck Project (2002)

clothes[135] is groundbreaking for the time. Donatello took naturalism to the extreme. His stark-naked young David,[136] clad only in a laurel-wreathed hat, stands casually on the severed head of the old Goliath. Almost shocking is another sculpture by Donatello. Centuries after its completion, the penitent Mary Magdalene[137] still had a disturbing effect on the poet Arndt. On his travels he saw Donatello's Mary in Florence, which aroused in him a "disgusting feeling" (from [248], p. 135):

> By this same master is a penitent Mary Magdalene of bronze, a beautiful work, if you want an anatomical lazaretto, toothless and fleshless, emaciated, sheathed in long flying curls.

Brunelleschi, like Donatello, also worked as a sculptor and had also made a penitent Magdalene.[138] More important, than this activity, which in any case has survived only in

135 e. g. in the representation of John the Evangelist or of the prophet Habbakuk, called "Zuccone" (in the Museo dell'Opera del Duomo, Florence, see also [247]).

136 Today at the Palazzo del Bargello in Florence.

137 Today in the Museo dell'Opera del Duomo in Florence.

138 Which, however, was of wood, and was therefore destroyed by fire in 1471 ([249], p. 19).

fragments, is that as an architect. Brunelleschi travelled to Rome in company with Donatello ([249], p. 29) to devote himself to the study of the ancient sites. His attention was directed to ancient architecture. While studying the Roman ruins, perhaps the desire ripened in him to revive the ancient architecture that had perished. On his return to Florence, the opportunity arose: Brunelleschi was commissioned to work with Ghiberti to build the dome of the Cathedral of Santa Maria del Fiore, the Florentine cathedral. The completion of this structure alone demonstrates Brunelleschi's mastery. More astonishing is what Manetti tells us about Brunelleschi from his own experience in "Vita di Filippo Brunelleschi" [250]. In Brunelleschi's time painting by mere imitation of nature had developed an astonishing realism, and led to a better understanding of scenery by sight. As an architect and connoisseur of geometry, Brunelleschi developed a practical presentation of perspective. Since this first description represents an important step in the development of mathematical perspective, it is reproduced here in translation of Fabriczy's German explanation (from [249], p. 46).

> In order to present the result of his observations and researches as vividly and clearly as possible, Brunelleschi produced two pictures in the most meticulous colorful execution. He depicted the Baptistery with its surroundings, from the main portal of the cathedral,—and the Palazzo de' Signori together with the square, seen from the corner of the Via dei Calimara in perspective. In order to achieve the intended illusion all the more perfectly, he covered the sky in the former panel with polished silver, on which the air and clouds were reflected. Then, at the point of the eye, he drilled a hole the size of a lentil in the panel, which widened conically towards the back, and placed it in front of a mirror in such a way that its distance from it, measured with the scale he had chosen for the pictorial representation, corresponded exactly to the distance at which he had been when he took the picture of the main object (the Baptistery). By thus prescribing the proper distance for the viewing of the image reproduced therein, he was sure that the truthfulness of his representation would be brought to bear and arouse astonishment.

Fabriczy's text is admittedly not easy to digest. What is clear, however, is that Brunelleschi painted two pictures (of which only the first is considered here). The picture must have been of excellent perspective quality. On it was depicted the Baptistery of San Giovanni (the baptistery of Florence Cathedral). There was a hole drilled in the picture through which one could look. This peephole was conical in shape so that the eye could be comfortably placed. By consulting a more modern translation of Manetti's text, it becomes clear how Brunelleschi's device was supposed to work (from [251], p. 116).

> Now, in order that no error might creep into the viewer's choice of point of view, by which the appearance of the eye might be distorted, he had made a hole in the picture-panel, which in the picture was exactly at the point of the church of San Giovanni opposite to the eye of him who looked from the place within the central door of Santa Maria del Fiore, on which the painter had placed himself when he painted. And this hole was as small as a lens on the painted side, and widened conically backwards, like a lady's straw hat, to the size of a ducat or a little more. And he desired the eye to lie against the back where the hole was large, through which he was to look, and he was to shield the eye with one hand, and with the other he was to hold a flat mirror to the other side, in which the image was to be reflected. The distance at which the mirror was stretched out with the hand was proportionally about the real distance from which

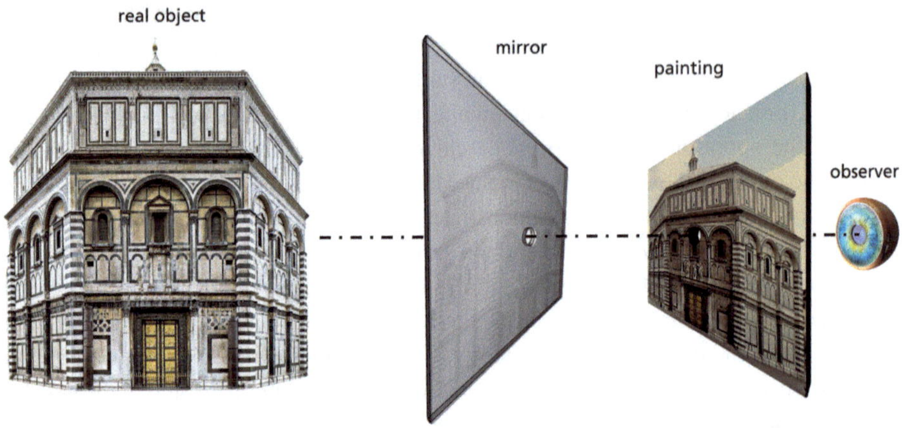

Fig. 2.40 Brunelleschi's experiment, Grasnick (2019)

> he had painted the church of San Giovanni. So that when one looked at it, it seemed—together with the circumstances already mentioned (the polished silver, the representation of the square, the accurate drawing, etc.)—as if one were seeing reality itself. And I have had it in my hand, and have seen it many times at that time, and can bear witness to it.

According to Manetti (see [251], p. 116), the picture panel had a size of half a "braccio"[139] in the square, which corresponds to about half an arm's length or the height of an DIN A4 sheet. The image could be held as convenient with one hand in front of the eye, while a mirror was held in front of the image with the other hand. Through the peepholes in the mirror and the painting, a viewer could simultaneously look at the real object and the reflection of the painting (see Fig. 2.40). If the viewer stood at the position in which Brunelleschi had made the painting, the overlap between reality and image was almost perfect, as Manetti testifies.

Since the pictorial plates have not survived, this early form of "Mixed Reality" was only transmitted in text form. It is not known how Brunelleschi made the paintings and whether he constructed his drawings or merely copied them precisely from nature. In any case, Brunelleschi is credited by his biographers Manetti and Vasari with the ability to determine the perspective representation of a building from its ground plan and elevation—which presupposes at least the geometric knowledge of how to generate an oblique image from the two-panel projection.

The first documented description of the construction of a perspective was made by Alberti in 1435, during Brunelleschi's lifetime. Alberti was a friend of Brunelleschi and Donatello and dedicated his three books on painting to the "admirable genius" Brunelleschi [253]. The first book is devoted to the nature of painting from a mathematical point of view.

[139] A "braccio" was the Florentine cubit of 58.36 centimeters (twice that of a Roman foot, from [249], p. 80, footnote 1) or (according to [252]) p. 581 55.06 cm (for canvas) and 58.21 cm (for wool).

The treatise begins with the definition of a point as the smallest representable sign. Several points join to form a line, and several lines, woven into one another as in a linen fabric,[140] make a surface. Then Alberti speaks to the surfaces that, by changing the boundary lines (hem), change not only the shape but also the name: "... what was called a triangle will now be called a quadrilateral or polygon." (from [253], p. 54). These surfaces do not necessarily have to be flat, but may well be spherical or concave and even composed of different surfaces. He calls the shape of the surface the "ridge" and can thus compose any object from different surfaces defined by the properties "ridge" and "border". Alberti recognizes that the perception of the shape of a surface is related to the "sight", that is, the shape or color of the surface altered by the change of the observer's position.

Alberti divides the colors according to the elements and thus comes up with four: red for fire, blue for air, green for water and grey for earth. By adding black or white and mixing the colors, one can produce all colors and create any gradations. This is amazingly well observed and in principle still valid today. A current screen produces all the colors of the image from the primary colors red, green and blue alone, plus the respective black value (the brightness).

When Alberti then mentions the visual rays of the visual pyramid, one immediately feels reminded of Euclid. And indeed, Alberti's library had contained a Latin translation of Euclid (after [254], p. 69). In general, Alberti then divides the whole Euclidean cone of vision into smaller pyramids of vision and establishes a fascinating idea (from [253], pp. 69–70):

> If this is so, as I have said, then in painting one sees nothing but a certain "cross-section of a pyramid. Painting, then, will be nothing other than the artistic representation, brought about on a surface by means of lines and colors, of a cross-section of the pyramid of vision according to a certain distance, a certain point of the eye, and a certain illumination.

Alberti subsequently describes how he creates a picture frame on the canvas and describes it with the metaphor of the open window. In Alberti's work, this "finestra aperta" becomes a virtual image of the real environment, the picture becomes a window.

Alberti does not provide any pictures in his instructions, thus leaving a little room for interpretation. In the 1804 edition [255], a sketch is provided (see Fig. 2.41).

On the basis of historical observation, Panofsky comes to the interpretation that Alberti's instruction is probably a "Costruzione legittima" [256] and provides a sketch (Fig. 2.42).

In della Francesca's "Prospectiva pingendi" [257] the rough description of Alberti's perspective construction becomes an exceedingly detailed and richly illustrated working instruction for the painter interested in perspective. And this can really be seen. Francesca first shows his understanding of perspective on simple bodies. Piece by piece he develops more and more complicated structures to finally be able to represent perspective projections

[140] Fabric made from the fibers of flax (linseed).

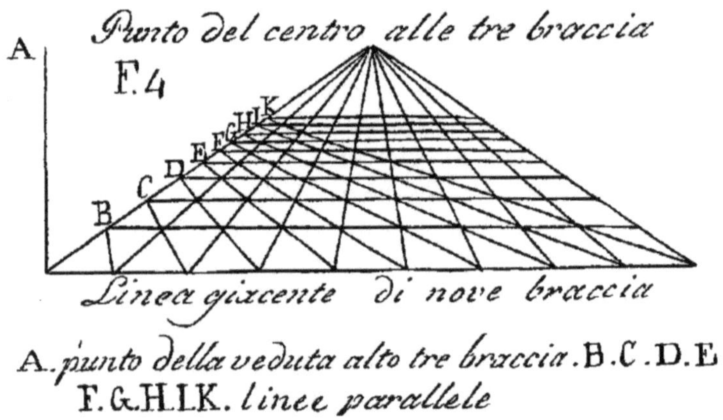

Fig. 2.41 Distance point construction of perspective, Ferrario Giusti (1804) from [255] Appendix, Fig. 4

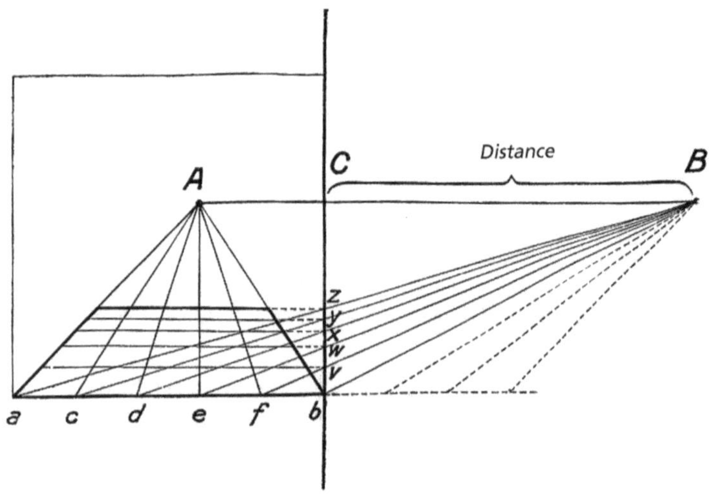

Fig. 2.42 Alberti's Costruzione legittima, Erwin Panofsky (1915) from [256], p. 514, Fig. V

also for the human body. From a precisely constructed human head in an extended three-panel projection (with additional viewing from below), a perspective view from another (arbitrary) viewing position is created by means of a split horizontal and vertical projection (see Fig. 2.43). This is remarkable. Even more than 500 years ago, 3D reconstructions were possible from a small number of 2D projections.

Della Francesca divides his geometrical perspective into five divisions (after [259] XI):

1. the sight (the eye)
2. the shape of the object seen

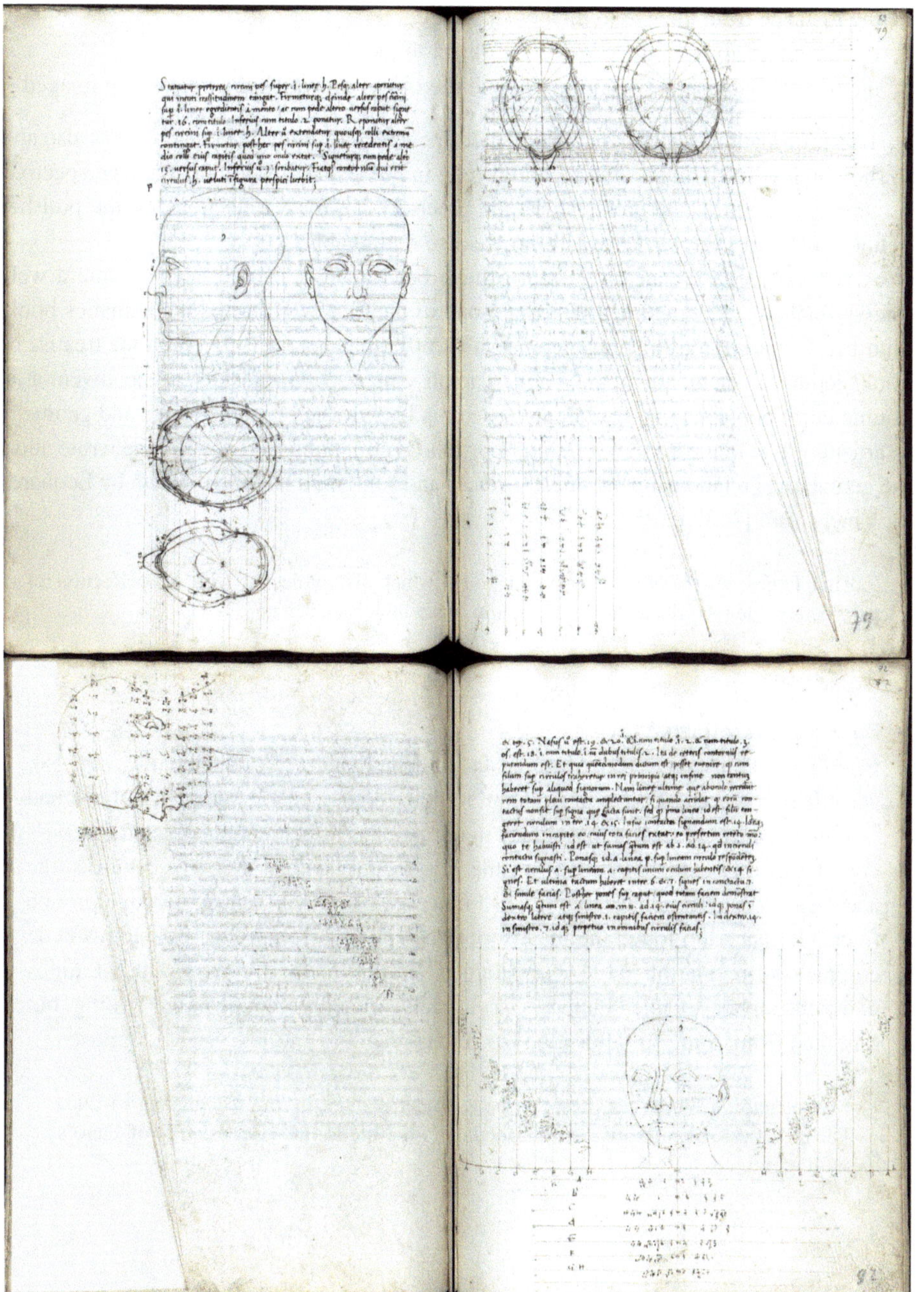

Fig. 2.43 3D reconstruction from perspectives, Piero della Francesca (1460), from [258] in manuscript 616, Bibliothèque de Bordeaux, fol. 74v, 79r, 80v, 82r

3. the distance from the eye to the object seen
4. the lines that go from the edge of the object to the eye
5. the boundary plane between the eye and the object on which the objects are imaged

This information is sufficient to construct a perspective, as della Francesca was also able to show impressively in his own paintings (e.g. in Fig. 2.44). In addition to the perspective-accurate depiction of the architecture, the precisely observed reflections on the polished helmet of the kneeling Duke of Urbino are also remarkable.

A pupil of della Francesca is said to have been Pacioli. Pacioli later became a well-known mathematician and had published one of the first ever printed mathematics books with his "Summa de arithmetica geometria" [260] published in 1494. With the treatise on bookkeeping[141] included in the book, Pacioli can also be considered the inventor of double-entry bookkeeping. Pacioli also used his knowledge of mathematics and geometry in artistic ways. In the book "Divina proportione" [262] published in 1509, he wrote about the geometrical relationships of lines, surfaces and solids and had them drawn by Leonardo da Vinci (from [263], p. 282):

> ...That figure will be that of the said body, which [is] projected with all perfection of perspective upon the plane, as our Leonardo Vinci understands it.

Background Information

Pacioli is not only said to have knowledge of della Francesca's mathematics, but also, since Vasari's accusation of plagiarism in della Francesca's biography, to have had an unfortunate tendency to inaccuracy in quotation. The accusation targeted Pacioli's copying of a Francesca treatise "on the five regular solids." For five centuries the accusation could not be cleared up, as Francesca's original manuscript could not be found. Jordan, who found the manuscript "De quinque corporibus regularibus et de corporibus irregularibus" [264] catalogued in 1880 in the Codex Urbinus Latinus under the author's name "Pictor[142]", originally wanted to republish the writing, but refrained from doing so (after [265], p. 115):

> My intention to publish it, however, was rendered moot by my recognizing in Luca Pacioli's treatise on the five regular bodies[143] the literal Italian translation of Piero's writing in Latin.

[141] Tractatus XI. Particularis de computis et scripturis, also in "Treatise on Accounting" [261].

[142] After the Latin name of Pietro della Francesca: Petrus Pictor Burgensis.

[143] With incipit "Libellus in tres partiales tractatus divisus quinque corporum regularium" as a supplement to the "Divina proportione" of 1509 (from Tractatus Primus in [262]).

Fig. 2.44 Pala Montefeltro (La Vergine con il Bambino e santi), Piero della Francesca, (c. 1470), Pinacoteca di Brera, Milan, photograph from "The Yorck Project"

The task of the perspective representation of the bodies mathematically described by Pacioli is masterly solved by da Vinci. Pacioli, like della Francesca before him, had given the explanation "why there cannot be more than five regular bodies" (from [263], p. 213 ff.). These are the Platonic solids, which, according to the philosopher Plato, are solids of the greatest possible symmetry. The symmetry of the polyhedron is defined by the shape of similar faces which together form the surface of these regular solids. There can in fact be only five Platonic solids. The names of the solids are always based on the number of faces:

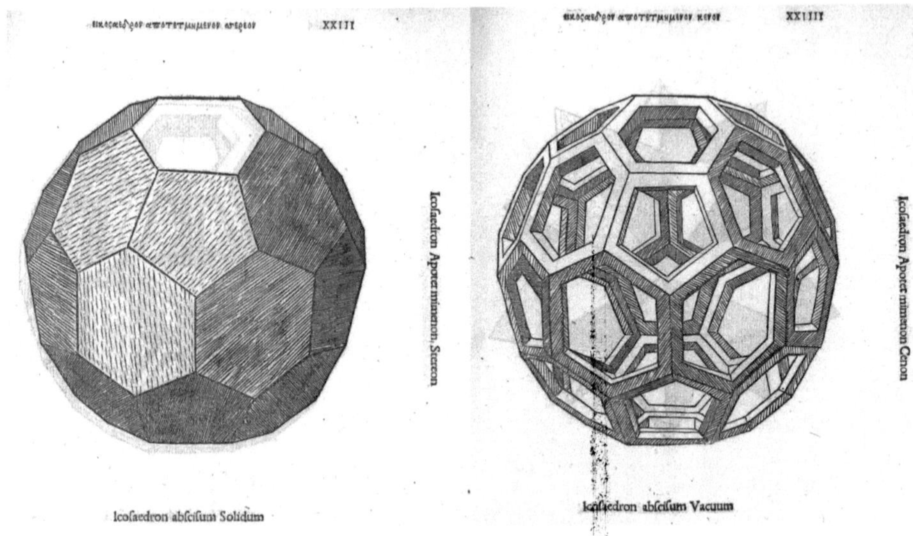

Fig. 2.45 Da Vinci's icosahedron, Leonardo da Vinci (1509) from [262] from XXIII and XXIIII

Tetrahedron (4 faces, pyramid of triangles), Hexahedron (6 faces, cube of squares), Octahedron (8 faces, double pyramid of triangles), Dodecahedron (12 faces of pentagons) and Icosahedron (20 faces of triangles).

In addition to the Platonic solids, da Vinci also drew some Archimedean solids. Archimedean solids are also made of different polygons, but they must meet at the corners in the same way. Da Vinci's truncated icosahedron corresponds to today's football, which consists of 12 pentagons and 20 hexagons. To illustrate da Vinci's understanding of perspective, an openwork "football" icosahedron is also shown here (Fig. 2.45).

In fact, the da Vinci drawings in Pacioli's writing are an extremely rare record of a publication authorized by Leonardo himself. In addition to his important pictorial work, his many scientific and engineering works have survived only through his numerous sketches and notebooks. It is due to his pupil and heir Melzi that some of da Vinci's thoughts on painting were summarized in the writing "Trattato della pittura" [266]. In addition to basic instructions on drawing, proportion, anatomy, ponderation,[144] light, shadow and reflection, the collection also contains teachings on the design of perspective including the aerial perspective[145] used by da Vinci. A suitable example of da Vinci's instruction seems to be the instruction on drawing a figure from the perspective of a frog (Fig. 2.46).

[144]Composition of paintings with regard to optimal representation of weight distribution (realistic representation of posture and movement).

[145]Da Vinci also uses the term "sfumato," [266] p. 105.

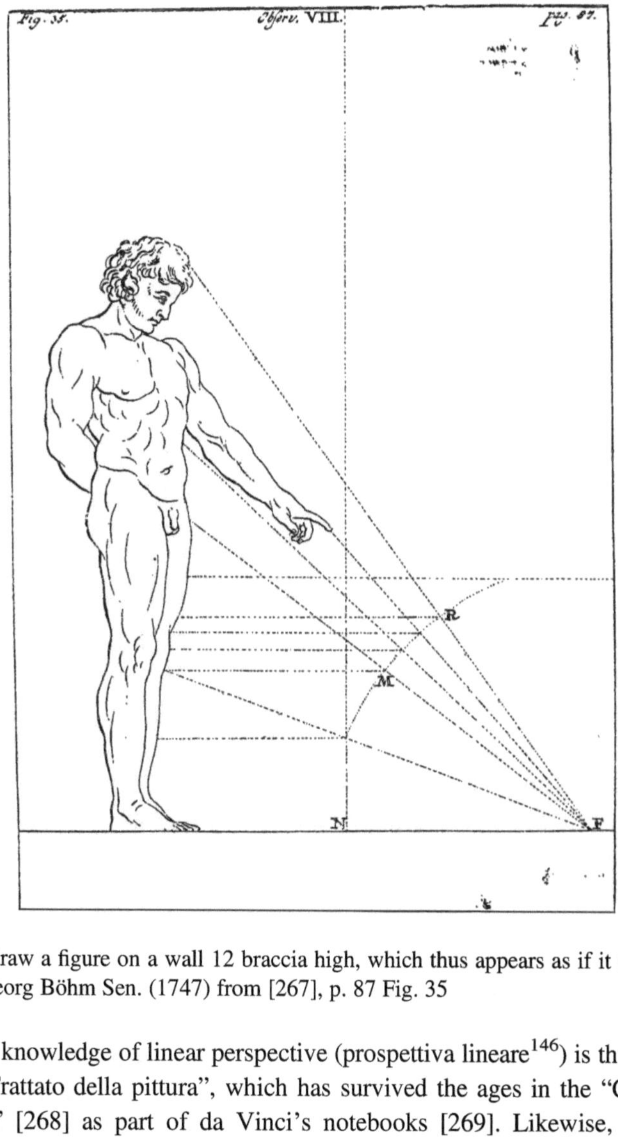

Fig. 2.46 To draw a figure on a wall 12 braccia high, which thus appears as if it stood 24 braccia high. Johann Georg Böhm Sen. (1747) from [267], p. 87 Fig. 35

Da Vinci's knowledge of linear perspective (prospettiva lineare[146]) is thus vouched for by Melzi's "Trattato della pittura", which has survived the ages in the "Codex Urbinus Latinus 1270" [268] as part of da Vinci's notebooks [269]. Likewise, knowledge of Vitruvius' ten books can be attested by the drawing of the Vitruvian Man according to Vitruvius' instruction.[147] Da Vinci succeeded so well in depicting the proportions of a

[146]Ludwig, however, uses the term "prospettiua lineale," [268] p. 218.

[147]"The center of the body further is by nature the navel. For if a man is laid on his back with his hands and feet stretched out, and the centre of the compass is placed in his navel, when the line of a circle is described, the fingers and toes of the two hands and feet are touched by the line. Just as the figure of a circle is represented on the body, so that of a square is found on it. For if you measure from

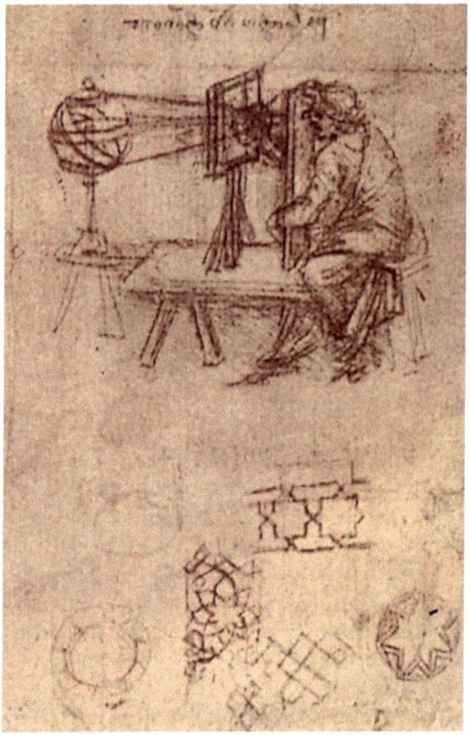

Fig. 2.47 Prospettografo, Leonardo da Vinci (ca. 1480) from Codex Atlanticus Fol. 5 r, Biblioteca Ambrosiana Milan

human body from Vitruvius' text that the drawing is still a popular subject today and is used exactly as da Vinci had drawn it—for example, on the Italian euro coin. Da Vinci had also sketched in the Codex Atlantico the 'prospettografo'(perspectograph) (Fig. 2.47), which is used to draw a more complicated object in perspective.

Although the picture is certainly one of his weaker works, it clearly shows the intended function. The draughtsman looks through a real eye point through a transparent drawing surface (presumably a coarser fabric) onto the object and can thus transfer it to the grid of the fabric.

In his time, however, da Vinci was better known as an artist and could hardly serve as a theorist of perspective to those who did not know him personally. At the same time as da Vinci was in Italy, the painter Dürer was living in Germany. Dürer traveled to Venice and studied the Italian masters there, who greatly influenced him. Dürer's painterly realism was at least as exceptional as his mathematical skills. Dürer knew the Latin edition of the "Elements of Euclid" and wrote the "Underweysung der Messung mit dem Zirckel und

the bottom of the feet to the top of the head, and transfer this measure to the hands stretched out, you will find the same breadth as height, as is the case with surfaces made square according to the angle measure." (from [201], pp. 131–133)

Fig. 2.48 The draughtsman of the seated man, Albrecht Dürer (1525) [270]

richtscheyt" (Instruction of measurement with the compass and straightedge) [270], the first German mathematics book.

Whether Dürer had seen or heard of da Vinci's "perspectograph" is not known. Certainly, however, a simpler device for recording perspective can be designated by this name to this day, which functions even more accurately in English as a "perspectograph" [271]. Such aids have been used since the Middle Ages to copy a scene correctly in perspective. Albrecht Dürer also sketched such a device in the woodcut "The Draughtsman of the Sitting Man" ("Der Zeichner des sitzenden Mannes") (Fig. 2.48).

Here a scene is viewed through a grid or transparent medium from a fixed point. In effect, this is a central projection from the master's eye point onto the drawing surface.

Today, the German word "Perspektograph" refers to a somewhat different device, albeit with a similar function. In 1883, the Berlin professor Hauck had presented a theory with which any third projection could be determined from two projections and proposed a mechanical apparatus for this purpose. The device was first called a "tricolograph" [272, 273] and was later marketed in a modified version as a perspectograph (also known as a perspektograph) by the Frankfurt architect Ritter [274, 275] (Fig. 2.49).

It is unquestionable that after da Vinci and Dürer, a period of natural realism and realistic perspective also began outside Italy and Germany. Without going into detail about every artist, only a few of the great Baroque painters are mentioned here. First and foremost the painter of light, color and the senses Rubens, Vermeer with his lively genre

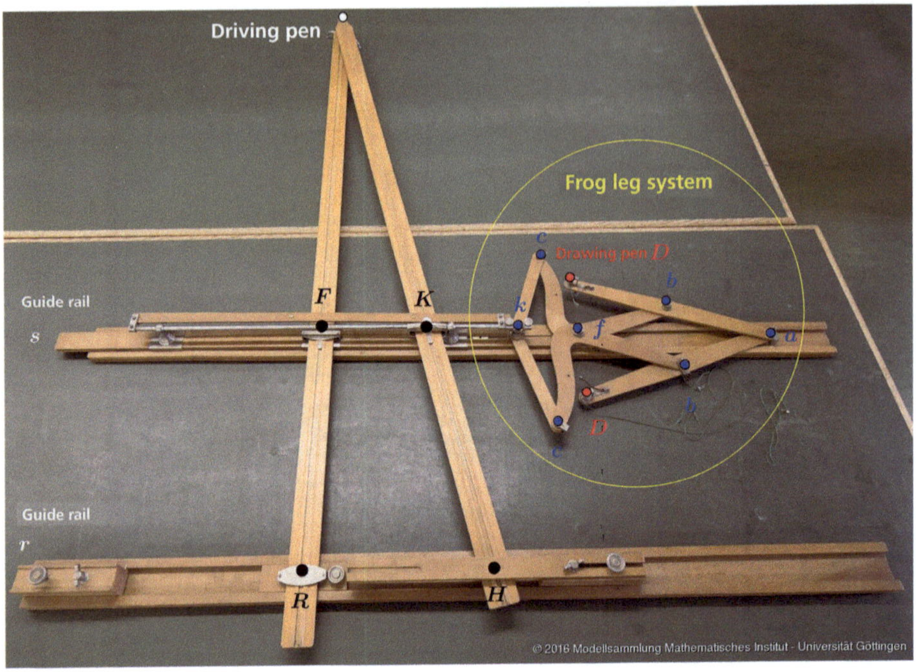

Fig. 2.49 Perspectograph by Ritter, model 834, Göttingen Collection of Mathematical Models and Instruments, courtesy of Prof. Bartholdi

scenes and finally the legendary portrait and landscape painter Rembrandt. These and many other artists have impressively demonstrated that realistic perspective images of real and virtual objects can be produced.

2.6 Multiple Perspectives

A Picture from Different Angles

Even a two-dimensional image can contain multiple perspectives. Some images contain elements that reveal themselves to their viewer only when observed from the correct perspective. These are often images with highly reshaped forms that can only be seen under extreme angles or a specially curved mirror. An early example of such anamorphoses[148] is Holbein's painting "The Ambassadors." In the picture (Fig. 2.50, left), an object is depicted in the foreground, which appears in the picture like an unknown foreign body and actually only seems to disturb the composition of the picture. Only an

[148]From Greek. Anamorphosis = transformation.

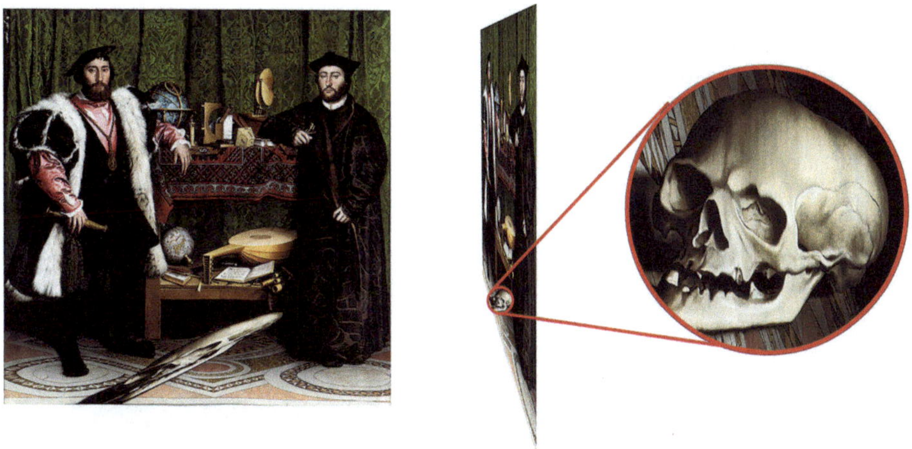

Fig. 2.50 The Ambassadors, Hans Holbein the Younger (1533), The National Gallery, London, Google Art Project, rectification Grasnick (2019)

extreme oblique view from the upper right reveals what Holbein had intended. In fact, the painter had depicted a skull, which only becomes perceptible as such in an unusual perspective (Fig. 2.50, centre and right, enlarged).

The depiction of such objects did not fundamentally serve the pure entertainment of the viewer, but quite particularly also the accentuation of the artist's abilities. Holbein had delivered with "The Ambassador" a strong illustration of his understanding of perspective contraction, a description of the way to the distorted result, however, not.

The second perspective does not necessarily have to be part of the same picture. The Italian painter Buti had already produced a painting in the sixteenth century depicting Charles III, the Duke of Lorraine and his daughter Christine. In this case, however, the background of the painting is not flat, but consists of prismatic strips. The Duke is painted on one side of the prism and Christine on the other. Seen from below, the duke appears, seen from above, his daughter (in Fig. 2.51 in the mirror above the painting).

The architect da Vignola was not only a user of perspective, but also a practitioner. In his "Rules of Practical Perspective" ("Le due regole di prospettiva pratica") [276] there are many manageable descriptions and small aids for creating usable perspective pictures. In it, Vignola now also describes the production of such trick pictures (see Fig. 2.52).

Such pictures were subsequently made not only in Italy, but were also known in Shakespeare's England [277].

In addition to the influences of antiquity and the insights of realistic perspective painting, the Renaissance of science and art was also strongly influenced by religious representation and biblical motifs. This was of course first and foremost due to the exclusive access of the clergy to the preserved knowledge in the libraries and monasteries, but of course also due to the wealth of the church, which was often also a client for the

Fig. 2.51 Gioco ottico, Ludovico Buti (1593), Museo Galileo Firenze—Istituto e Museo di Storia della Scienza, Sala I, Inv. 3197, courtesy of the Museo Galileo Florence

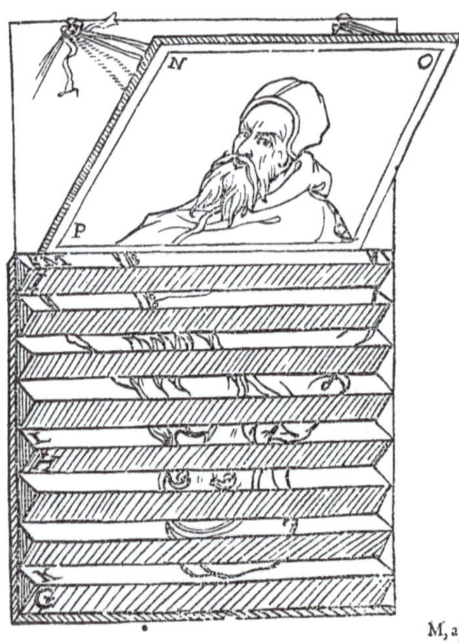

Fig. 2.52 Facetted image, Giacomo da Vignola (1583) from [276], p. 95

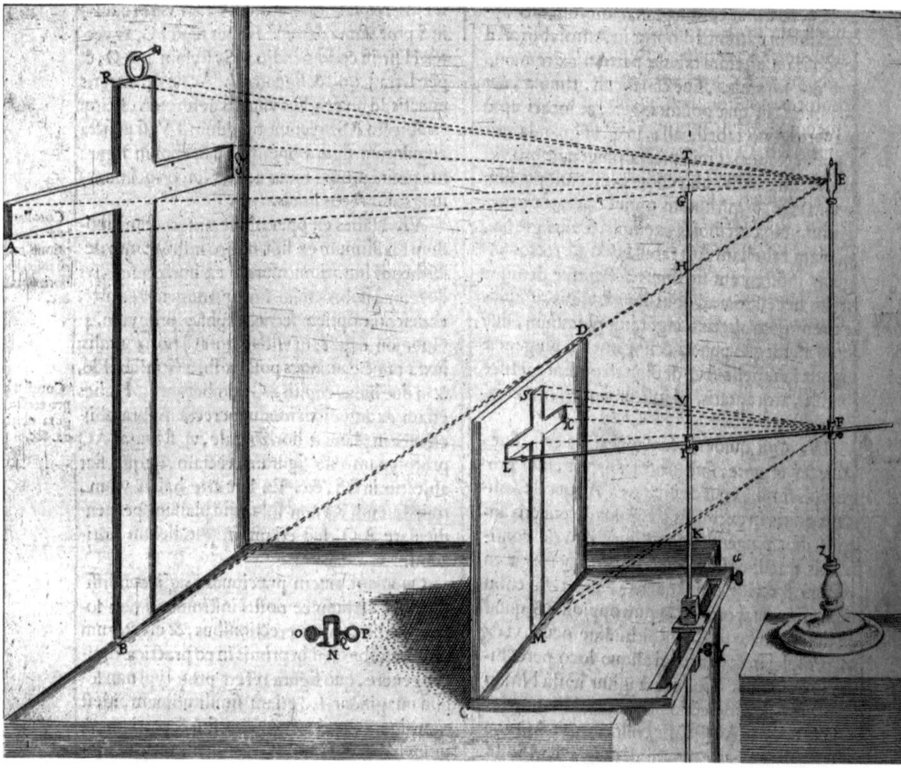

Fig. 2.53 Constructio instrumenti scenographici, Mario Bettini (1642) from [278] Apiar V p. 36

artists. In addition, the clergy were often also scholars, read and spoke Latin and Greek, and sometimes traveled widely.

It is therefore not surprising that all kinds of priests and monks were also concerned with the mathematics of representation.

The Jesuit Bettini collected numerous mathematical curiosities in the "Hive of the whole philosophy and mathematics" (Apiaria Universae Philosophiae Mathematicae) and described, among other things, the practical determination of distances, angles and heights with Jacob's staff. A large part of the book is devoted to geometry. However, it actually becomes exciting from the moment Bettini explains scientific projection. Bettini also describes a perspectograph, but he calls it a skenographic construction element, which describes it quite well. With this device, a view or an object can be transferred into another perspective (see Fig. 2.53).

The rod closer to the image is adjusted in height so that its upper end forms a line with the eye point and the observation point. The horizontal rod then determines the height in the image. Bettini designs other aids to projection with which he also achieves the desired result, namely the true-to-scale reproduction of an original (see Fig. 2.54).

Fig. 2.54 Instrumenti scenographici parallelo plano, Mario Bettini (1642) from [278] Apiar V p. 52

The knowledge of the projection facilitated Bettini the geometrical explanation of the anamorphosis. And finally, he also provides instructions on how to create an anamorphosis of Holbein's kind. What Bettini does here is first to divide the undistorted original image (as it is meant to be seen) into a grid. Each grid point is projected from point E onto the surface on which the image is then painted (Fig. 2.55). If you look carefully, you can already see the basic features of computer graphics illustrated here.

The knowledge of the projection on almost arbitrary surfaces (and even curved mirrors) and the necessary pre-distortion of the images for a certain perspective was not only present with Bettini. The Jesuit Schott, with the chapter "De magia anamorphotica" of his "Magia

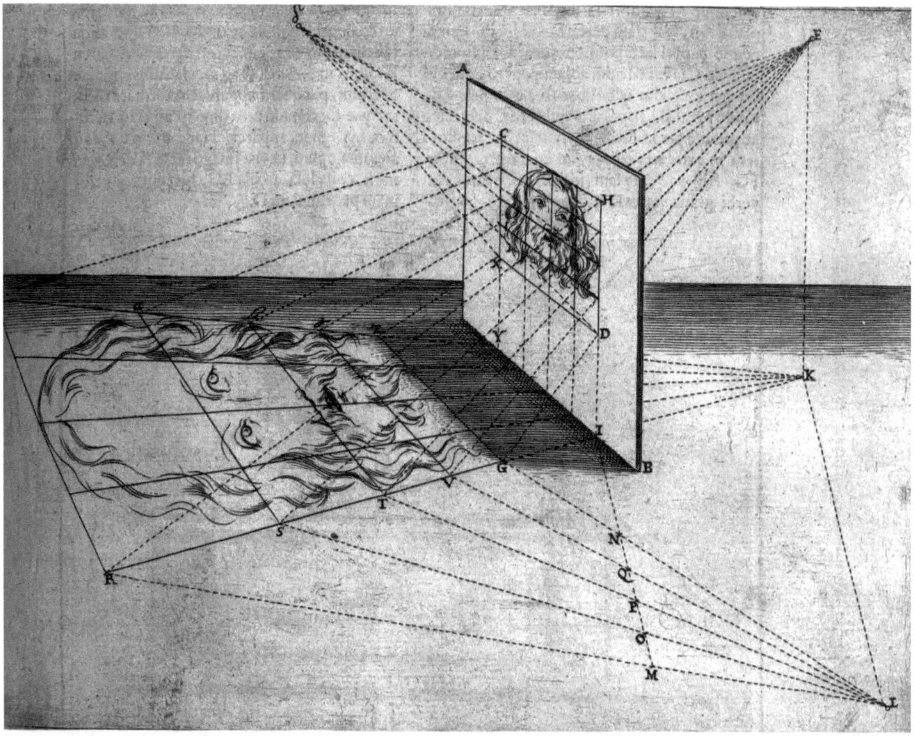

Fig. 2.55 De deformationibus è superficiebus planis, Mario Bettini (1642) from [278] Apiar V p. 25

universalis naturae et artis", had also included instructions for the production of anamorphoses [279]. This has already been discussed frequently[149] and will therefore not be the subject of investigation again here. However, both Bettini in the "Apiaria universae" and Schott in the "Magia anamorphotica" describe a method that leads to an image with two equivalent perspectives. The method, later called two-way paintings (in German "Riefelbilder"—groove paintings), basically requires the prisms (grooves) for the image montage (see Fig. 2.56 left), as well as the two images that are applied to the grooves (in frontal view with both images in Fig. 2.56 right). Due to the higher quality of the drawings, Bettini's design is shown here.

The lines starting from a point L in Fig. 2.56 on the left also show well the shading of the image contents by the prisms when viewed from point L. Today this would be called vignetting.

Such pictures were of course not only described by learned Jesuits, but actually also painted by artists. The two-way painting of Pastor Pauli and his wife from Brekum [281][150]

[149] e.g. in [280], p. 102.

[150] see in [280], p. 100.

Fig. 2.56 Bettini's two-way painting, Mario Bettini (1642) from [278] Apiar V p. 29

Fig. 2.57 King Christian V and Queen Charlotta Amalia of Denmark and Norway; two portraits on triangular facets, Gaspar Antoine de Bois-Clair (1692), Old Masters Picture Gallery, courtesy of the Museumslandschaft Hessen Kassel, Wilhelmshöhe Castle

was made by an unknown artist. There exist, however, other works besides, whose painter is quite well known. The painter de Bois-Clair had painted a double portrait of the royal couple Christian V. and Charlotta Amalia of Denmark and Norway on the triangular facets sketched by Bettini and Schott. This painting is now in the collections of the Gemäldegalerie Alte Meister (Old Masters Picture Gallery) of the Museumslandschaft Hessen Kassel (Fig. 2.57).

Fig. 2.58 Souvenir icon; lamellar image, Ident No. N (32 S) 1/1994,8301, Berlin State Museums, photograph Ute Franz-Scarciglia, by courtesy of bpk Bildagentur, Museum Europäischer Kulturen SMB

In Fig. 2.57, the different perspectives can be seen quite well. If one moves in front of the picture from left to right, first King Christian appears, then a mixed picture of the royal couple is created and finally, when looking from the right side, the queen reveals herself.

Another double portrait of the Danish King Frederik IV, the son of Christian V and Charlotta Amalia, and Frederik's wife Louise zu Mecklenburg-Güstrow was made by Bois-Clair, which was offered by the New York gallery "Robert Simon Fine Art" some years ago and has meanwhile presumably disappeared in a private collection.

However, this at least shows that such pictures were made from time to time, especially if the representation of a couple was connected with it, which is well described by the term "double portrait".

Viewing a two-way painting from the central position directly in front of the image necessarily leads to a deterioration of the image impression. It is therefore obvious to assign a third image to this viewing position. Three perspectives at the same time can be represented in a so-called lamella or harp image. The terms "trisceneorama" or "three-way-paintings", which are sometimes used, are smoothly descriptive of the situation. Each picture is composed of three individual images (scenes or perspectives). Such pictures were therefore excellently suited for the religious representation of the Trinity, the Holy Trinity or Trinity of Father, Son and Holy Spirit [282] and revealed the essential unity of the three divine hypostases[151] to the observer moving in front of the picture. A religious lamellar image with Madonna is shown in Fig. 2.58.

[151] Hypostasis—personification of divine attributes or religious ideas into a distinct divine being, e.g., in Christian theology, the three persons of the Trinity.

The mechanical construction is somewhat more difficult to produce compared to a grooved image. The narrowest possible lamellae with a small base area must be attached vertically to a flat substrate and the lateral perspective images must also be mounted on the side surfaces.

As a rule, the perspective views of the alternating image do not correspond to the real central projections of a scene from different positions. The term perspective is not always of a purely geometric nature, but sometimes also describes the observation of a certain thing. For example, the previously mentioned alternating image representation of the divine Trinity is essentially a philosophical perspective.

It all depends on the point of view. The German bureaucracy uses terms such as "community of need" or "fiscal unit" to describe life partnerships or units of interest. These units consist of single individuals (or legal entities) and can be considered individually or jointly, depending on the perspective chosen.

Literature

1. Darwin C. Über die Entstehung der Arten im Thier- und Pflanzenreich durch natürliche Züchtung. Stuttgart: E. Schweizerbart'sche Verlagsbuchhandlung; 1863.
2. Sedgwick A. On the classification and nomenclature of the Lower Palaeozoic rocks of England and Wales. Q J Geol Soc. 1852;8:136–68.
3. Sedgwick A, Murchison RI. On the Silurian and Cambrian systems, exhibiting the order in which the older sedimentary strata succeed each other in England and Wales. Report of the Fifth Meeting of the British Association for the Advancement of Science; Held in Dublin in 1835; Notices of Communications; 1836. pp. 59–61.
4. Sprigg RC. Early Cambrian (?) jellyfishes from the flinders ranges, South Australia. Trans R Soc S Aust. 1947;71:212–24.
5. Xiao S, Shen B, Zhou C, Xie G, Yuan X. A uniquely preserved Ediacaran fossil with direct evidence for a quilted bodyplan. Proc Natl Acad Sci. 2005;102:10227–32.
6. Gould SJ. Zufall Mensch: das Wunder des Lebens als Spiel der Natur. München: Hanser; 1991.
7. Nash M. Evolution's big bang: when life exploded. Time Mag. 1995;146(23):66–74.
8. Holmes B. Evolution's big bang: how life on Earth took off. New Sci. 2015;(Ausg. 3010);38–41.
9. Herloßsohn C, editor. Damen Conversationslexikon. Adorf: Verlags-Bureau; 1835.
10. Ehrenberg CG. Organisation, Systematik und geographisches Verhältniß der Infusionsthierchen. Berlin: Königliche Akademie der Wissenschaften; 1830.
11. Ehrenberg C. Die Infusionsthierchen als vollkommene Organismen: ein Blick in das tiefere organische Leben der Natur: nebst einem Atlas von vierundsechszig colorirten Kupfertafeln. Leipzig: Leopold Voss; 1838.
12. Grell KG. Protozoologie. 2. Berlin: Springer; 1968.
13. Iseki M, Matsunaga S, Murakami A, Ohno K, Shiga K, Yoshida K, et al. A blue-light-activated adenylyl cyclase mediates photoavoidance in Euglena gracilis. Nature. 2002;415:1047–51.
14. Garm A, Coates MM, Gad R, Seymour J, Nilsson D-E. The lens eyes of the box jellyfish Tripedalia cystophora and Chiropsalmus sp. are slow and color-blind. J Comp Physiol A. 2007;193:547–57.

15. Berger EW, Conant FS. Physiology and histology of the Cubomedusae, including Dr. F.-S. Conant's notes on the physiology. Baltimore: The Johns Hopkins Press; 1900.
16. Garm A, Oskarsson M, Nilsson D-E. Box jellyfish use terrestrial visual cues for navigation. Curr Biol. 2011;21:798–803.
17. Haeckel E. Berg- und Seefahrten 1857/1883. Leipzig: R. F. Koehler; 1923.
18. Haeckel E. Das System der Siphonophoren auf phylogenetischer Grundlage entworfen. Jenaische Z Naturwissenschaft. 1888;22:1–46.
19. Haeckel E. Kunstformen der Natur. Leipzig: Bibliographisches Institut; 1899.
20. Young GA, Hagadorn JW. The fossil record of cnidarian medusae. Palaeoworld. 2010;19:212–21.
21. Kutschera U. Evolutionsbiologie: Ursprung und Stammesentwicklung der Organismen. 4., vollständig überarbeitete und erweiterte Aufl. Stuttgart: Verlag Eugen Ulmer; 2015.
22. Linné C. Systema naturae per regna tria naturae, secundum classes, ordines, genera, species, cum characteribus, differentiis, synonymis, locis. Stockholm: Laurentius Salvius; 1766.
23. Herrn LC. Carl von Linnés Reisen durch Oeland und Gothland 1741. Halle: Johann Jacob Curt; 1764.
24. Woltersdorff JL. Systema minerale in quo regni mineralis producta omnia systematice per classes, ordines, genera et species proponuntur. Mineral-System worin alle zum Mineral-Reich gehörige Körper in ordentlichem Zusammenhange nach ihren Classen, Ordnungen, Geschlechtern und Arten vorgetragen werden. Berlin: Christian Ludewig Kunst; 1748.
25. Walch JEI. Die Naturgeschichte der Versteinerungen zur Erläuterung der Knorrischen Sammlung von Merkwürdigkeiten der Natur. Nürnberg: Paul Jonathan Felßecker. 1771;
26. Clarkson E, Levi-Setti R, Horváth G. The eyes of trilobites: the oldest preserved visual system. Arthropod Struct Dev. 2006;35:247–59.
27. Portlock JE. Report on the geology of the county of Londonderry, and of parts of Tyrone and Fermanagh: examined and described under the authority of the master general and board of ordnance. Dublin: Andrew Milliken; 1843.
28. Clarke JM. The structure and development of the visual area in the trilobite, Phacops Rana. Green J Morphol. 1888;2:253–70.
29. Jell PA. The abathochroal eye of Pagetia, a new type of trilobite eye. In: Martinsson A, editor. Evolution and morphology of the Trilobita, Trilobitoidea and Merostomata: proceedings of a NATO Advanced Study Institute. Oslo: Univ.-Forl; 1975.
30. Kepler J. Strena seu de Nive Sexangula. In: Dornau C, editor. Amphitheatrum Sapientiae Socraticae Joco-seriae. Hannover: Schleichius; 1619. p. 751–7.
31. Lagrange J-L, d'Arithmetique R. Nouveaux Memoires de l'Academie Royale des Sciences et Belles-Lettres. Berlin: Frederic Voss; 1775. p. 265–312.
32. Gauß CF. Recension der "Untersuchungen über die Eigenschaften der positiven ternären quadratischen Formen von Ludwig August Seeber, Dr. der Philosophie, ordentl. Professor an der Universität in Freiburg. 1831. 248 S. in 4." J für die reine und angewandte Mathematik (Crelles J). 1840;1840:312–20.
33. Hsiang W-Y. On the sphere packing problem and the proof of Kepler's conjecture. Int J Math. 1993;04:739–831.
34. Hales TC. An overview of the Kepler conjecture. arXiv:math/9811071 [Internet]. 1998. http://arxiv.org/abs/math/9811071. Accessed on 18. Aug. 2018.
35. Hales T. A proof of the Kepler conjecture. Ann Math. 2005;162:1065–185.
36. Hales T, Adams M, Bauer G, Dang TD, Harrison J, Hoang LT, et al. A formal proof of the Kepler conjecture. Forum of Mathematics, Pi [Internet]. 2017; 5. https://www.cambridge.org/core/product/identifier/S2050508617000014/type/journal_article. Accessed on 18. Aug. 2018.

37. v. Frisch K. Der Farbensinn und Formensinn der Biene. In: Spengel JW, editor. Zoologische Jahrbücher, Abteilung für allgemeine Zoologie und Physiologie der Tiere. Jena: Gustav Fischer; 1915. p. 1–179.
38. Schroeder TBH, Houghtaling J, Wilts BD, Mayer M. It's not a bug, it's a feature: functional materials in insects. Adv Mater. 2018;30:1705322.
39. Stange G, Howard J. An ocellar dorsal light response in a dragonfly. J Exp Biol. 1979;83:351–5.
40. Brückner A, Duparré J, Dannberg P, Bräuer A, Tünnermann A. Artificial neural superposition eye. Opt Express. 2007;15:11922.
41. Medathati NVK, Neumann H, Masson GS, Kornprobst P. Bio-inspired computer vision: towards a synergistic approach of artificial and biological vision. Comput Vis Image Underst. 2016;150:1–30.
42. Cristóbal G, Perrinet L, Keil MS, editors. Biologically inspired computer vision: fundamentals and applications. Weinheim: Wiley-VCH Verlag GmbH & Co. KGaA; 2016.
43. Exner S. Die Physiologie der facettirten Augen von Krebsen und Insekten. Leipzig: Franz Deuticke; 1891.
44. Mesnik B. Detection, recognition, identification: the new criteria [Internet]. Kintronics. 2017; https://kintronics.com/detection-recognition-identification-new-criteria/. Accessed on 19. Aug. 2018
45. Nilsson D-E, Pelger S. A pessimistic estimate of the time required for an eye to evolve. Proc R Soc B Biol Sci. 1994;256:53–8.
46. Gaarder J. Sofies Welt. Roman über die Geschichte der Philosophie. München: Hanser; 1993.
47. v. Goethe JW. Goethes Werke – Gedichte Dritter Theil. Weimar: Hermann Böhlau; 1890.
48. Schlange-Schöningen H. Die römische Gesellschaft bei Galen: Biographie und Sozialgeschichte. Berlin: De Gruyter; 2003.
49. Galen. Sieben Bücher Anatomie des Galen. Leipzig: J C Hinrichssche Buchhandlung. 1906;
50. Hunain Ibn-Isḥāq. De oculis. In: Africanus C, Editor. Omnia opera. Lyon: Andrea Turini; 1515.
51. Hunain Ibn-Isḥāq. The book of the ten treatises on the eye ascribed to Ḥunain ibn Isḥâq (809–877 A. D.). Kairo: Government Press; 1928.
52. Alhazen, Witelo. Opticae thesaurus. Risner F, Editor. Basel: Episcopios; 1572.
53. Witelo. Witelonis Perspectiva. Biblioteca Apostolica Vaticana: Manuscript; 1201.
54. Witelo. Vitellionis mathematici doctissimi περὶ ὀπτικῆς, id est de natura, ratione et proiectione radiorum visus, luminum, colorum atque formarum, quam vulgo Perspectivam vocant, libri X. Tannstetter G, Apian P, Editor. Nürnberg: Johannes Petreius; 1535.
55. Vesalius A. De humani corporis fabrica libri septem. Baseline. 1543;
56. Stolberg M. Eine anatomische Inszenierung. Felix Platter (1536–1614) und das Skelett der Frau. In: Schwarte L, Lazardzig J, Editor. Spuren der Avantgarde: Theatrum anatomicum. Berlin: De Gruyter; 2011.
57. Platter F. De corporis humani structura et usu Felicis Plateri libri III: tabulis methodicè explicati, iconibus accuratè illustrati. Basel: Froben; 1583.
58. Koelbing HM. Felix Platter (1536–1614) als Augenarzt. 1990. https://www.e-periodica.ch/digbib/view?pid=ges-001:1990:47::446. Accessed on 26. Aug. 2018.
59. Platter F. Praxeos Seu de cognoscendis, praedicendis, praecavendis, curandisq[ue] affectibus homini incommodantibus Tractatus. Basel: Conrad Waldkirch; 1609.
60. Scheiner C. Oculus hoc est fundamentum opticum. Innsbruck: Daniel Agricola. 1619;
61. Scheiner C. Rosa Ursina sive Sol. Bracciani [Bracciano]: pud Andream Phaeum Typographum Ducalem; 1630.
62. Huygens C. Opuscula postuma, quae continent dioptricam. Leiden: Cornelis Boutesteyn; 1703.
63. Muth HL. Anatomia Oculi Humani, Una Cum Structura Artificiali. Oder: Zergliederung Des Menschlichen Auges, Wie auch Eine genaue Darstellung aller desselbigen Theile woselbsten

alles accurat vorgebildet, und durch Kunst verfertiget, bestehend aus Horn, Helffenbein, Crystall und Glas. Kassel: Henr. Harmes; 1730.
64. Listing JB. Beitrag zur physiologischen Optik. Göttinger Studien Göttingen: Vandenhoeck und Ruprecht. 1845:52–110.
65. v. Helmholtz H. Handbuch der physiologischen Optik. Leipzig: Leopold Voss; 1867.
66. Schmidt RF, Lang F, Heckmann M. Editor. Physiologie des Menschen: mit Pathophysiologie: mit Online-Repetitorium. Sonderausgabe der 31. Aufl. Berlin: Springer; 2017.
67. v. Helmholtz H. Handbuch der physiologischen Optik. Gullstrand A, v. Kries J, Nagel W, Editor. Hamburg: Leopold Voss; 1909.
68. Lachenmayr B, Friedburg D, Buser A. Auge – Brille – Refraktion: Schober-Kurs: verstehen – lernen – anwenden. 5., vollständig überarbeitete Aufl. Stuttgart: Georg Thieme; 2016.
69. Haigis W. Zur Herkunft der Hornhautdaten des Gullstrand-Auges. Klin Monatsbl Augenheilkd. 2010;227:69–74.
70. Lotmar W. Theoretical eye model with aspherics*. J Opt Soc Am. 1971;61:1522.
71. Kooijman AC. Light distribution on the retina of a wide-angle theoretical eye. J Opt Soc Am. 1983;73:1544.
72. Liou HL, Brennan NA. Anatomically accurate, finite model eye for optical modeling. J Opt Soc Am A Opt Image Sci Vis. 1997;14:1684–95.
73. Navarro R. Adaptive model of the aging emmetropic eye and its changes with accommodation. J Vis. 2014;14(21):1–17.
74. Atchison DA, Thibos LN. Optical models of the human eye: optical models of the human eye. Clin Exp Optom. 2016;99:99–106.
75. Celsus AC. Aulus Cornelius Celsus über die Arzneiwissenschaft in 8 Büchern. 2nd. ed. Braunschweig: Friedrich Vieweg und Sohn; 1906.
76. Celsus AC. A. Cornelii Celsi de medicinae. Daremberg C, Editor. Leipzig: Teubner; 1859.
77. Celsus AC. De medicina libri VIII. Florenz: Manuscript. 1465;
78. von Amida A. Aetii Medici Graeci Contractae ex veteribus medicinae sermones XVI. Venedig: Johannes Gryphius. 1553;
79. Richter IA, Wells T, Kemp M. Leonardo da Vinci Notebooks. New York: Oxford University Press; 2008.
80. Clark K. The drawings of Leonardo Da Vinci in the collection of her majesty the queen at Windsor Castle; anatomical manuscripts A, B, C. 2. Aufl. Edinburgh: R. & R. Clark; 1935.
81. Conard NJ. A female figurine from the basal Aurignacian of Hohle Fels Cave in southwestern Germany. Nature. 2009;459:248–52.
82. Conard NJ, Kind C-J, Kretschmann W, Wolf C, Engler B. Als der Mensch die Kunst erfand: Eiszeithöhlen der Schwäbischen Alb. Darmstadt: Theiss; 2017.
83. UNESCO World Heritage Centre. UNESCO World Heritage Centre – Decision – 41 COM 8B.24 [Internet]. UNESCO World Heritage Centre. 2017. http://whc.unesco.org/en/decisions/6896. Accessed on 6. Jan. 2018.
84. Urgeschichtliches Museum Blaubeuren|Venus [Internet]. https://www.urmu.de/de/Forschung%2BArch%C3%A4ologie/Eiszeitkunst/Venus. Accessed on 6. Jan. 2018.
85. Heidenreich SM. Virtuelle Archäologie in Baden-Württemberg. Von der wissenschaftlichen Dokumentation zur öffentlichen Web-Präsentation von 3-D-Modellen archäologischer Denkmale. Denkmalpflege in Baden-Württemberg. 2014;43:261–4.
86. Landesamt für Denkmalpflege Baden-Württemberg. Iceageart: 3D [Internet]. High-Tech trifft Eiszeit. 2018. https://www.iceageart.de/3d/. Accessed on 6. Jan. 2018.
87. Antl-Weiser W. Venus-forschung [Internet]. Naturhistorisches Museum Wien. 2017; http://www.nhm-wien.ac.at/forschung/praehistorie/forschungen/venus-forschung. Accessed on 6. Jan. 2018

88. Nigst PR, Haesaerts P, Damblon F, Frank-Fellner C, Mallol C, Viola B, et al. Early modern human settlement of Europe north of the Alps occurred 43,500 years ago in a cold steppe-type environment. Proc Natl Acad Sci. 2014;111:14394–9.
89. Teschler-Nicola M. Willendorf – die anthropologischen Quellen. Mitteilungen der Anthropologischen Ges in Wien. 2008;138:89–100.
90. Scan The World. 3D Printable Venus of Willendorf at The Naturhistorisches Museum, Vienna [Internet]. MyMiniFactory. 2015. https://www.myminifactory.com/object/venus-of-willendorf-at-the-naturhistorisches-museum-vienna-austria-11455. Accessed on 6. Jan. 2018.
91. Snow D. Sexual dimorphism in European Upper Paleolithic cave art. Am Antiq. 2013;78:746–61.
92. Pettitt P, Castillejo AM, Arias P, Peredo RO, Harrison R. New views on old hands: the context of stencils in El Castillo and La Garma caves (Cantabria, Spain). Antiquity. 2014;88:47–63.
93. Leroi-Gourhan A. Gesture and Speech. Cambrige: MIT Press; 1993.
94. Clottes J. Kunst im Morgenlicht der Menschheit. Spektrum der Wissenschaft. 2003;
95. v. Helmholtz H. Populäre wissenschaftliche Vorträge. Braunschweig: Friedrich Vieweg und Sohn; 1871.
96. Platon. Platons Werke. 2. Aufl. Berlin: Realschulbuchhandlung; 1817.
97. Platon. Timaeus und Kritias. Breslau: Georg Philipp Aderholz; 1841.
98. Ebers G, editor. Das Hermetische Buch über die Arzeneimittel der alten Ägypter in hieratischer Schrift – Einleitung und Text. Leipzig: Wilhelm Engelmann; 1875.
99. Ebers G, editor. Das Hermetische Buch über die Arzeneimittel der alten Ägypter in hieratischer Schrift – Glossar und Text. Leipzig: Wilhelm Engelmann; 1875.
100. Alexandrinus C. Clementis Alexandrini: Stromatum Liber Quintus. In: Potter J, editor. Clementis Alexandrini Opera quae extant. Venedig: Antonius Zatta; 1757.
101. Hirschberg J. Geschichte der Augenheilkunde. In: Seamisch T, Graefe A, Editor. Handbuch der gesamten Augenheilkunde. 2. Aufl. Leipzig: W. Engelmann; 1899.
102. Plinius GS. Arzneimittel vom Menschen, vom Wasser und von den Thieren. Leipzig: Gressner & Schramm; 1882.
103. Paullin KF. Heilsame Dreck-Apotheke. Friedrich Knochen: Frankfurt; 1697.
104. Lewis W, Rotheram J, Lavoisier AL. Neues Edinburger Dispensatorium (The Edinburgh new dispensatory). Black J, Editor. Leipzig: Gerhard Fleischer der Jüngere; 1793.
105. Hornung E. Der ägyptische Mythos von der Himmelskuh. 3. Aufl. Göttingen: Vandenhoeck & Ruprecht; 1982.
106. Maul SM, Editor. Das Gilgamesch-Epos. 7. Aufl. München: Beck; 2017.
107. Kratz RG, Spieckermann H, editors. Das erste Buch Mose Genesis: die Urgeschichte Gen 1–11. Göttingen: Vandenhoeck & Ruprecht; 2018.
108. Aristoteles. Die Metaphysik des Aristoteles. Berlin: L. Heimann; 1871.
109. Homer. Homers Ilias. 2. Aufl. Königsberg: Friedrich Nicolovius; 1802.
110. Diels H. Die Fragmente der Vorsokratiker. 2. Aufl. Berlin: Weidmannsche Buchhandlung; 1906.
111. Nietzsche F. Friedrich Nietzsche Gesammelte Werke. München: Musarion; 1921.
112. Prantl C, editor. Aristoteles. Vier Bücher über das Himmelsgebäude und Zwei Bücher über Entstehen und Vergehen. Leipzig: Wilhelm Engelmann; 1857.
113. Huygens C. Traite de la lumiere. Leiden: Pierre van der Aa. 1690;
114. Descartes R. Discours de la methode pour bien conduire sa raison, & chercher la verité dans les sciences: plus la dioptrique et les meteores. In: qui sont des essais de cette methode. Paris: Theodore Girard; 1657.
115. Descartes R. Entwurf der Methode: mit der Dioptrik, den Meteoren und der Geometrie. Hamburg: Felix Meiner; 2013.

116. Huygens C. Abhandlung über das Licht. Lommel E, Editor. Leipzig: Wilhelm Engelmann; 1890.
117. Newton I. Opticks: or, a treatise of the reflexions, refractions, inflexions and colours of light. 4. Aufl. London: William Innys; 1730.
118. Hermann A. Einstein auf der Salzburger Naturforscherversammlung 1909. Phys J. 1969;25: 433–6.
119. Einstein A. Über die Entwicklung unserer Anschauungen über das Wesen und die Konstitution der Strahlung. Phys Z. 1909;10(22):817–25.
120. Einstein A. Äther und Relativitätstheorie: Rede gehalten am 5. Mai 1920 an der Reichs-Universität zu Leiden von Albert Einstein. In: Janssen M, Schulmann R, Illy J, Lehner C, Buchwald DK, Editor. The Berlin years: writings, 1918.–1921. Berlin: Julius Springer; 1920.
121. Einstein A. Über den Äther. 1924. https://www.e-periodica.ch/digbib/view?pid=sng-005:1924:105::623. Accessed on 5. Oct. 2018.
122. Bohr N. Discussion with Einstein on epistemological problems in atomic physics. In: Schilpp PA, editor. Albert Einstein, philosopher-scientist. Evanston: Library of Living Philosophers; 1949.
123. Einstein A, Podolsky B, Rosen N. Can quantum-mechanical description of physical reality be considered complete? Phys Rev. 1935;47:777–80.
124. Nobel Media AB. The nobel prize in physics 1954 [Internet]. Nobelprice.org. 2018. https://www.nobelprize.org/prizes/physics/1954/summary/. Accessed on 5. Oct. 2018.
125. Planck M. Zur Theorie des Gesetzes der Energieverteilung im Normalspectrum. In: König A, editor. Verhandlungen der Deutschen Physikalischen Gesellschaft im Jahre 1900. Leipzig: Johann Ambrosius Barth; 1900. p. 237–45.
126. Nestle W, editor. Die Vorsokratiker in Auswahl übersetzt und herausgegeben. Eugen Diederichs: Jena; 1908.
127. Mansfield DF, Wildberger NJ. Plimpton 322 is Babylonian exact sexagesimal trigonometry. Hist Math. 2017;44:395–419.
128. Freudenthal J. Alexander Polyhistor und die von ihm erhaltenen Reste jüdischer und samaritanischer Geschichtswerkce. Jahresbericht des jüdisch-theologischen Seminars Fraenkel'scher Stiftung Breslau: Grass, Barth und Comp. 1874;
129. Apelt O, editor. Diogenes Laertius: Leben und Meinungen berühmter Philosophen; Buch VII–X. Leipzig: Felix Meiner Verlag; 1921.
130. Kittler FA, Berz P, Strauss J, Weibel P, Scharbert G. Götter und Schriften rund ums Mittelmeer. 2017.
131. Phillipson L. Hylē anthrōpinē: Pars I. De internarum humani corporis partium cognitione Aristotelis cum Platonis sententiis comparata. Pars II. Philosophorum veterum usque ad Theophrastum doctrina de sensu. Theophrasti de sensu et sensilibus fragmentum … Aristotelis Doctrina de sensibus. Theophrasti fragmenta de sensu, phantasia et intellectu e Prisciani metaphrasi primum excerpta. Berlin: A. List; 1831.
132. Magnus H. Die Augenheilkunde der Alten. Breslau: J U Kern's (Max Müller). 1901;
133. Wendt A. H. Ritter, Geschichte der Philosophie 1.-3- Theil (2. Artikel). Jahrbücher für wissenschaftliche Kritik 1832;45:345–352.
134. Baylens P. Peter Baylens Philosophisches Wörterbuch. Jakob LH, Editor. Johann Gottfried Ruff: Halle; 1797.
135. Ritter H. Geschichte der Philosophie. 2. Aufl. Hamburg: Friedrich Perthes; 1836.
136. Haas AE. Antike Lichttheorien. In: Stein L, editor. Archiv für Philosophie. Berlin: Georg Reimer; 1907. p. 345–86.
137. Diels H. Die Fragmente der Vorsokratiker. 4. Aufl. Berlin: Weidmannsche Buchhandlung; 1922.

138. Bischoff M. Interview: Das Universum als zellurärer Automat. Spektrum – Die Woche. Ausgabe 47;2018:20–23.
139. Theophrast. Pars II. Philosophorum veterum usque ad Theophrastum doctrina de sensu. In: Phillipson L, Editor. Hylē anthrōpinē Scripsit et edidit Ludovicus Philippson. Berlin: J. A. List; 1831.
140. Marx K. Über die Differenz der demokritischen und epikureischen Naturphilosophie. Ergänzungsband: Schriften bis 1844. Berlin: Dietz; 1968. p. 264.
141. Lucretius CT. De Rerum Natura. Berlin: Weidmannsche Buchhandlung; 1923.
142. von Goethe JW. Goethe's saemmtliche Werke. Paris: Tetot Freres; 1836.
143. Lucretius CT. Von der Natur der Dinge. 2. Aufl. Leipzig: Georg Joachim Göchen; 1831.
144. Xenophon. Isokrates Panegyrikos. Xenofons Gastmahl. Xenofons sokratische Gespräche. Wien: Anton Dell; 1813.
145. Xenophon. Oikonomikos oder Xenophon vom Haus-Wesen. Hamburg: König und Richter; 1734.
146. Platon. Apologie des Sokrates und Kriton. Leipzig: Felix Meiner; 1919.
147. Platon. Platons Dialog Phaidon. Leipzig: Felix Meiner; 1913.
148. Platon. Der Staat. Schleiermacher F, Editor. Berlin: G. Reimer; 1828.
149. Aristoteles. Aristoteles über die Seele. Aus dem Griechischen übersetzt und mit Anmerkungen begleitet. Frankfurt: Johann Herrl; 1794.
150. Aristoteles. Aristotelis Stagiritae Parua quae vocant naturalia. De sensu & sensili. De memoria & reminiscentia. De somno & vigilia. De Insomnijs. De diuinatione per somnia. De animalium motione. De animalium incessu. De extensione & breuitate vitae. De iuuentute & senectute, morte & vita, & de spiratione. Omnia in latinum conuersa, & antiquorum more explicata a Nicolao Leonico Thomaeo. Eiusdem opuscula nuper in lucem edita. Item eiusdem dialogi, quotquot extant. de Colines S, Editor. Paris: Simonem Colinaeum; 1530.
151. Koelbing HM. Zur Sehtheorie im Altertum: Alkmeon und Aristoteles. Sauerländer. 1968;
152. Euklid. Euklid's Elemente. Fünfzehn Bücher aus dem Griechischen übersetzt. 5. Aufl. Mollweide CB, Editor. Halle: Buchhandlung des Waisenhauses; 1824.
153. Meschkowski H. Denkweisen großer Mathematiker: Ein Weg zur Geschichte der Mathematik [Internet]. Wiesbaden: Vieweg + Teubner; 1990. http://nbn-resolving.de/urn:nbn:de:1111-2012041134. Accessed on 16. Febr. 2019.
154. Euklid. Euclidis Optica & Catroptrica è Greco. Paris: Andreas Wecheli; 1557.
155. Euklid. Euclides Latinus. Ps.-Robertus Grosseteste. Alfraganus. Geber. Thüringer Universitäts- und Landesbibliothek Jena: Manuscript; 14. Jhdt.
156. Ṭūsī N al-DM ibn M. Arabic versions of seven Greek treatises on mathematics. British Library: Oriental Manuscripts: Manuscript; 1198.
157. Lindberg DC. Theories of vision. From Al-Kindi to Kepler. Chicago: The University of Chicago Press; 1976.
158. Euklid. Euclidis Optica, Opticorum Recensio Theonis, Catoptrica. In: Heiberg JL, Editor. Euclidis opera omnia. Leipzig: B. G. Teubner; 1895.
159. Eco U. Die geheimnisvolle Flamme der Königin Loana: illustrierter Roman. Ungekürzte Ausg München: dtv. 2006;
160. Lobatschewski NI. Zwei Geometrische Abhandlungen. Leipzig: B. G. Teubner; 1898.
161. Reichardt H. Gauß und die Anfänge der nicht-euklidischen Geometrie. 1. Aufl. Leipzig: Teubner; 1985.
162. Gauss CF. Allgemeine Flächentheorie (Disquisitiones generales circa superficies curvas) 1827. Leipzig: Wilhelm Engelmann; 1889.
163. Riemann B. Über die Hypothesen, welche der Geometrie zu Grunde liegen. Berlin: Springer; 1919.

164. Cranz C. Gemeinverständliches über die sogenannte vierte Dimension. Stuttgart: Verlagsanstalt und Druckerei A.G. (cvormals J. F. Richter); 1890.
165. Minkowski H. Raum und Zeit. (Vortrag gehalten auf der 80. Naturforscher-Versammlung zu Köln am 21. Sept. 1908). In: Hilbert D, Editor. Gesammelte Abhandlungen von Herbert Minkowski. Leipzig: B. G. Teubner; 1911. pp. 431–444.
166. Archimedes, Sturm JC. Des unvergleichlichen Archimedis Sand-Rechnung/oder tiefsinnige Erfindung einer, mit verwunderlicher Leichtigkeit aussprechlichen, Zahl, welche er unfehlbar beweiset grösser zu seyn als die Anzahl aller Sandkörnlein, mit welchen die Höhle der ganzen Welt könnte ausgefüllet werden. Nürnberg: Paul Fürstens; 1667.
167. Ptolemäus C. Ptolemy's theory of visual perception: an english translation of the optics. Philadelphia: The American Philosophical Society; 1996.
168. Ptolemäus C. Liber de optica. Staatsbibliothek zu Berlin: Manuscript; 14. Jhdt.
169. Ptolemäus C, Delambre J-BJ. Die Optik des Ptolemäus, verglichen mit der Euclid's Alhazen's und Vitellio's, von Delambre. In: Gilbert LW, editor. Annalen der Physik. Leipzig: Johann Ambrosius Barth; 1812. p. 371–88.
170. Ptolemäus C. L'Ottica Di Claudio Tolomeo. Govi G, Editor. Turin: G. B. Paravia; 1885.
171. Cicero MT. Cicero's philosophische Schriften. In deutschen Uebertragungen. Klotz R, Editor. Leipzig: Carl Focke; 1840.
172. Ingenkamp HG. Zur stoischen Lehre vom Sehen. Rhein Mus Philol (NF). 1971;114:240–6.
173. Haeckel E. Die Welträthsel. Gemeinverständliche Studien über monistische Philosophie. Bonn: Emil Strauss; 1899.
174. de Lavoisier AL, de Laplace P-S. Expériences sur la respiration des animaux: mémoire sur la chaleur, altérations qu'éprouve l'air respiré.
175. Galen. On the Doctrines of Hippocrates and Plato. – De placitis Hippocratis et Platonis. Akademie. 2. Aufl. Berlin; 2005.
176. Alhazen. MXI. (Cod. 201 Gol.) Commentarius ad Ibno-'l-Haithami. Catalogus codicum orientalium bibliothecae Academiae Lugduno-Batavae. Leiden: E J Brill; 1865. pp. 61–62.
177. Wiedemann E. Ueber "Die Darlegung der Abhandlung über das Licht" von Ibn al Haitam. In: Annalen der Physik und Chemie. Leipzig: Johann Ambrosius Barth; 1883. p. 337–45.
178. Ibn-al-Haiṯam al-Ḥasan I-Ḥasan. Abhandlungen zur Optik 1246.
179. Baarmann J. Abhandlung über das Licht von Ibn al-Haitam. Z der Dtsch Morgenl Ges. 1882;36 (2):195–237.
180. Fehrenbach F. Der oszillierende Blick. Sfumato und die Optik des späten Leonardo. Z Kunstgesch 2002;65:522–544.
181. Herzfeld M. Leonardo da Vinci: Der Denker, Forscher und Poet. 2. Aufl. Jena: Eugen Diederichs; 1906.
182. da Vinci L. Pariser Manuskript D [Internet]. Paris: Manuscript; 1508.
183. Isaacson W. Leonardo da Vinci: the biography. 2017.
184. Noest AJ, Kemp M. No refraction in Leonardo's orb. Nature. 2011;480:457.
185. Kepler J. Ad Vitellionem Paralipomena: Astronomiae pars Optica Traditur. Frankfurt a. M: Claud Marne & Johannes Aubers; 1604.
186. Kepler J, Hammer F [Hg. /Red.], Dyck W von [Hg. /Red.], Caspar M [Hg. /Red.]. Gesammelte Werke. Astronomiae pars optica. München; 1939.
187. Kepler J. Dioptrice seu demonstratio eorum quae visui & visibilibus propter Conspicilla ita pridem inventa accidunt. Augsburg: Davidis Franci; 1611.
188. Kepler J. Dioptrik oder Schilderung der Folgen, die sich aus der unlängst gemachten Erfindung der Fernrohre für das Sehen und die sichtbaren Gegenstände ergeben. Leipzig: Wilhelm Engelmann; 1904.

189. Descartes R, v. Kirchmann JH. René Descartes' philosophische Werke. Erste Abtheilung. Lebensbeschreibung. Abhandlung über die Methode. Berlin: L. Heimann; 1870.
190. Newton I. Sir Isaac Newton's Optik oder Abhandlung über Spiegelungen, Brechungen, Beugungen und Farben des Lichts. Leipzig: Wilhelm Engelmann; 1898.
191. Le Stierlin H. Buste de Néfertiti: Une imposture de l'égyptologie? Gollion: Infolio; 2009.
192. Borchardt L. Porträts der Königin Nofret-Ete. Leipzig: J. C. Hinrichssche Buchhandlung; 1923.
193. Illerhaus B, Staude A, Meinel D. Das zweite Gesicht der Nofretete. ZfP-Zeitung. 2009;
194. Schelper T. Vom Original zum Malmodell: Die Bemalung der neuen Replik der Nofretete-Büste der Gipsformerei, Staatliche Museen zu Berlin. Berliner Beiträge zur Archäometrie, Kunsttechnologie und Konservierungswissenschaft. Berlin: Rathgen-Forschungslabor, Staatliche Museen zu Berlin – Preußischer Kulturbesitz; 2016.
195. Merz M. Das neue Abbild der Schönen: 3D-Scantechniken ermöglichen objektive Reproduktion der Nofretete-Büste. Restauro. 2009;115:154–5.
196. Bürja A. Der mathematische Maler oder gründliche Anweisung zur Perspektive. Berlin: Christian Gottfried Schöne; 1795.
197. Wilde E. Geschichte der Optik, vom Ursprunge dieser Wissenschaft bis auf die gegenwärtige Zeit. Berlin: Rücker und Püchler; 1838.
198. Wilde E. Ueber die Optik der Griechen. In: Köpke GGS, editor. Ladung zu der öffentlichen Prüfung der Zöglinge des Berlinischen Gymnasiums zum Grauen Kloster. Berlin: Dieterleische Buchhandlung; 1832.
199. President's Challenge. MythBusters. USA: Discovery Channel; 2010.
200. Plutarch. Biografien des Plutarchs. Berlin: George Jacob Decker; 1777.
201. Vitruv, Reber F, Schneider G. Zehn Bücher über Architektur: De architectura libri decem. 3. Aufl. Reber 2015. Rom/Wiesbaden: Matrixverlag; um 20 v. Chr.
202. Aristoteles. The Poetics of Aristotle. 3. Aufl. London: MacMillan and Co.; 1902.
203. Dörpfeld W, Reisch E. Das griechische Theater: Beiträge zur Geschichte des Dionysos-Theaters in Athen und anderer griechischer Theater. Athen: Barth & von Hirst; 1896.
204. Stangl G. Antike Populationen in Zahlen. Frankfurt a. M.: Lang; 2008.
205. Cicero MT. Von dem Redner. Drei Gespräche Altona: Johann Friedrich Hammerich; 1801.
206. Lukian von Samosata. Lucian's Werke. Tafel GLF, Osiander CN, Schwab G, Editor Stuttgart: J B Metzlersche Buchhandlung; 1830.
207. Polybios von Megalopolis. Geschichte des Polybius. Berlin: Georg Ludewig Winter; 1762.
208. Klein HA. Science of measurement: a historical survey. New York: Dover Publications; 2012.
209. Plinius GS. Die Naturgeschichte des Cajus Plinius Secundus. Leipzig: Gressner & Schramm; 1882.
210. Zöllner F. Sandro Botticelli als Apelles und als Grisaillemaler. Petronius' "Monocremos" als Inspiration für die Selbstreferentialität der Malerei? In: Lehmann C, Gramaccini N, Rößler J, Dittelbach T, Editor. Chiaroscuro als ästhetisches Prinzip Kunst und Theorie des Helldunkels 1300–1550. Berlin: De Gruyter; 2018. pp. 91–104, 400.
211. von Sandrart J. Teutsche Academie der Bau-Bildhauer- und Maler-Kunst, Dritter Theil. Volkmann JJ, Editor. Nürnberg: Johann Anreas Endter; 1774.
212. Schürer M. Die Enzyklopädie der berühmten Männer und Frauen: Domenico Bandini, sein Fons memorabilium universi und die kompilatorische Biographik der Renaissance. Tübingen: Mohr Siebeck; 2017.
213. Böttinger CA. Ideen zur Archäologie der Malerei. Erster Theil. Dresden: Waltersche Hofhandlung; 1811.
214. Mau A. Geschichte der decorativen Wandmalerei in Pompeji. Berlin: G. Reimer; 1882.
215. Mau A. Führer durch Pompeji. 2. Aufl. Leipzig: Wilhelm Engelmann und F. Furchheim; 1896.

216. Pugliese Carratelli G. Pompei: pitture e mosaici. Documentazione nell'opera di disegnatori e pittori dei secoli XVIII e XIX. 1990. Aufl. Rom: Istituto della Enciclopedia Italiana; 1911.
217. Beacham R, Denard H, Baker D, Blazeby M. Skenographia Homepage [Internet]. The Skenographia Project. 2002; http://www.skenographia.cch.kcl.ac.uk/index.html. Accessed on 30. Jul. 2016
218. Blazeby M. House of Apollo [Internet]. Perspective Analysis. 2007; http://www.kvl.cch.kcl.ac.uk/wall_paintings/apollo/perspective/default.htm. Accessed on 30. Jul. 2016
219. Panofsky E. In: Oberer H, Verheyen E, editors. Aufsätze zu Grundfragen der Kunstwissenschaft. Berlin: Volker Spiess; 1980.
220. Zahn W. Die schönsten Ornamente und merkwürdigsten Gemälde aus Pompeji, Herculanum und Stabiae nach den an Ort und Stelle gemachten Originalzeichnungen. Berlin: G. Reimer; 1842.
221. Bayerische Verwaltung der staatlichen Schlösser, Gärten und Seen. Bayerische Schlösserverwaltung I Schlösser I Pompejanum [Internet]. https://www.schloesser.bayern.de/deutsch/schloss/objekte/as_pom.htm. Accessed on 19. Mar. 2019.
222. von Larissa D. Damianos Schrift über Optik. Mit Auszügen aus Geminos, Griechisch und Deutsch. Schöne R, Editor Berlin: Reichsdruckerei. 1897;
223. Drecker J. Das Planisphaerium des Claudius Ptolemaeus Isis. 1927;9:255–78.
224. Ptolemäus C. Planisphaerium. Commandino F, Editor Venedig: Aldus; 1558.
225. Schnyder R. Zur Entdeckung der wissenschaftlichen Perspektive in der Antike. Z Schweiz Archäol Kunstgesch. 1962;22:143–57.
226. Neugart T. Episcopatus Constantiensis Alemannicus. St. Blasien: Typis S. Blasii; 1803.
227. Bayerische Akademie der Wissenschaften. Mittelalterliche Bibliotheken (ca. 500–1500 n.Chr.) [Internet]. Mittelalterliche Bibliothekskataloge Deutschlands und der Schweiz (MBK). http://www.mbk.badw.de/bibliothekskataloge.html. Accessed on 23. Mar. 2019.
228. Schmidt D. UNESCO-Welterbe Klosterinsel Reichenau in Baden-Württemberg. Landesamt für Denkmalpflege im Regierungspräsidium Stuttgart. Editor Landesamt für Denkmalpflege im Regierungspräsidium Stuttgart. 2013;
229. Vasari G. Le vite de piu eccellenti architetti, pittori, et scultori italiani, da Cimabue insino a' tempi nostri. Florenz: Torrentino. 1550;
230. Vasari G, Nova A, Lorini V. Die Edition Giorgio Vasari: Giorgio Vasari Lebensläufe der hervorragendsten Künstler: Gesamtausgabe. Berlin: Klaus Wagenbach; 2004.
231. Florenski PA. Die umgekehrte Perspektive: Texte zur Kunst. München: Matthes & Seitz; 1989.
232. von Phourna Dionysios. Das Handbuch der Malerei vom Berge Athos. Trier: Verlag d. Fr. Lintz'schen Buchhandlung; 1855.
233. Tauchert S. Form und Funktion von Perspektive in der englischen Lyrik des 16. und 17. Jahrhunderts. Norderstedt: Books on Demand; 2007.
234. Peckham J. Jo. Archiepiscopi Cantuariensis Perspectiva communis. Gauricus L, Editor Venedig: Io Baptistam Sessam; 1504.
235. Bacon R. Rogerii Bacconis Angli, Viri Eminentissimi Perspectiva. Combach J, Editor Frankfurt: Wolffgang Richter; 1614.
236. Ghiberti L. Lorenzo Ghibertis Denkwürdigkeiten (I Commentarii). Zum ersten Mal nach der Handschrift der Biblioteca Nazionale. In: von Schlosser J, editor. Florenz vollständig herausgegeben und erläutert. Berlin: Julius Bard; 1912.
237. Schwarz MV, Theis P. Giottus pictor. Wien: Böhlau; 2004.
238. Alighieri D. Die göttliche Komödie. Berlin: Enslin'sche Buchhandlung; 1842.
239. Boccaccio G. Das Dekameron. Zweiter Theil. 2. Aufl Leipzig: F A Brockhaus.

240. Comune di Siena. Sala dei Nove – Museo Civico – Strutture Museali – Cultura – La Città – Comune di Siena [Internet]. http://www.comune.siena.it/La-Citta/Cultura/Strutture-Museali/Museo-Civico/Sala-dei-Nove. Accessed on 31. Mar. 2019.
241. Schmidt D. Der Freskenzyklus von Ambrogio Lorenzetti über die gute und schlechte Regierung [Elektronisches Medium]: eine danteske Vision im Palazzo Pubblico von Siena/[Dissertation]. St. Gallen: Universität St.Gallen; 2003.
242. Zoege von Manteuffel K. Die Gemälde und Zeichnungen des Antonio Pisano aus Verona. Halle a. S: Buchdruckerei der "Rev Ztg"; 1909.
243. Vallardi G. Desegni di Leonardo da Vinci. Mailand: Pietro Agnelli; 1855.
244. Pisanello. Pisanello: le peintre aux sept vertus; Musée du Louvre, Paris, 6 mai – 5 août 1996. Réunion des Musées Nationaux, Editor. Paris: Éd. de la Réunion des Musées Nationaux; 1996.
245. Galleries U. Adoration of the Magi I Artworks I Uffizi Galleries [Internet]. Gli Uffizi. https://www.uffizi.it/en/artworks/adoration-of-the-magi. Accessed on 31. Mar. 2019
246. Dvořák M. Geschichte der italienischen Kunst: Das 14. und 15. Jahrhundert München: R Piper & Co. 1927;
247. Verdon T, Zolli DM, Museum of Biblical Art, Opera di Santa Maria del Fiore, Exhibition Sculpture in the Age of Donatello: Renaissance Masterpieces from Florence Cathedral, Editor. Sculpture in the age of Donatello: Renaissance masterpieces from Florence Cathedral; [... accompanies the Exhibition Sculpture in the Age of Donatello: Renaissance Masterpieces from Florence Cathedral ... on display at the Museum of Biblical Art from February 20 to June 14, 2015]. London: Giles [u.a.]; 2015.
248. Arndt EM. Ernst Moritz Arndts Reisen durch einen Theil Teutschlands, Ungarns, Italiens und Frankreichs in den Jahren 1798 und 1799. 2. Leipzig: Heinrich Gräff; 1804.
249. von Fabriczy C, Brunelleschi F. Sein Leben und seine Werke. Stuttgart: Cotta'sche Buchhandlung; 1892.
250. Manetti A, Perrone C. Vita di Filippo Brunelleschi. Roma: Salerno Editrice; 1992.
251. Cramer T, editor. Wege in die Neuzeit. München: W. Fink; 1988.
252. Gernrath JohK. Abhandlung der Bauwissenschaften oder Theoretisch-praktischer Unterricht in der bürgerlichen Bauskunst, dem Strassenbau, und in der Hydrotechnik oder Wasserbaukunst für angehende Ingenieurs, Hydrotekten, Bauleute, Brunnenmeister und Müller, dann zum Gebrauch der Wirthschaftsämter. Brünn: J. G. Gastl; 1825.
253. Alberti LB. Drei Bücher über die Malerei (Della Pictura libri tre). Leone Battista Alberti's kleinere kunsttheoretische Schriften Im Originaltext herausgegeben. Wien: Wilhelm Braunmüller; 1877.
254. Schäffner W. Euklids Zeichen. Zur Genese des analogen Codes in der Frühen Neuzeit. In: Bredekamp H, Bruhn M, Werner G, Editor. Mathematische Forme(l)n. Berlin: Akademie; 2017. pp. 62–73.
255. Alberti LB. Della Pittura e della Statua di Leonbattista Alberti. Giusti F, Editor. Dalla Societa Tipografica de Classici Italiani: Mailand; 1804.
256. Panofsky E. Das perspektivische Verfahren des Leone Battista Albertis. Kunstchronik. 1914;26: 506–16.
257. Della Francesca P. De prospectiva pingendi. Gizzi C, Editor Venedig: Edizioni Ca' Foscari – Digital Publishing; 2016.
258. Della Francesca P. Petrus Pictor Burgensis. De prospectiva pingendi. Bordeaux: Manuscript ms. 616; 1460.
259. Burgensis PP. De Perspectiva Pingendi. Strassburg: Heitz & Mündel; 1899.
260. Pacioli L. Summa de arithmetica geometria. Proportioni: et proportionalita. 2. Aufl. Toscolano: Paganino; 1523.

261. Pacioli L, Penndorf B. Abhandlung über die Buchhaltung 1494. 3., unveränd. Repr. der Ausg. von 1933. Stuttgart: Schäffer-Poeschel; 2009.
262. Pacioli L, Da Vinci L. Divina proportione. Venedig: A Paganius Paganinus. 1509;
263. Pacioli L. Fra Luca Pacioli Divina Proportione. Die Lehre vom Goldenen Schnitt. Eitelberger v. Edelberg R, Ilg A, Editor. Carl Graeser: Wien; 1889.
264. Della FP. De quinque corporibus regularibus et de corporibus irregularibus. Codex Urbinus Latinus Rom: Manuscript Urblat632. 1476:1r–68r.
265. Jordan M. Der vermisste Traktat des Piero della Francesca über die fünf regelmäßigen Körper. In: Bode W, Dohme R, Grimm H, Jordan M, Lippmann Fr, Editor. Jahrbuch der Königlich Preussischen Kunstsamlungen. Berlin: Weidmannsche Buchhandlung; 1880. pp. 112–118.
266. Da Vinci L, Della Bella S. Trattato della pittura. 1792. Aufl. Florenz: Pagani & Grazioli; 1792.
267. Da Vinci L. Des vortreflichen Florentinischen Mahlers Lionardo da Vinci höchst-nützlicher Tractat von der Mahlerey. Nürnberg: Christoph Weigel. 1747;
268. Da Vinci L. Das Buch von der Malerei. Nach dem Codex vaticanus (Urbinas) 1270. Ausgabe 1882. Melzi F, Ludwig H, Editor. Wien: Wilhelm Braunmüller; 1519.
269. Da Vinci L. The Notebooks of Leonarda da Vinci. 1955. Aufl. MacCurdy E, Editor. Florenz: George Braziller; 1955.
270. Dürer A. Underweysung der Messung, mit dem Zirckel und Richtscheyt, in Linien ebnen unnd gantzen Corporen. Nürnberg: Hieronymus Formschneyder. 1538;
271. Encyclopædia Britannica. Definition of PERSPECTOGRAPH Merriam-Webster [Internet]. Merriam-Webster 2017. https://www.merriam-webster.com/dictionary/perspectograph. Accessed on 1. Jan. 2018.
272. Schiffner F. Über Photogrammetrie und ihre Anwendung bei Terrainaufnahmen. Mittheilungen aus dem Gebiete des Seewesens. Wien: Druck und Commission-Verlag Carl Gerolds Sohn.
273. Hauck G. Mein perspektivischer Apparat. Fortschritte der Physik im Jahre 1884 Dargestellt von der physikalischen Gesellschaft zu Berlin. Berlin; 1890.
274. Photogrammetrie und Luftbildwesen. Handbuch der wissenschaftlichen und angewandten Photographie. Wien: Julius Springer; 1930.
275. Ritter H. Perspectograph: Apparat zur Herstellung der Perspective aus geometrischen Figuren. 2. Frankfurt a. M: J Maubach & Co; 1884.
276. da Vignola GB. Le due regole di prospettiva pratica. Danti E, Editor Rom: Francesco Zannetti. 1583;
277. Shickman A. "Turning Pictures" in Shakespeare's England. Art Bull 1977;59:67.
278. Bettini M. Apiaria Universae Philosophiae Mathematicae, Bd. 1. Bolognia; 1642.
279. Schott C. Magia Universalis Naturae et Artis. Frankfurt: Schönwetter; 1657.
280. Grasnick A. 3D ohne 3D-Brille: Handbuch der Autostereoskopie. Springer; 2016.
281. Nommensen K-U. Pastorenehepaar nach 350 Jahren auf Reisen [Internet]. Kirchenkreis Nordfriesland. 2013; http://www.nordfriesland-evangelisch.de/42.0.html?&tx_ttnews%5Btt_news%5D=24&cHash=7e4e0fd2478b14a6e3077dd215a3f71a. Accessed on 26. Aug. 2016
282. Ucke C, Schlichting H-J. Morphing, Zoom und 3D. Lamellen- und Linsenrasterbilder Phys Unserer Zeit. 2010;41:43–6.

Images from Light

3

The Recording and Reproduction of Reality

Summary

Since ancient times, the imaging of bright objects through small apertures onto random surfaces has been described. This simplest kind of projection works quite satisfactorily in a darkened room and became known by the Latin name "camera obscura". The improvement of the camera obscura with lenses or mirrors led not only to what we know today as a camera, but also to an opposite consideration. Couldn't the images inside the darkroom be displayed outside the chamber? This consideration had a magical consequence, the magic lantern. Its original use for spreading fear and terror was quickly forgotten when the suitability of the magic lamp for scientific purposes became apparent. With the advent of photography, techniques for recording and projecting moving scenes developed. The advantages of motion pictures were matched by few disadvantages. To see a motion picture, one had to go to the cinema and the images reproduced were always shots from the past. Seeing at a distance, television, only became possible with the discovery of light-sensitive materials. While the first televisions were still large devices with a small field of view, the practical application of liquid crystals also made the production of large-format flat screens a reality.

3.1 Camera Obscura

Pictures from the Darkroom

Already Aristotle was credited with observing the multiple imaging of the sun or moon through a foliage or sieve [1]. A "camera obscura" was first and foremost only a dark room in which a bright object (e.g. sun or moon) was projected through a small hole in the wall onto the opposite wall for observation. The term "camera obscura" is merely the Latin

A. Grasnick, *Basics of Virtual Reality*,
https://doi.org/10.1007/978-3-662-64201-6_3

translation of the Arabic expression "al-bayt al-muzlim" for a dark room. This simple description of how a pinhole camera works is found in Alhazen's book of optics [2], although Alhazen is not yet concerned with the display of an image on the wall [3]. Alhazen's "camera obscura" was thus only a dark chamber for observing the propagation of light.

Example

Aristotle's observation can easily be repeated on a sunny summer day in a beer garden. A nice description for this experiment can be found in the second "Book of Inventions" ("Buch der Erfindungen") [4] in the section "From the light. Essence and reproduction of light" ("Vom Lichte. Wesen und Fortpflanzung des Lichtes", p. 351–352):

> Which of our readers, sitting under a shady tree, through the gaps in whose leaves the rays of the sun fell on the white surface of a tablecloth or on the light gravel floor, would not yet have remarked in amazement that all the individual spots of light have a circular shape . . .?
>
> They are small solar images, conditioned in their shape merely by the outer form of the light-radiating body of light. . ."

The corresponding illustration (Fig. 3.1) shows the well-known phenomenon.

Much more rarely one will perceive the following light phenomenon (Fig. 3.2). To do this, one would have to be in the above-mentioned beer garden during a partial solar eclipse and could then perceive the crescent-like projections of the celestial phenomenon.

The natural philosopher Roger Bacon was demonstrably not only a connoisseur of the writings of Aristotle, but also those of Alhazen [5] and thus familiar with the basic features of optics. In his Letter "On the Secret Workings of Art and Nature and on the Vanity of Magic" (Epistola de Secretis Operibus Artis et Naturae et de Nullitate Magiae), in the chapter "On Artificial Appearances", he described optical imaging in such a way that not only a camera appears interpretable, but even a microscope or telescope (from the German translation [6]):

> Indeed, there are found quite marvelous natural ideas of the rays. For transparent things and mirrors can be made in such a way that a single thing appears as if there were many of them, and a single man as if he were a whole army, indeed that many, and as many as we wish, suns and moons become visible. For nature sometimes gives the vapors such a shape that two suns and two moons, even sometimes three suns, can be seen at the same time in the air, as Pliny tells us in his Natural History. In the same way, many and even innumerable things can appear as a single thing; for if a thing is more perfect than is required of it, it no longer has a definite number, as Aristotle suggests in the chapter on space. . . . For the visible can take on such a form that what is very far away seems as if it were very near, and so also vice versa. So that at an incredible distance we can read the smallest letters, and count the smallest things, and imagine stars as visible where we will.

Fig. 3.1 Solar images with uncovered sun, A. Dressel (1898) from [4], p. 352

Fig. 3.2 Solar images during partial solar eclipse, A. Dressel (1898) from [4], p. 353

One can see that Bacon's writing is particularly concerned with multiple imaging through innumerable pinholes in the sense of an Aristotelian crown of leaves, but also clearly refers to optical magnification (or diminution).

In the philosophical work "Milchamot Ha-shem" [7] ("Wars of the Lord"), in German "Die Kämpfe Gottes" [8], the religious scholar, mathematician and astronomer Levi Ben Gershon deals from 1317 with the great questions of his time:

1. The immortality of the soul
2. The revelation of the future
3. The divine knowledge of things
4. The divine providence
5. The heavenly bodies, their movers, their mutual ranking, their relation to God
6. The genesis of the world

For the exploration of the celestial bodies Gershon used a simple device called "Revelator secretorum[1]", which is described in detail in the writing "De sinibus, chordis et arcubus"[2] from the year 1342. The tool, later known as Jacob's staff, was used to determine the angle of elevation of a celestial body by means of triangulation. A vertical plate was moved on a rod until the star could be aimed at via the distance AB (line of sight) (Fig. 3.3). The angle α is now obtained from the read division of the rod. Gershon calculated the angle via the sine and added a sine table for the purpose of conversion.

In the same work, Gershon also discusses the camera obscura. Goldstein describes the construction as follows (from [14], p. 156):

> The instrument described is a combination of the Jacob's staff and the camera obscura, the rod being 16 or more spans[3] (i.e. greater than about 3 m). At one end of the rod is a plate with a circular hole, the diameter of which is equal to[4] one or two units through which the light of the luminous body enters. At the other end of the rod is a screen parallel to the plate on which the image of the star can be seen.

---

[1] Revealer of secrets.

[2] The writing was handed down in the Latin translation by Petrus de Alexandria [9]. A work with the identical title "De sinibus, chordis et arcubus" was written by Johannes von Gmunden in Vienna in 1437. According to Busard, who published the work again in 1971 with commentary [10], von Gmunden refers, however, among other things to the "Canones Tabularum primi mobilis" of Johannes de Lineriis from 1322, in particular to his sinus tables (see s. [11]) and fraction calculus (see also [12]).

[3] A span is: "…the distance between the extended thumb and the equally extended little finger, and is estimated to be 9 inches…" (from [15], p. 248).

[4] Since each unit is 1/8 of a span, the value for one unit is 25 mm. So the hole had a diameter between 25 and 50 mm.

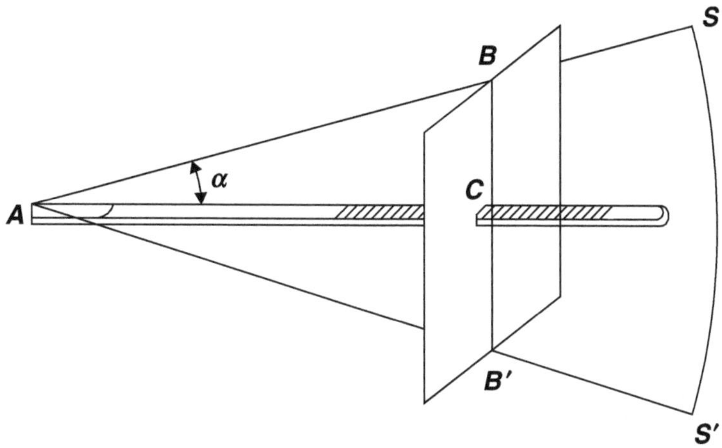

Fig. 3.3 Revelator secretorum, Anton von Braunmühl (1900) from "Geschichte der Trigonometrie" (History of trigonometry) [13], p. 104

An interesting recourse to the original meaning of the term camera obscura is made by Curtze in his article "Die Dunkelkammer" (The darkroom), who analyzed Gershon's Latin text of "De sinibus, chordis et arcubus[5] "in the translation of Petrus de Alexandria in the Codex Vindobonensis Palatinus. The description of the third chapter already reveals the basic function of the darkroom [17]:

> The third chapter develops a method by which the proportions . . . of the sun and moon to that of the circle which they pass trough outside their deferent at the time of observation may be found from the size of the image which they sketch after passing through the windows of a house.

The Epicyclic Hypothesis—Epicycle and Deferent

According to Ptolemy's theory, the planets should move in circular orbits around the Earth. However, more precise astronomical observations showed a clear deviation from this ideal conception. Thus, in the "Amalgest" [18], Ptolemy used the epicyclic hypothesis of two simultaneous circular motions to describe a planetary orbit around the Earth. The planet rotates around a small circular orbit (epicycle) whose center moves on a larger circular orbit (deferent). Copernicus in his revolutionary work "De revolutionibus orbium coelestium" ("On the Revolutions of the Heavenly Spheres")

(continued)

[5] Curtze, an acknowledged connoisseur of Gershon's writings, had referred to the camera obscura some years earlier in a treatise on Gershon's trigonometry [16].

[19], moved the sun into the center of the system, but still used the epicyclic hypothesis to describe the track of the stars. With opposite directions of rotation on epicycle and deferent, an ellipse can thus also be represented with some effort. Only with Kepler's 1609 in the "Astronomia Nova" [20, 21] presented idea of an elliptical planetary orbit could be dispensed with the epicycle theory.

In the following, Gershon develops a detailed description of the darkroom, of which only the "main sentence" is reproduced here: (from [17], pp. 226–227)

> When rays from the sun, moon, or other sources of light enter through any window or hole and strike an object behind the aforesaid opening, the image of the same is on all sides wider than the aperture according to the angle formed by the radius of the luminous body at the entrance.

A little later in the text follow two observations worth mentioning:

> ...and the smaller the opening in the window, the more accurate will be the result of the measurement
>
> Further, it follows that when the eclipse appears on one side of the luminous body, the arch is on the opposite side of the image on the screen. For it is clear from the foregoing proof that the lower half of the celestial body sends light to the upper part of the image, and always the opposite side of the celestial body gives rise to the opposite part of the image.

With the "main theorem of the darkroom", his precise description of the influence of the hole diameter on the image quality and the description of side- and height-inverted imaging, Gershon proves that he had understood the basic principle of optical imaging and was able to apply it (even if he did not use the term camera obscura). Since it is known from the text of 1342 that Gershon also gained his knowledge by observing lunar and solar eclipses, he could well have observed the lunar eclipse of October 3, 1335 and the solar eclipse of March 3, 1337 during his lifetime (1288–1344).

It is only natural to use this type of representation for the purposes of drawing. The use of perspective in painting, however, requires a camera in the modern sense, i.e. a device for projecting the three-dimensional environment onto a surface. Brunelleschi understood and used the rules of perspective to graphically represent his architectural designs. Interestingly, however, he also developed the device described earlier that allowed the painted perspective to be compared to the original via an image on a mirror [22].[6] By imaging via the mirror, the image could now be perceived virtually floating in space at twice the distance from the mirror.

[6]However, a stereoscopic effect, as claimed by Scott ([23], p. 9), will hardly have occurred with one eye.

Brunelleschi had not developed a camera obscura in this way, but he had developed an amazing device with which one could compare the drawn perspective with the real view by superimposition. Most certainly a helpful apparatus for the master builder Brunelleschi. With it he had found a way to prove the perspective correctness of his reduced projections of the great buildings impressively and beyond doubt.

A reduction in the size of objects also takes place when they are depicted in the eye. It is not surprising that Leonardo da Vinci, who had worked intensively on perspective, representation and human anatomy, also wanted to fathom how vision works.

A glance at Codex Atlanticus [24] reveals his view of the eye. In a drawing he compares the rays path in the eye with the image in a camera obscura, where the image is made on the back wall of a sphere and is seen there by the observer (Fig. 3.4).

In his essay on the darkroom, Curtze translates a passage from da Vinci's Paris manuscript D [25] on sheet 8 [17] as follows:

> As soon as the images of the illuminated objects enter a very dark room through some small round opening, and then one catches these images within this room on a white sheet of paper brought near the opening, one will see all the aforementioned objects on the sheet with their real shape and colors, but they will be smaller and reversed...

Da Vinci's Manuscript D examines the optical function of the eye quite intensively and draws some simple setups for simulating the imaging, of which the one described above comes closest to a camera obscura. That da Vinci built this or another pinhole camera and used it for his experiments is probable if one looks a little more closely at the description of it. Da Vinci proposes not only a metal plate with a hole to cover the window aperture (presumably to obtain the most stable and small aperture possible for better imaging), but he proposes projection onto a thin paper viewed from behind [26].[7] We may thus assume that da Vinci had experimented with his apparatuses and improved their shortcomings. A small sketch (Fig. 3.5) explains the function of da Vinci's darkroom. Let us have the structure of the "Abitatione Oscura[8]"explained by the master himself (after [27], p. 45):

> Let **a b c d e** be the object illuminated by the sun, and **o r** be the front of the dark chamber in which is the aforesaid hole at **n m**. Let **s t** be the sheet of paper, which turns the rays of the images of these objects upside down, because the rays are straight, **a** on the right hand becomes **k** on the left, and **e** on the left becomes **f** on the right; and the same takes place within the pupil.

Moreover, there are further references to a pinhole camera as well as simple descriptions in the various works[9] attributed to da Vinci (see also [27–30]). The knowledge of this simple

[7] From the description of folio 8 (recto).

[8] Italian original designation of da Vinci, to dt. approximately "dark apartment".

[9] E.g. in the various editions of the Codex Atlanticus or in the Codex Leicester.

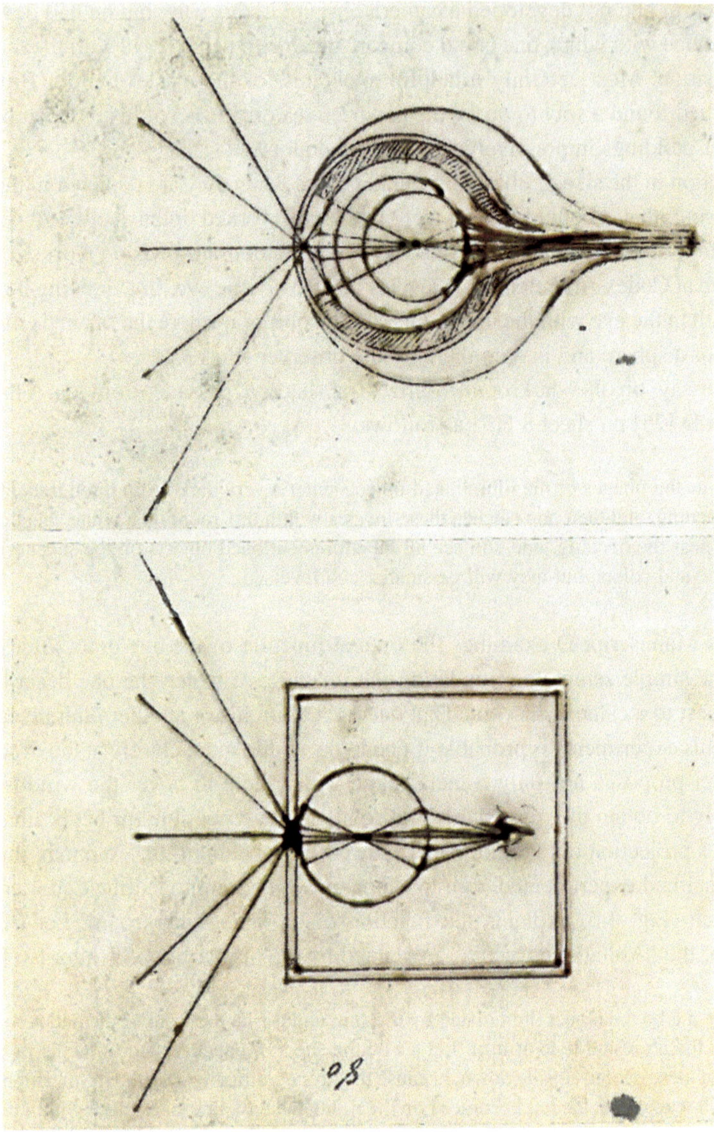

Fig. 3.4 Eye and camera obscura, Leonardo da Vinci (1478–1519), from Codex Atlanticus [24], folio 921r

form of imaging, however, seems to have become known and common knowledge only in the years after da Vinci's death.

Evidence for the practical use of a camera obscura is then handed down by Gemma Frisius, who in his writing "De radio astronomico & geometrico" [31] reports on the solar eclipse of the year 1544 and without further ado adds a drawing of his observation device (Fig. 3.6).

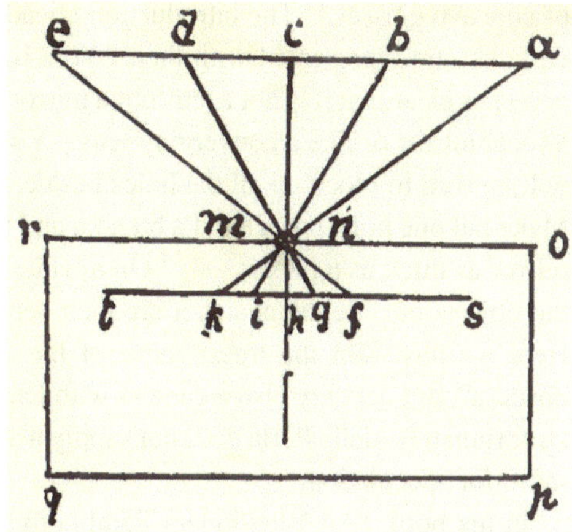

Fig. 3.5 "An experiment showing how objects transmit their images. . .", Leonardo da Vinci (1508) [25] Foglio 8 recto, reproduction Jean Paul Richter (1883) [27], p. 45

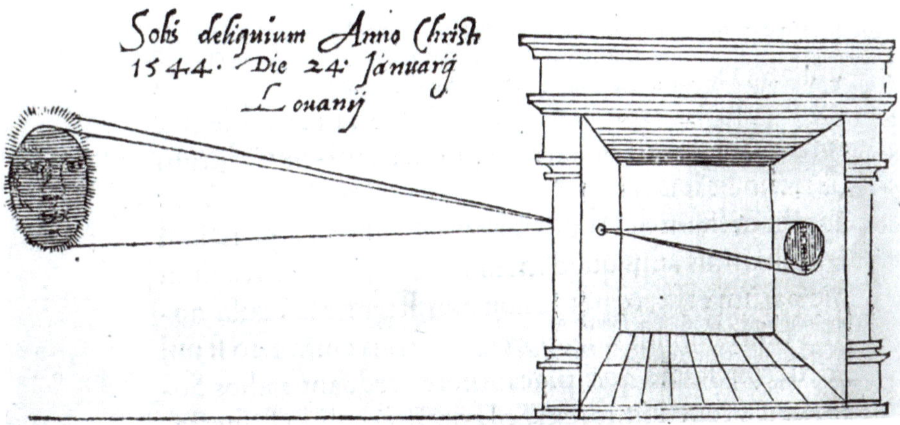

Fig. 3.6 Solis deliquium, Rainer Gemma Frisius (1545) from [31], p. 31, second sheet

One could interpret that the casualness of the mention of this apparatus indicates a high degree of awareness of the camera obscura at that time. Perhaps, however, the construction of a simple pinhole camera was obvious to the experienced device maker Frisius and not worth mentioning.

Giovanni della Porta devoted himself to the study of nature in Naples. In the twenty-volume work "Magia naturalis" [32] in the seventeenth book "in which the burning mirrors and their wonderful phenomena are presented",[10] there is a chapter on the effects of the glass lens.[11] An astonishing application of such a lens is described in the chapter on the use

[10] In quo ustoria specula, & mirabiles eorum visiones proponuntur.

[11] De crystallinae lentis effectibus.

of concave glasses.[12] The introduction already hints at a camera: "In this way all things in darkness are represented with their colors by the sun outside."[13] In the rest of the text it becomes clear that it is not a lens but a mirror, which has a concave surface and thus acts as a focal mirror or like a converging lens: "You must close all the chamber windows, and it will be well to block up all the holes beside, any light that comes in will spoil everything. Make but one hole, it is a hand's breadth and length; above it is a small lead or brass table, a mirror as thick as a paper;. . .".[14] On a white surface of (paper or cloth) opposite the hole, the objects outside the chamber are then depicted with their sides and heights reversed.[15] Here we have—in the truest sense of the word—not only a dark chamber, a "camera obscura", but actually also a camera with lens and projection surface in the modern sense. Unfortunately, della Porta does not supply a drawing to this observation, so that there is still room for speculation.

In his book "Ad Vitellionem Paralipomena" Johannes Kepler describes a visit to the "Finstergemach" [34] of the Dresden Kunstkammer in 1604 (from [35], p. 182).

> A disc, thicker in the middle, or a ocular lens , . . ., stood at the entrance of a closed chamber against a small window, which was the only one open, ... As the sight traveled through the dark emptiness, accidentally it hit the place of the image. . .

Kepler was quite familiar with the preliminary works of Euclid, Alhazen, Witelo, Porta and of course those of his mentor Brahe. Since Kepler dealt very intensively with the propagation of light in the chapter "On the figures of light", he recognized the connection between the object and its image (from [36], p. 19):

> If the window could be a mathematical point, the light on the opposite wall would perfectly reproduce the figure of the illuminated surface, but inverted, and there would be the same proportion of the diameters of the illuminated surface to those of the illuminated wall, as between the distances on either side from the point of the window.

This then essentially describes the image on a pinhole camera, the function of which Kepler also drew (Fig. 3.7).

The art of "bringing to representation in a closed chamber . . . everything that is outside the chamber . . ." has been handed down to Kepler by Porta and is known to him in the description from the "Magia naturalis". Kepler also investigates the influence of the hole diameter on the imaging. A hole that is too small gives a faint but clear image, while one that is too large gives a raw and blurred image. Kepler's chamber is also dark, but he uses

[12] Alia speculi concavi operationes.

[13] Veluti ut omnia in tenebris, qua foris a sole illustrantur cum suis coloribus (from [32], p. 545).

[14] From the English edition of 1658 [33], p. 363.

[15] ". . .and you will see everything even without the sun, and those who walk on the road will appear as antipodes, and what was on the right is now on the left,. . . (from [33], p. 363, own translation).

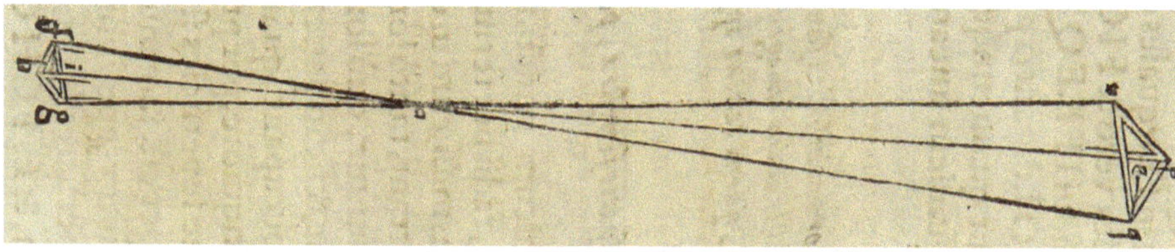

Fig. 3.7 Sinestra punctum, Johannes Kepler (1604) from [35], p. 43

the term "closed", i.e. "camera clausa" ([35], p. 51). The term "camera obscura" is also coined by Johannes Kepler: it is found in the index entry "Camera obscura res foris representans" as a reference to the art of the "camera clausa".

Kepler investigated in detail the optical properties of lenses, mirrors, prisms and even lens systems [37] and put optics on a scientific basis. The use of a lens for imaging was thus known to subsequent researchers as well as the pinhole camera. And so it is not surprising that the combination of both systems in the seventeenth century led to the technical development of some useful examples of the camera obscura and to the development of the camera in today's sense.

Christoph Scheiner observed the sunspots from 1610 in Ingolstadt by means of projection, camera obscura and astronomical telescope, an invention of Kepler. The publication of his findings under the pseudonym "Apelles latens post tabulam" [38] leads to the decades-long dispute with Galileo Galilei about the actual discoverer of the sunspots. Of importance for the understanding of the visual process was his work "Oculus" [39] about the eye, in which he also describes, among other things, the pupil constriction with accommodation and thus anticipates the near triad.

In his "Ars Magna" Athanasius Kircher was particularly concerned with the practical application of optics for the representation of images. The simple construction of the darkroom had inspired Kircher to some interesting uses. It is significant that Kircher not only depicted self-illuminating objects, but also reproduced drawings and paintings produced in advance (Fig. 3.8).

An outstanding student of Kircher was Kaspar Schott, who, parallel to Kircher, was concerned with mathematical and perspective issues and intensively investigated the possibilities of imaging. From Schott comes a design of a multi-sided camera obscura, inside of which was another chamber made of projection walls. Through the openings in the outer chamber, images were projected onto the inner, translucent, screens. A draughtsman placed in the middle of the inner chamber could thus observe and record a 360° view of the scene (Fig. 3.9).

Johann Zahn was a student of Caspar Schott and, like his teacher, was also interested in optical imaging. His proposal of a camera with mirror deflection and lens from 1685 is highly topical. The drawing (Fig. 3.10) shows the ray path in such a system. An object (here the arrow **AB**) is imaged via a lens **CD** and a plane mirror **EF** onto a translucent

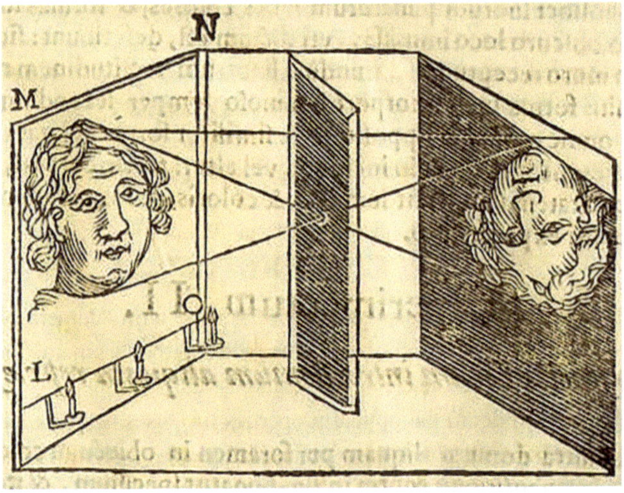

Fig. 3.8 Illustration of a picture by means of camera obscura, Athanasius Kircher (1646) [40], p. 121

Fig. 3.9 360° camera obscura, Caspar Schott (1657) from [41], p. 194, Fig. II

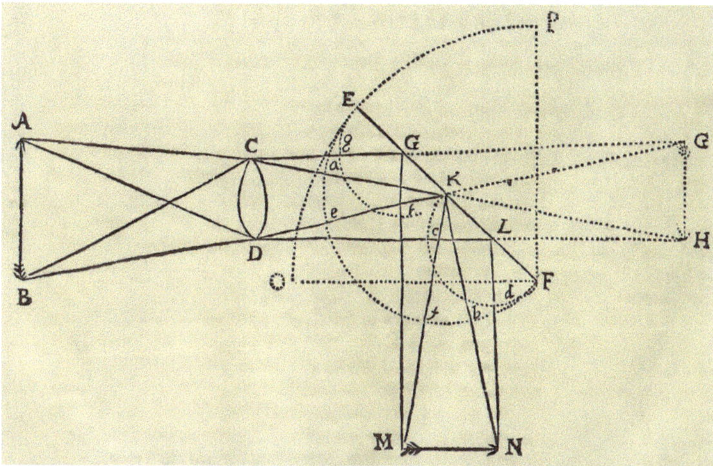

Fig. 3.10 Imaging by means of a mirror, Johann Zahn (1685) from [42], p. 178

surface at the position **MN.** The mirror can be rotated around the point **F.** In position **OF** it would be removed from the optical path and the image would appear at position **GH.**

This describes the optical construction of a single lens reflex camera (SLR) quite well. Unlike his predecessors, however, Zahn went one step further. He designed a practicable and actually mobile camera (Fig. 3.11).

On the ground-glass screen of the camera, tilted in a suitable position, the images of the environment appeared and could thus be easily traced by the observer on a translucent paper. This type of camera with lens, mirror and ground glass was used for a long time in a similar form.

Only a few years later (1694), Robert Hooke[16] gave a lecture to the Royal Society on his picture box[17] [45]. The thing that he uses is quite remarkable and needs to be illustrated (Fig. 3.12).

The artist carries the entire camera obscura on his shoulders and is inside the device with his head and hands. Since the picture box is attached to the back of the artist's head, one can certainly speak of a head-mounted display here.

Even in the nineteenth century, the camera obscura, now also called the "camera clara" (light chamber), and the smaller camera lucida were obviously still in use. One of the most famous users was Johann Wolfgang von Goethe, who used this device for his scientific studies [46]. Goethe wrote to Charlotte Schiller on April 18, 1788 [47]

[16]Hooke was known to his contemporaries through his work "Micrographia" [43]. The drawings shown in it were based on observations with his reflected-light microscope and allowed a glimpse into the unknown microscopic world. Today, however, Hooke's name is more commonly associated with Hooke's law of spring force "Ut tensio, sic vis" (as the elongation, so the force) [44].

[17]"An Instrument of Use to take the Draught, or Picture of any Thing." [45], p. 292.

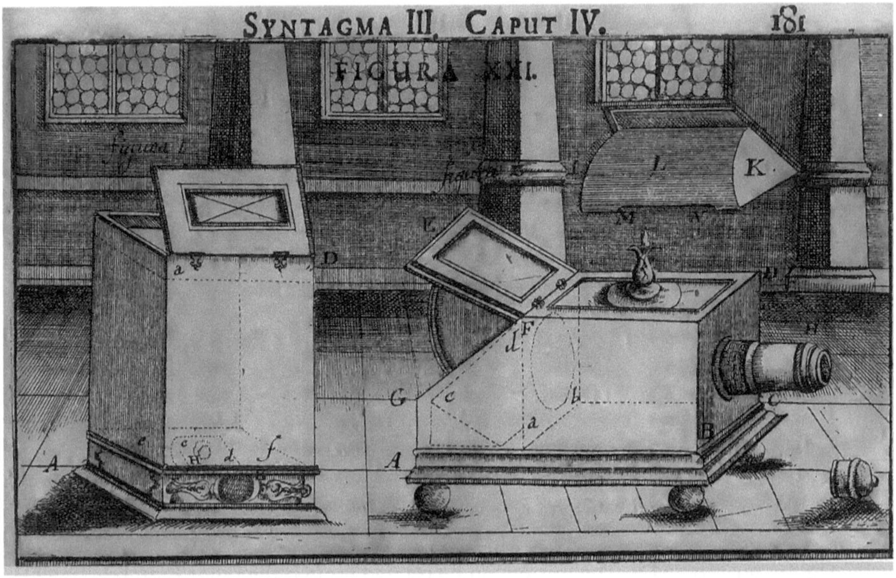

Fig. 3.11 Portable camera obscura, Johann Zahn (1685) from [42], p. 181

> Yesterday I prepared my camera obscura and, while looking at the apparatus, characterized my way in this part of physical science. You can see quite clearly the detours you have made when you can still see before your eyes all the means and tools you have used for that purpose.

Goethe's original camera obscura is kept in the Goethe National Museum in Weimar. (Fig. 3.13). Two things in particular catch the eye in this model: the sliding lens for focusing and the flap to protect the ground-glass screen during transport.

Of course, this construction was not the only one used for recording. The camera lucida[18] was proposed in 1807 by the London physician and chemist William Wollaston. Wollaston was an avid explorer[19] who was also engaged in sketching landscapes. A conventional camera obscura seemed to him unsuitable for sketching, if only because the single reflection on the mirror made things appear laterally reversed. It is obvious that an additional reflection can be used to revert the image again. Wollaston uses a four-sided prism (Fig. 3.14, Fig. 2), with the vertices **abcd** to reflect the view onto the drawing paper twice. In the figure, this ray of vision is represented by the distance **ehgf.** At the same time (or with roving gaze) the drawer looks across the prism at the scene to be sketched (here the visual ray **a**), so that there is a superposition of the image on the drawing paper with reality [50].

---

[18]From Latin lucidus: clear, distinct.

[19]Wollaston is probably better known for the discovery and description of the elements rhodium and palladium [48] and is familiar to opticians for his refractometer [49] and—most especially—for Wollaston's birefringent prism, used to split unpolarized light into its polarized portions (the ordinary and extraordinary rays).

Fig. 3.12 Picture box, Robert Hooke (1726) from [45], p. 295

3.2 Magic Lantern

Magic Lantern and Fright Lamp

The reversal of the light path turns the darkroom into a magic lamp. The difficulty is really only due to the fact that the image to be projected must have a very high brightness in order to still be seen with the naked eye after a pinhole. It is not surprising that projection lanterns only came into the limelight when the manufacture of lenses and mirrors also reached a quality sufficient to focus the light and bright lamps could be produced. Equally unsurprising is the fact that those researchers who were engaged in the study of optics and the camera obscura were also among the fathers of the magic lantern.

Fig. 3.13 Portable camera obscura from Goethe's private collection, Goethe National Museum, Hajotthu (2015), Wikimedia Commons

An early depiction of the magic lantern (ca. 1420) is found in a manuscript by Johannes de Fontana in his collection Bellicorum instrumentorum liber cum figuris [51]. On the last page, a lady is depicted holding a lantern (Fig. 3.15). In the original to the right, the projection of the small drawing from inside the lantern is shown and probably represents an inhabitant of Hell. The title of the drawing is therefore "Nocturnal Apparition to the Horror of the Beholder" (Apparentia nocturna ad terrorem videntium). To this the coded text says: "You know the method with the lantern and have seen with your own eyes [the lantern, author's note] made by myself according to my own idea."[20] The lantern is shown again enlarged in greyscale on the left side of the picture in Fig. 3.15. It is easy to see in the enlargement that Johannes de Fontana has correctly reproduced the projected motif inside the lantern at a smaller scale. This could indicate that Fontana had practical experience in using the lantern and had actually made it himself.

The "Uomo universale"[21] Alberti described in 1436 in his work "Della pittura" [54] not only the use of perspective. As Tiraboschi tells us in his monumental work "Storia della Letteratura Italiana" [55], he possibly used something similar to a peep-box ("Camera

[20] Lat.: "Habes modum cum lanterna, quam propriis oculis vidisti exmea manu fabricatam et proprio ingenio.", decipherment and translation by Horst Kranz [52].

[21] Ital.: the "all-round man", ideal image of man from the time of the Renaissance. Alberti is praised as a universal man, who was versed in all arts and sciences and could masterfully bring them to bear, coupled with extraordinary physical skill. The most famous universal genius, however, is certainly Leonardo da Vinci (according to [53]).

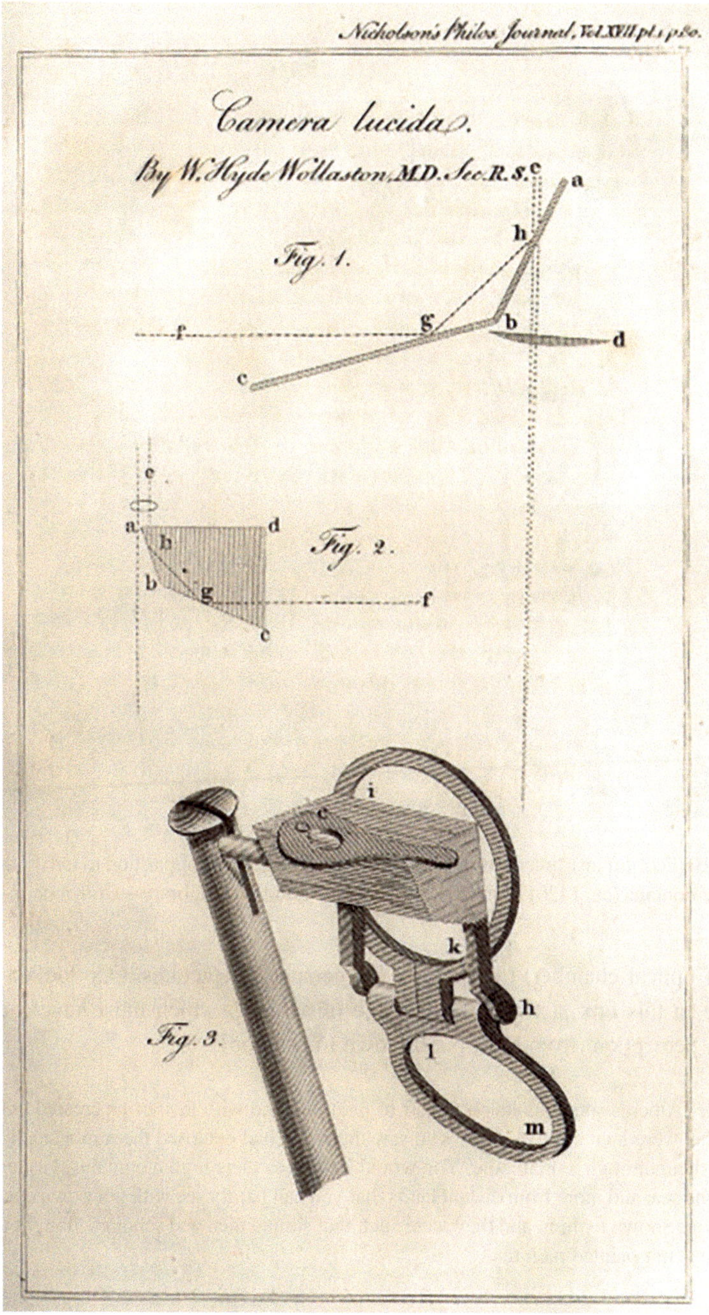

Fig. 3.14 Camera lucida, William Hyde Wollaston (1807) from [50]

Fig. 3.15 Bellicorum instrumentorum liber cum figuris, Nocturnal apparition to terrify the beholder, Johannes de Fontana (ca. 1420) from [51], Munich, Bavarian State Library—Cod. icon. 242, folio 70r

ottica", an optical chamber) to represent his perspective pictures. One looked through a small hole in this box at the painted image of a scene, which must have seemed very realistic in perception (from [55], p. 368, own translation):

> He wrote (Alberti) some books on the art of painting, and with this art he created unheard-of works that were incredible to those who saw them. He had enclosed them in a small box and showed them through a small hole. You would have seen there high mountains, vast provinces around the sea, and more from distant lands that you can hardly see with your own eyes. These things were shown by him, and they were such that simple men and scholars thought they saw real things, not painted pictures.

It is probable that Alberti's optical chamber was already a device for image reproduction, a kind of early slide viewer, in which the hole took over the function of the lens.

However, a projection of the image was not yet possible with Alberti's device.

Fig. 3.16 Lantern with lens, Leonardo da Vinci (between 1487 and 1519) from [24] Folio 34 r

Of course, Leonardo da Vinci is also credited with a magic lantern. Indeed, the Codex Atlantico [24] also contains a drawing of a box with a lens from which light seems to emanate (Fig. 3.16).

A bright lamp is certainly an important prerequisite for a projector. In Schwenter's "Mathematische und Philosophische Erquickstunden" ("Mathematical and Philosophical Refreshment Lessons") from 1636 [56] the reader can enjoy numerous mathematical, geometrical and physical tasks. In the sixth part, some tasks and questions on the "Catropticam or mirror art" are posed. For the solution of the 31st task a bright lantern is needed:

> Now I will uncover another excellent invention, namely, how to read a text at a distance by means of a mirror and a light: First, one must find in the mirror **punctum ustionis**, that is, the focal point, …, and hold to it a lighted candle: thus the candle jumps far back through the parallel lines and thus makes a bright light.

With the candle at the focal point of the concave mirror, Schwenter had created an improved light source that was now also suitable for the illumination beam path of a projector (Fig. 3.17).

In Bettini's richly illustrated treatises of 1642 [57], in addition to drawings of various projections, there are also representations of different illumination concepts. Bettini knows the illumination beam path with reflector and condenser lens (Fig. 3.18) and is aware of the imaging problems with spherical lenses, which are represented here in the figure by the different focal points.

The aspherical, hyperbolic curvature proposed by him leads almost to an ideal lens and improves the image decisively. In the sixth part (Dioptrica) of the "Apiaria Universae

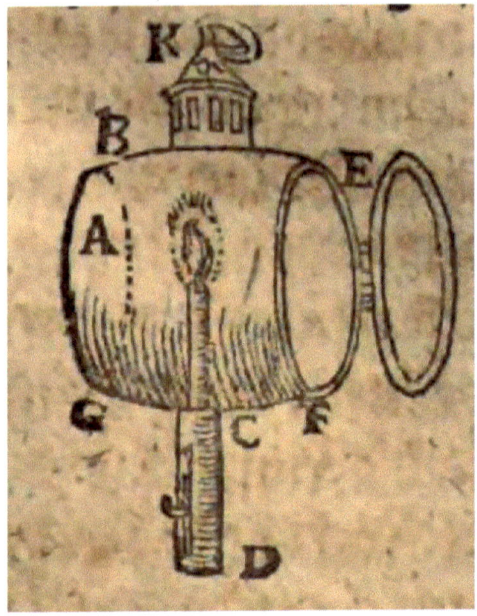

Fig. 3.17 Hand lantern, Daniel Schwenter (1636) from [56], p. 301

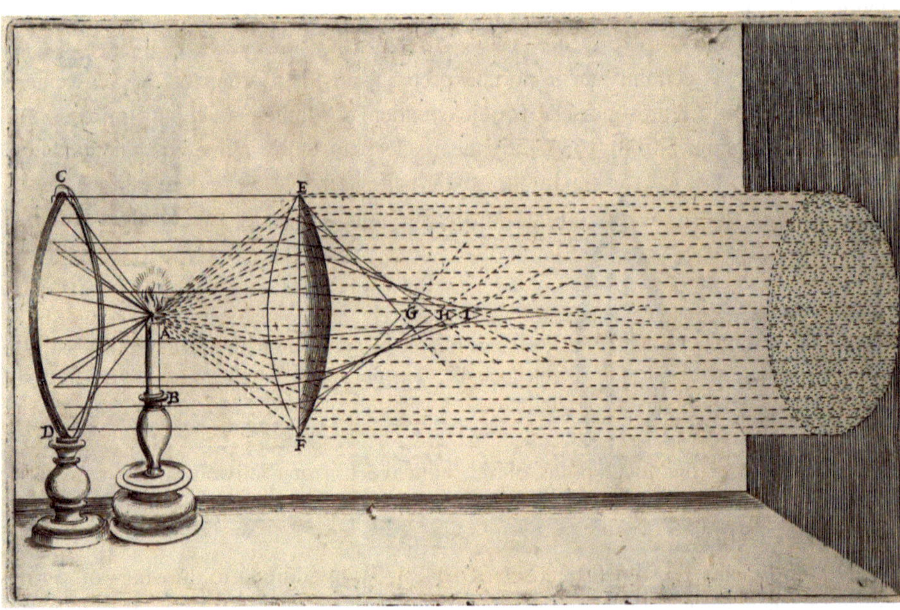

Fig. 3.18 Illumination beam path, Mario Bettini (1642) from [57], Dioptrica p. 26

Fig. 3.19 Projection with aspherical lens, Mario Bettini (1642) from [57], Dioptrica p. 27

Philosophiae Mathematicae" this lens is then even used to project an image onto a wall (Fig. 3.19).

Bettini also uses a lens to reverse the image in the camera obscura and illustrates the effects of different image distances (Fig. 3.20).

This idea is taken up in the second part of the "Erquickstunden", which was compiled by Harsdörffer after Schwenter's death [58], with reference to Bettini. In the 15th task of the fifth part (of the art of seeing), a beautiful drawing is shown, which also contains an imaging ray path with image reversal and is depicted as a mixture of camera obscura and magic lantern (Figs. 3.17 and 3.21).

This means that all parts of a projector are known: Mirror, illumination lens and imaging lens. The ray paths for imaging have been sufficiently explained and the processes involved in projecting an object or image have been comprehensibly illustrated. Now all that is missing is the scientist who can piece together a solution from the extensive preliminary work. And this scientist seems to be Athanasius Kircher.

The learned Jesuit Kircher, lecturer at the papal university "Collegio Romano" in Rome, dealt in his optical textbook "The great art of light and shadow" (Ars magna lucis et umbrae) in 1646 [59] with a variety of scientific topics of his time, but especially with all forms of light propagation and its application. As already stated, he is familiar with the

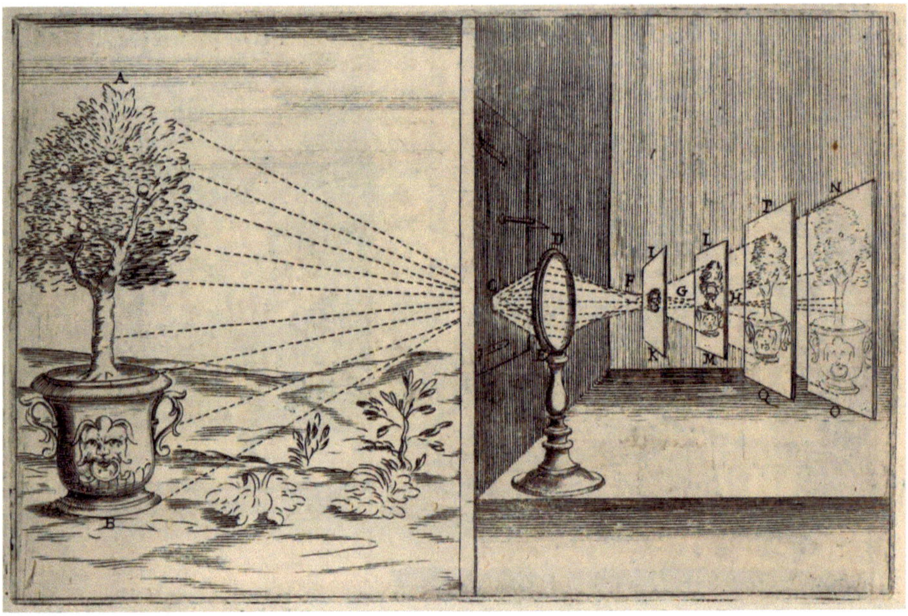

Fig. 3.20 Camera obscura as magic lantern, Mario Bettini (1642) from [57], Dioptrica p. 27

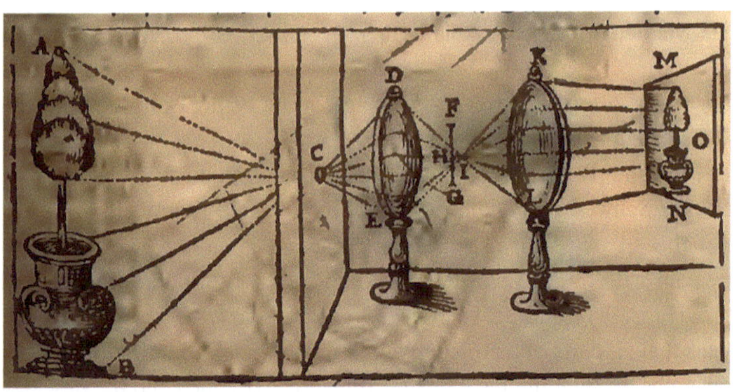

Fig. 3.21 Camera obscura as magic lantern, Georg Philipp Harsdörffer (1651) from [58], p. 209

construction of a camera obscura and so it is obvious that the reverse direction of light leads to a projection by means of a "magic lamp" (Lucerna magica). The book contains several illustrations of image generation in a camera obscura with solar illumination, of which the representation with multiple image was selected as an example (Fig. 3.22).

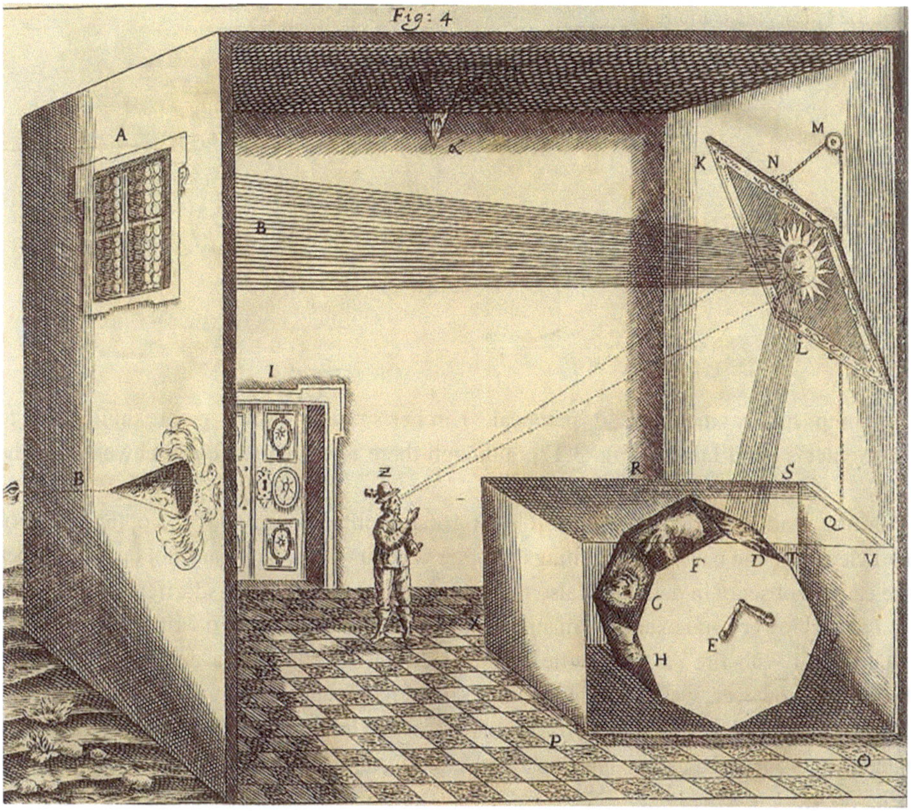

Fig. 3.22 Multiple image projection in a camera obscura, Athansius Kircher (1646) from [59] Folio 901

Such demonstrations were indeed already held as a kind of "slide show", as Schott relates in 1671. He reports in the Magia optica [60] of a performance by Andreas Tacquet from Louvain in Flanders, who reproduced Martin Martini[22]'s journey to China with Kircher's "mirror speech and mirror writing" ("Spiegelred- und Spiegelschreibkunst")[23] in a pictorial story.

The use of a polygon wheel for the image holder certainly brought a speed advantage when changing images, with the help of which it would in principle also have been possible to depict moving scenes. From today's point of view, however, the reflecting lantern "Lucerna catopirica"(Fig. 3.23) was of greater importance. The lantern is characterized by a concave mirror **AB**, a candle **F**, a smoke vent **C**, and a window **D**. A handle **E** makes

[22]Martino Martini was a student of Kircher and traveled China as a Jesuit missionary [61] and geographer [62].

[23]Ibid, p. 398.

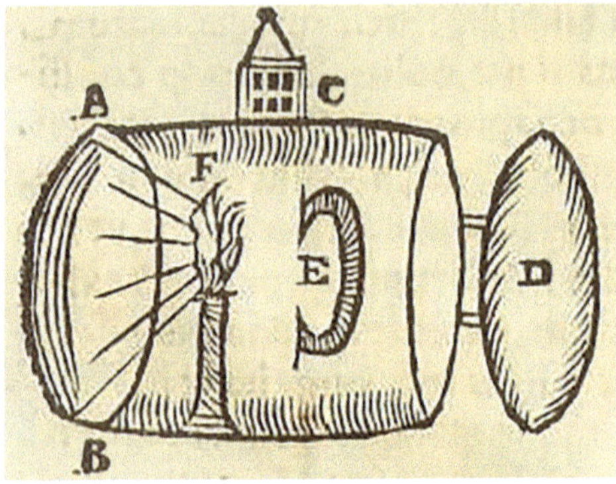

Fig. 3.23 Lucerna catopirica, Athanasius Kircher (1646) from [59], p. 887

the lamp highly mobile and practical. Kircher's lamp shows great similarities to Schwenter's hand lantern (Fig. 3.17), although there is no reference to Schwenter in the text.

In the work "Tractatus opticus" [63] of 1663, Kohlhans compiles the optical state of knowledge on the basis of the writings of other scholars. In it, Kohlhans not only describes the camera obscura in detail, but also the various applications of Harsdörffer's mirror art[24] in a rather flowery and extensive manner. It contains such tasks as "To write in the air or on a dark wall with fire" or "To write several pictures or letters on a dark wall at night". Kohlhans, however, was not yet familiar with the magic lantern.

Only in a later edition of the Ars Magna Lucis Et Umbrae from 1671 [40] is the magic lantern mentioned by name and illustrated in addition to the previous projection devices (Fig. 3.24).

This page also mentions the Danish mathematician Thomas Walgenstein as the inventor of the magic lantern. Walgenstein is repeatedly mentioned in connection with the first applications of the magic lantern[25] and presentations in Europe are attributed to him from 1662 [68]. However, since no records of Walgenstein's own are known, the information about his magic lantern is transmitted only by witnesses of his presentations.

A careful observer of such an event in 1665 was probably the mathematician Claude Dechales. In his 1674 "Cursus seu Mondus Matematicus" [69], geometry and mathematics are treated as well as their practical application in architecture, hydraulics, navigation and optics. Optical imaging is repeatedly explained on the eye, and there are even treatises on binocular vision "Why both eyes do not see objects twice" (pp. 380, 415), in which disparity is already pointed out. The parallel reflection of the rays of light from a candle at the focus of a concave mirror is outlined, as is its use in a lantern (even with an additional

[24] Harsdörffer had understood under the name mirror art (s. e. g. [64]) in particular Kircher's depictive catoptrics (doctrine of reflection) and published it in German.

[25] See also "The magic of the magic lantern" [65], "The magic lantern of literature" ("Die Laterna magica der Literatur") [66], "Scientific applications of photography" ("Wissenschaftliche Anwendungen der Photographie") [67].

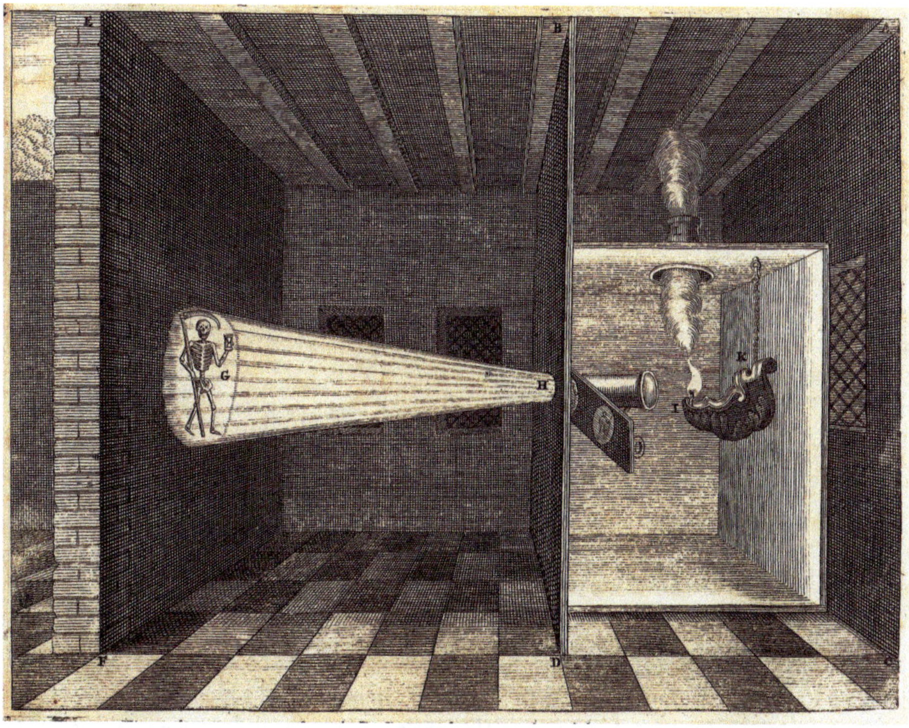

Fig. 3.24 De Lucerna Magicae seu Thaumaturgae constructione, Athanasius Kircher (1671) from [40], p. 769

converging lens to focus the light, p. 586). After a treatise on imaging by lens systems, Dechales turns to the task of projecting onto a wall. Dechales tells of a Danish scholar (who is presumed to have been Walgenstein) who, while passing through Lyon, demonstrated his projection lamp to him. Dechales now also uses the name "Laterna magica"to describe his prototype for enlarging images for display on walls: "Here we present the Lyon dioptric machine under the name of Laterna Magica...".[26]

Dechales made a drawing of the "Lyon dioptric machine", which now had all the elements of a projector: light source, reflector, imaging lens (Fig. 3.25).

Another hint for the existence of a magic lantern comes from Juncker, who translated the travelogue "Voyage d'Angleterre, Pais-Bas, Allemagne, & Italie" by the Frenchman

[26] In the original, "Vidimus hic Lugduni dioptricam machinam, sub nomine Laternae magicae" ([69], p. 655).

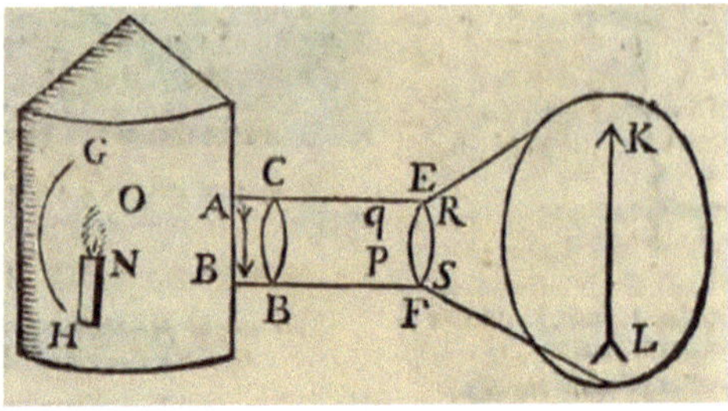

Fig. 3.25 Magic lantern, Claude Dechales (1674) from [69], p. 666

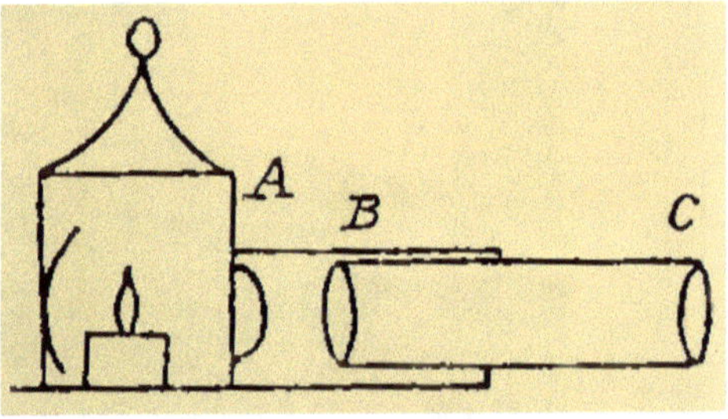

Fig. 3.26 Lanterne de peur, Pierre Petit (1662) from [74], p. 269

Monconys [70] into German. On his journey to England in 1663, Monconys visited the optician Reeves[27] in London, who demonstrated a simple optical lantern to Monconys [72].

The invention of the magic lantern is also attributed to a well-known personality of his time, the physicist Huygens. In addition to Huygens' drawings of a dancing skeleton, which was intended as a model for a projection with a lamp with convex glasses,[28] his extensive correspondence with the Frenchman Pierre Petit is especially credited for this. In this, the projection lantern appears again and again. In a letter to his letter of November 28, 1662, Petit describes the "lantern of fear" ("Lanterne de peur", [74], p. 269) and also adds a simple sketch (Fig. 3.26).

[27]Richard Reeves, whose son John was also acquainted with Huygens (see also [71], pp. 45, 271, 279, 314, 445), manufactured optical instruments in James Street.

[28]"Pour des representations par le moyen de verres convexes à la lampe" (from [73], p. 197).

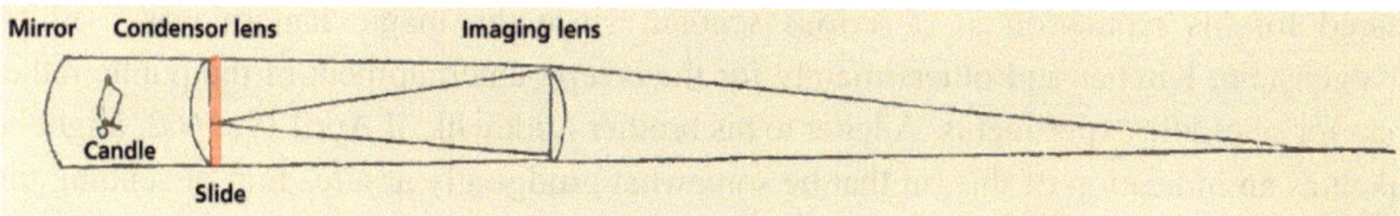

Fig. 3.27 Magic lantern, Christiaan Huygens (1664) from [77], p. 162

Background Information

In 1696, in a description of the collections in the Royal Danish Copenhagen Kunstkammer, Holger Jacobsen also describes, among other things, the "Lucerna magica"([75], p. 67). In a German translation of the Latin text in the "Contributions to Natural History" from 1752 for the entry "...also a magic lantern", an associated anecdote is reproduced, which perhaps explains Petit's designation "lantern of fear" ([76], p. 154):

> Jacobaeus thinks that the excellent Danish glass artist (opticus), Thomas Walgenstein, has surpassed all others in the production of this [the magic lantern, author's note]. He showed, shortly before King Frederick the Third[29]'s death, among others, the image of death through it; all others were frightened, but the king punished their fearfulness, said he could not see this image often enough, and ordered it to be shown more than once.

Another design comes from Huygens himself, who explains the placement of the lenses and the concave mirror in a letter on December 11, 1664 to Petit, ([77], p. 162). Even if the sketch in Fig. 3.27 does not appear particularly impressive at first glance, we are dealing here in principle with the construction of a projector still in use today. The labels and the position of the slide (in red) have been added for explanatory purposes only. There is no labeling in the original. In the intertwined beam path, the light source (here a candle) is located in front of a concave mirror (reflector), ideally at its focal point. This reflector serves to redirect the light emitted by the candle in the opposite direction to the projection direction, which would otherwise be wasted, back into the correct direction. With this "light recycling" Huygens obtains almost twice the brightness. An illuminating lens (condenser) is placed in front of the light source. Again, the light source should be near the focal point of the condenser. The film stage for the translucent object to be imaged is attached to the flat surface of the condenser. Through this arrangement, this object (slide) is illuminated by an almost parallel beam of light. Finally, the slide is imaged onto the projection screen via the imaging lens (today usually a lens), whereby the object must be located in the object distance and the projection screen in the image distance according to the lens equation.

In contrast to his groundbreaking optical discoveries, Huygens had not published his ideas on projection, but only exchanged them with his pen pals. It is possible that Huygens

[29] Here, of course, Frederick III of Denmark and Norway.

feared for his reputation as a serious scholar, since the magic lantern was used by Walgenstein, Kircher and others mainly for the creepy entertainment of the public rather than for scientific experiments. A letter to his brother Lodewijk of April 19, 1662, might be taken as an indication of this, in that he somewhat grudgingly assures him of sending the promised lanterns, and actually only does so because he could find no reasonable excuse for not doing so ([74], p. 111).

Background Information

Huygens later (1690) formulated in his book "Traite de la lumiere" [78] the statement that every point of a wave front can be regarded as the starting point of a new spherical wave; what is known today as Huygens' principle. Huygens himself explains it this way (from the German edition by Lommel [79], p. 21):

> I have thus shown in what way the gradual propagation of light by spherical waves may be conceived......In the first place, namely, it follows from the above considerations that every small spot of a luminous body, such as the sun, a candle, or a glowing coal, produces its waves, the centre of which is that spot.

These spherical waves have a common line of contact, which can be understood as a common wavefront (from [79], p. 24).

> ... they nevertheless meet nowhere else but ... in their common line of contact, ... to form a line delimiting the movement.

After the magic lantern became known to a wider public with the writings of Kircher and Dechales, other versions of the lamp appeared in quick succession.

In the "Collegium Curiosum" of 1676 [80], Johannes Christoph Sturm also provided a beautiful drawing of a projection lantern in addition to all kinds of physical experiments (Fig. 3.28):

In a drawing of the beam path you can see how it works (Fig. 3.29) and also a problem: If the two lenses are the lense system for imaging, the system lacks the condenser lens. If the slide is located far from the first lens, the projected image will be less bright than it would have been with a condenser. This is unfortunate, especially when projecting with a candle as the light source.

Possibly the lantern drawn by Sturm is a device from the workshop of the optician Griendel,[30] who obviously already built such optical devices some years before Sturm's publication.[31] Kohlhans then describes Griendel's lantern in 1677 ([83], p. 318) thus:

[30]Known especially for his richly illustrated microscopy book "Micrographia nova" [43].

[31]At least Griendel seems to mention the magic lantern on the occasion of a letter to Leibniz on December 28, 1671 ([81], pp. 316–319) in an advertising supplement "Specificatio, was Johann Frantz Griendl von Ach, Mathematicus und Opticus in Nürnberg, von optischen Raritäten machen

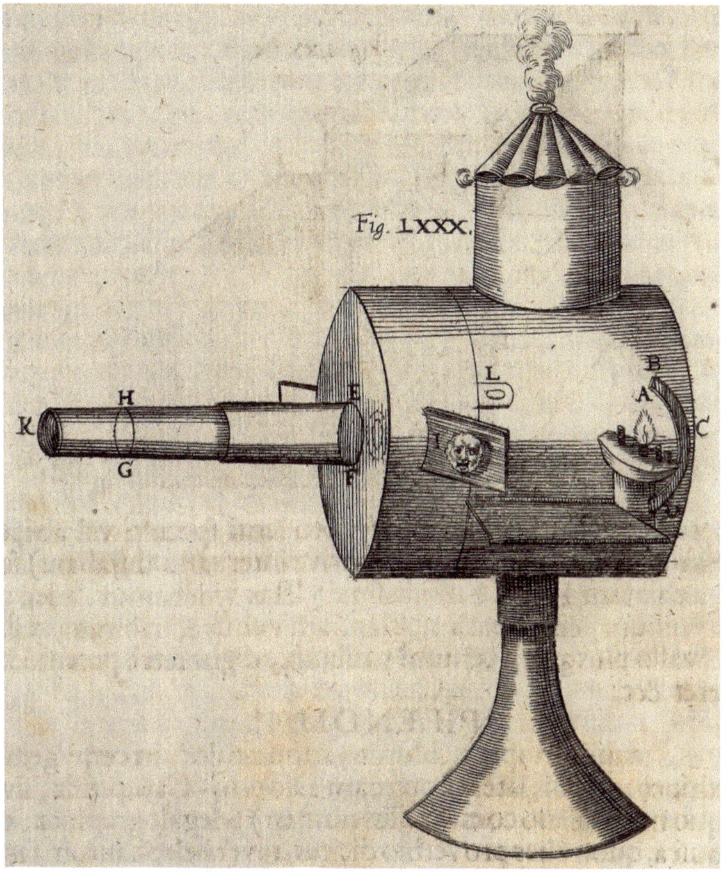

Fig. 3.28 Laterna magica, Johannes Sturm (1676) from [80], p. 164

> Of an optical lantern/so that all kinds of paintings/in a dark room/are presented small and large before the eyes: Such a lantern has been invented by Mr. Johann Frantz Gründel von Ach auf Wankhausen/a distinguished optician in Nuremberg/All kinds of pictures/that one wants to present/are painted on glasses and pushed and pulled through the lantern.

In 1685, Johann Zahn presented a mobile camera with mirror deflection, the predecessor of all modern reflex cameras. In a later, fundamental treatise on optics, the second edition of the "Oculus artificialis"of 1702 [84], a slide projector for the reproduction of an image sequence is shown on a double page (see Fig. 3.30).

thut" ("Specificatio, what Johann Frantz Griendl von Ach, Mathematicus and Opticus in Nuremberg, does of optical rarities") (see [82], footnote 6).

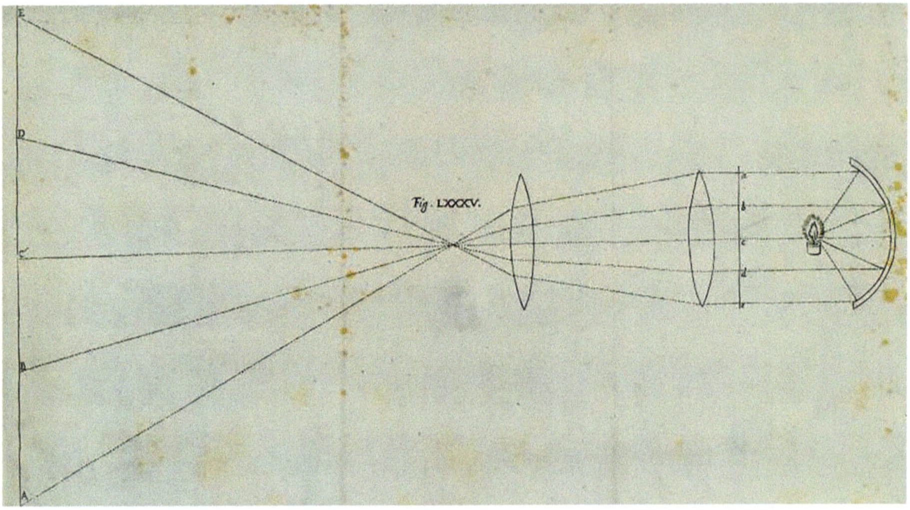

Fig. 3.29 Ray path in Sturm's lantern, Johannes Sturm (1676) [80] Drawing to p. 168

Fig. 3.30 Slide projector, Johann Zahn (1702) from [84], p. 398

The corresponding beam path is also shown on the same page. The illustration in the lantern is illustrated below using the example of his "Laterna megalographica" (Fig. 3.31). The term "megalography" stands for "large painting".

The term "Laterna thaumaturga", also used by Zahn, indicates the use of the lamp to produce miracles, since the thaumaturge is a "miracle worker". This is not surprising when

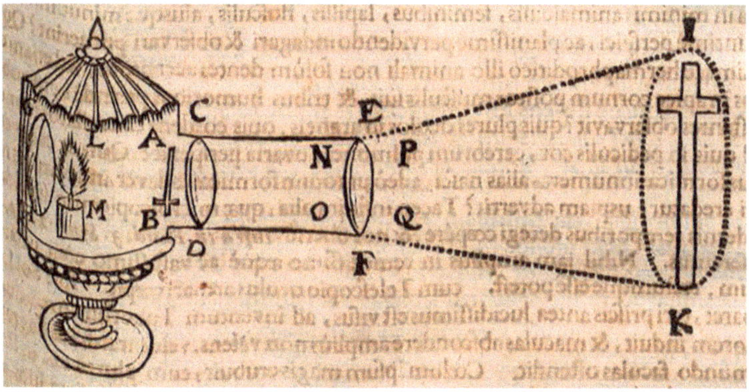

Fig. 3.31 Laterna megalographica, Johann Zahn (1702, p. 401)

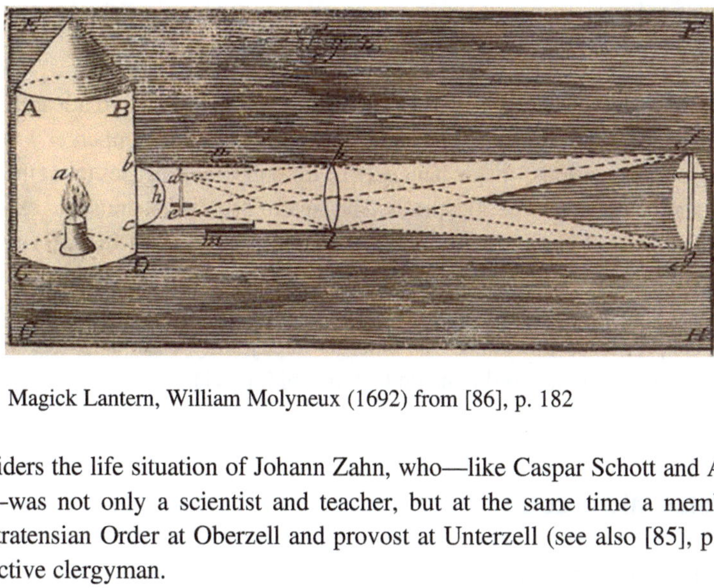

Fig. 3.32 Magick Lantern, William Molyneux (1692) from [86], p. 182

one considers the life situation of Johann Zahn, who—like Caspar Schott and Athanasius Kircher—was not only a scientist and teacher, but at the same time a member of the Premonstratensian Order at Oberzell and provost at Unterzell (see also [85], p. 402) and thus an active clergyman.

Like Sturm, Zahn had also dispensed with a condenser lens, but (like Sturm) had moved the slide close to the lens. William Molyneux in his "Dioptrica nova" [86] 1692 uses the name "Magick Lantern", but also mentions the name "Laterna megalographica" used by Zahn. The drawing (Fig. 3.32) now shows the correct use of a condenser lens (although the reflector is missing for this).

The Molyneux Problem

The Irish naturalist William Molyneux is known less for his magic lamp than for the formulation of the "Molyneux problem", with which he confronted the English philosopher John Locke in 1693: Can a man born blind recognize the objects, whose shape he knows only by feeling them, later (when he suddenly can see) only by sight? Locke, in his in his Essay on Human Perception, gives the answer "No." ([87], p. 67) Besides Locke, Leibniz, Voltaire, and Helmholtz, among others, were also engaged in the search for the answer to this question. Apart from these philosophical investigations, there is also an experimental answer.

In an Indian aid project called Prakash (which means "light" in Sanskrit) (see [88, 89]), blind children are to be helped to regain their sight through various treatments and operations. In 2011, these blind children had been given various objects to touch before surgery. After the surgery, the now sighted children were to try to recognize the objects. As expected, this was not immediately successful, but could be learned very quickly [90].

Through these and other publications, the subject became known to a learned minority and became the subject of scientific analysis. In Jena, Bonifatius Ehrenberger wrote in 1713 about the "novel and curious" magic lantern [91]. Ehrenberger proposes to set the images in mechanical motion by means of cogwheels. The lantern itself is not unusual - except for the telescopic lens tube (Fig. 3.33). However, Ehrenberger's reference (p. 7) to the Jena mathematician Weigel's previous involvement with the magic lantern is interesting. The statement seems credible, since one of Weigel's students was also the already mentioned Johann Christoph Sturm. Sturm and Ehrenberger later jointly wrote a textbook of mathematics [92] with a reference to the magic lantern (ibid. p. 41).

3.3 Solar Microscope

Projection with Solar Power

With the increasing spread of the "fright lantern" towards the end of the seventeenth century, scholars tried to use the now well-known physical effect of magnifying an object beneficially for science. The idea of a projection microscope can be assumed to have been obvious to those who had knowledge of the function of a microscope, the camera obscura and the magic lantern at the same time.

Until then, a serious problem of large image projection by means of a candle was the low brightness. The solar microscope now used the radiant power of the sun as illumination and thus produced an incomparably brighter image. Due to the distance of the sun, the sun's rays hit the earth almost parallel. With solar illumination, therefore, one saves not only the

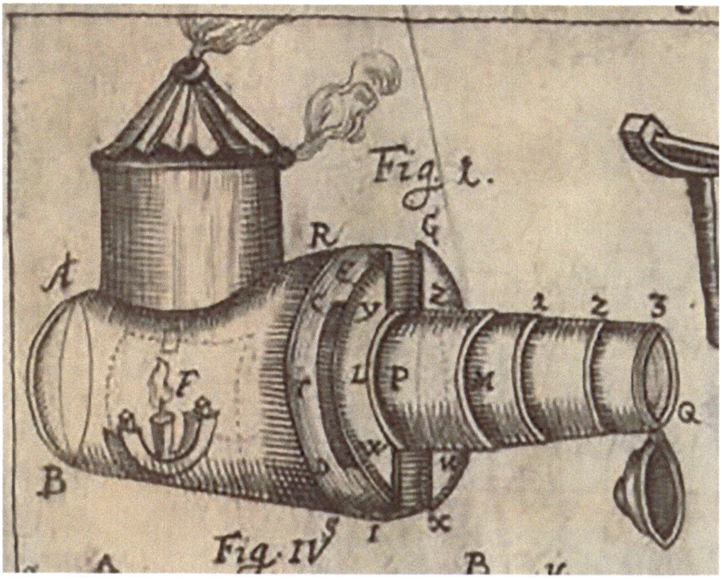

Fig. 3.33 Novum et Curiosum Laternae Magicae, Bonifacius Ehrenberger (1713) from [91] inlaid picture

rear collecting mirror, but also the condenser lens, and thus obtains an extremely simple instrument for microscopic observation.

The first description of such a device is sometimes considered to be Theodor Balthasar. In 1710, Balthasar compares the projection of an object onto a paper with the image in a magic lantern (in [93], p. 119). The part-time microscopist Ledermüller[32] defends in his book "Versuch zu einer gründlichen Verteidigung derer Saamenthiergen[33]" ("Attempt at a thorough defense of the spermatozoon") from 1758 [96] not only the "the dignified life of the seed worms", but also proclaims Balthasar as the true inventor of the solar microscope.

At this time, the solar microscope was generally associated with the name Lieberkühn. Lieberkühn had presented to the Royal Society a microscope whose illumination depended

[32] At this time according to his own statement notary at Nuremberg. Ledermüller is also author of the "Mikroskopische[n] Gemühts- und Augen-Ergötzung" ("Microscopic mind and eye entertainment") [94].

[33] A book by the professor of medicine Nicolai from 1746 explains what the spermatozoa are all about: "Man is, according to his origin, a worm, and all men, before they were born, were nothing else than small spermatozoa. Every father carried all his children with him in the form of spermatozoa before the mother became pregnant with them, and still contains within himself all his future children in just this way." (from [95], p. 101).

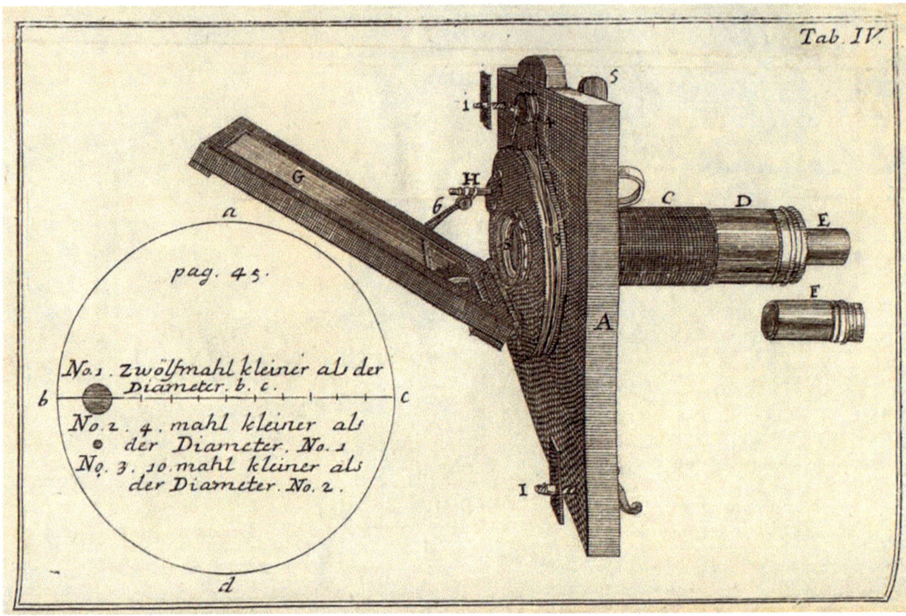

Fig. 3.34 Solar or Microscopio in Camera obscura, Johann Ludwig Steiner (1756) from [99] Table IV

solely on the sun. Baker[34] describes the device in 1743 as "The Solar, or Camera obscura Microscope" (in [97], p. 22) and at this point also mentions the improvements introduced by the optician Cuff. In the German translation by the optician Steiner[35] [99], a drawing (Table IV, Fig. 3.34) and a description (VI. Chapter, p. 24) are given. Steiner explains the principle in great detail

> **A** is a square wooden board through which two screws **I i** pass, with which the board can be conveniently fastened to a shutter. In the middle of the board is a round hole, into which a sleeve is inserted, and on one side of which comes a wooden shaft, which surrounds a gut string, in the middle above this shaft is another smaller shaft of brass, which at the same time has two surrounds from the sides, both ends of which are made on the larger string. On the other side of the small shaft there is a handle which makes a slight movement to drive the wooden shaft **B** with all its accessories around.

Somewhat awkwardly, but precisely, Steiner describes the turning device of the mirror in this. A brass tube **C** (interestingly, it is recommended to cover this with seal skin) takes

[34]Henry Baker, published under the pseudonym "Henry Stonecastle of Northumberland" together with his father-in-law Daniel Defoe (author of the novel "Robinson Crusoe") the magazine "The Universal Spectator".

[35]Besides also translator of Euler's color theory (from [98], p. 97).

another brass tube **D** (without seal skin) is pushed into tube **C** and now forms with a short (soldered) tube **E** the receptacle for tube F (screw-on point for the imaging optics, the "microscopium"). The room is darkened and the apparatus is placed in the shutter so that the mirror **G** is on the outside and the tube **CDEF** is on the inside. The plane mirror can be adjusted in angle by means of the hinge and the adjusting screw **H.** The mirror is set up in this way ("put in the right order") so that the sunlight hits the converging lens **5** (somewhat difficult to see in the drawing, of course in the middle of the shaft). Steiner: "..., that is, when the rays of the sun are directed through the glass **5** onto the white wall, and thus present a perfectly round glow."

The object to be microscoped is located on a glass slide which is inserted between the illumination beam path (mirror and lens 5) and the imaging beam path (in the objective tube F) between two brass plates. This is shown in the drawing by the two beads on tube D. Tube D can now be pushed in or pulled out so that an object lies more or less in the focal point of the lens. The sharpness of the image on the wall is adjusted by moving the tube F.

Steiner's explanation shows the similarity between the camera obscura and the solar microscope. In his "Treatise on the Solar Microscope" ("Abhandlung vom Sonnenmikroskop") Rußwurm[36] aptly writes in 1781: "The step from the magic lantern to the solar microscope was thus very short. Not much more was needed than to use the first lens glass of the magic lantern as a collecting glass and to attach a more magnifying one instead of the second one" (from [100], p. 4).

The mathematician Euler developed an improved Laterna magica as a kind of solar microscope from the idea of the camera obscura [101]. In a letter[37] to the Princess of Prussia (Friederike Charlotte von Brandenburg-Schwedt) in 1762, he once again describes his intention (in [102], p. 197): "It is therefore important to find a means of illuminating the object in the camera obscura as much as possible without the light penetrating the camera.

Euler had solved this problem by means of incident light illumination (Fig. 3.35). In the picture there is also no sun illumination, but only a candle illumination. Today, such magic cameras[38] are called reflected-light projectors or episcopes.

---

[36] Wilhelm Friedrich von Gleichen-Rußwurm.

[37] In the 196th letter, in the original in french, here from the German translation of 1780.

[38] The term camera magica ("Wunderkamera") is used at least by Liesegang for the episcope (see [67]) Perhaps from the composition of **camera** obscura and Laterna **magica** to **camera magica.**

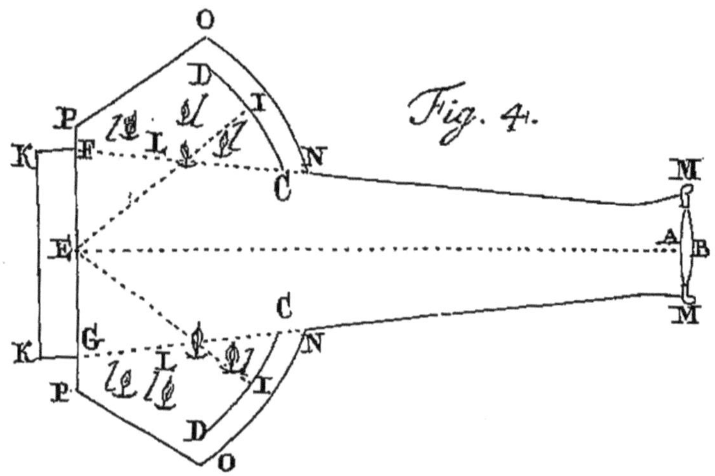

Fig. 3.35 Camera Magica, Leonhard Euler (1753) from [101], p. 381

3.4 Moving Images

Wonder Wheel and High-Speed Viewer

The reproduction of an image turns the camera obscura into a magic lantern. Through a superimposition of individual images, the perspectives merge into an impression of movement. But it is only through the permanent recording of the latent image that the pure viewing device of the camera obscura becomes a photographic camera. By reproducing the recorded images in a sequence, the magic lantern finally becomes a film projector.

A limited form of projection of movement as a sequence of images or by means of image rotation had already been described by Kircher, Zahn or Ehrenberger. However, a realistic impression is created when the images are cross-faded. The term "dissolving views"describes not only the cross-fading of two subsequently played individual images, but also the sometimes desired blurring of the image projection during the image change (by adjusting the projection lens into the blurred area).

The effect goes back to the end of the eighteenth century and the illusionist Phylidor.[39] Little is known about Phylidor's origins, but numerous of his performances, especially in

[39]Presumably Phyldor is identical with Philidor/Philidore/Phillidore and also with Paul de Philipsthal, as reported by the magicians Schirmer and Scholl in 1805 [103]: "...Mr. Phillidore, alias Philipsthal..." (p. 15), "...Mr. Philidore, alias Philipsthal..." (p. 16) and "...Mr. Philidore, alias Philipsthal..." (p. 19).

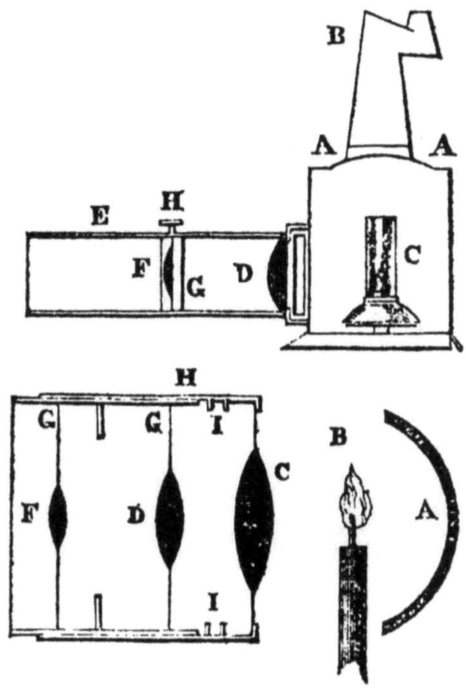

Fig. 3.36 Phylidor's phantasmagorical lantern, William Charlton Wright (1825) from [105], p. 314

necromancy and black magic, have been handed down by contemporary witnesses.[40] In London, he—now under the name Philipsthal—extended his performances in London by the phantasmagoria. In an explanation of his magic tricks from 1825 [105], his magic lantern is also shown (Fig. 3.36).

Philipsthal cross-faded two different images to achieve amazing effects. A well-known demonstration was the cross-fading of the Witch of Endor[41] with the ghost of Samuel (by means of two "Laterna magica"). However, Philipsthal later added the cross-fading of landscapes to these presentations [106]. In England, these dissolves were commonly referred to as dissolving views [107].

Philipsthal received a royal patent on his phantasmagoria on January 26, 1802 ([108], pp. 303–305):

> Specification of the Patent granted to Mr. Paul de Philipsthal, of the Lyceum, in the Strand, in the County of Middlesex, Gentleman; for his Invention of representing, in a dark Space or Scene, the human Figures, in various Characters, Proportions, and Sizes, and by which Means

[40] E.g. "The unmasked Phylidor as a blackmagic artist, ghost-citirer, and charlatan" (in [104], p.

[41] A fortune teller and necromancer from biblical history who invoked the spirit of the deceased prophet Samuel, read in the First Book of Samuel, Chapter 28.

Fig. 3.37 Phantasmagoria, Étienne-Gaspard Robertson (1831) from [110]

> Painters and other Artists may accurately enlarge or diminish any painting with more certainty and facility than has hitherto been known or done*.
>
> Dated January 26, 1802.This invention is now publickly exhibited under the name of the Phantasmagoria.

At the same time as Philipsthal, the Belgian magician Robertson also appeared with his phantasmagorias [109]. According to his own information, Robertson was initially a professor of physics[42] at the College of the Ourthedepartement[43] (College des Ourthedepartements), but quickly turned to phantasmagoria. His presentations were particularly intended to frighten the audience rather than to edify it, as a drawing on the inside cover of the first volume of his memoirs clearly illustrates [110] (Fig. 3.37).

In the picture one sees what Robertson had hoped his audience would see: a spectator lies gripped by fear on the floor, a brave soldier draws his sabre, a lady needs to be calmed down, others slap their hands in front of their faces, turn away or praise the apparitions. The room is filled with smoke, upon which the illusionist maps his visions of fear. The device Robertson uses to project the images onto the smoke is called a "fantascope". For this

[42]A professor of physics at a college in France is, according to the German view, usually a physics teacher at a secondary school. The official title high school professor was also common in Germany.

[43]The Département de l'Ourthe (named after the river Ourthe) was a Belgian department of the French Republic from 1795 to 1814.

device, Robertson obtained in the turmoil of the French Revolution a patent N°. 2755 (from [111], p. 14[44]):

> On the 27th Ventôse of the year VII, he obtained a patent of invention for five years, on this day, to Étienne-Gaspard C. Robert, professor of physics, residing at Paris, rue de Provence, n. 24, for the purpose of designing, constructing, and selling throughout the republic the same, an apparatus which he called fantascope, and which he intended to perfect Kircher's lantern, commonly called magic lantern;. . .

After the French Revolution of 1789, the "Décret de la Convention nationale portant sur la création du calendrier républicain" [112] introduced the French Revolutionary Calendar from the year 1792, which placed the beginning of Year I (Year One of Liberty) on the day after the storming of the Bastille (i.e. on July 15, 1789). However, the count begins again with Year I (Year I of the Republic) after the abolition of the monarchy on September 22, 1792 by the National Convention. Thereby not only a decimal division of time was invented (a day had 10 h of 100 min at 100 s each), but also completely new month names. The 27th Ventôse (wind month) of the year VII in the republican revolutionary calendar is therefore the 17th March of the year 1799 in the Gregorian calendar (conversion table from [113], p. 3).

In Robertson's memoirs [110], in addition to many illustrations and explanations of his magic tricks, there is also a drawing of the "Fantascope"in the chapter on "Fantasmagoria" (see Fig. 3.38). In principle, it is also a standard magic lantern with light source, reflector mirror and condenser lens.

The spread of phantasmagoria and the various types of magic tricks challenged scientific analysis. No less a person than David Brewster devoted himself to this elucidation in the book "Letters on Natural Magic" [114] and described the workings of magic tricks; besides the illusion of Kircher, Robertson and Philipstal also other optical, mechanical or chemical tricks.

Robertson wasn't just an entertainer, though. In addition to his necromancy, he was also a pioneer of balloon flight. In an essay he gives a detailed account of his ascent to 3600 toises[45] (about 7000 m) [115]. The balloon flight with the Hamburg merchant Lhoest (also L'Hoest) on July 18, 1803 [116] almost ended in disaster [117]:

---

[44] Own translation.., in the original: "LE 27 ventôse de l'an VII, il a été délivré un brevet d'invention pour cinq années, à compter dudit jour, au C. [en] Étienne-Gaspard Robert, professeur de physique, demeurant à Paris, rue de Provence, n.° 24, à l'effet de construire, faire construire, et vendre dans toute l'étendue de la République, un appareil qu'il a nommé *fantascope,* èt qu'il a annoncé être le perfectionnement de la lanterne de Kircher, vulgairement appelée *lanterne magique;* à la charge de suivre les procédés dont il a fourni les dessins et le mémoire explicatif."

[45] A toise de Paris was equivalent to 6 feet at 12 thumb widths each, about 1.96 m today.

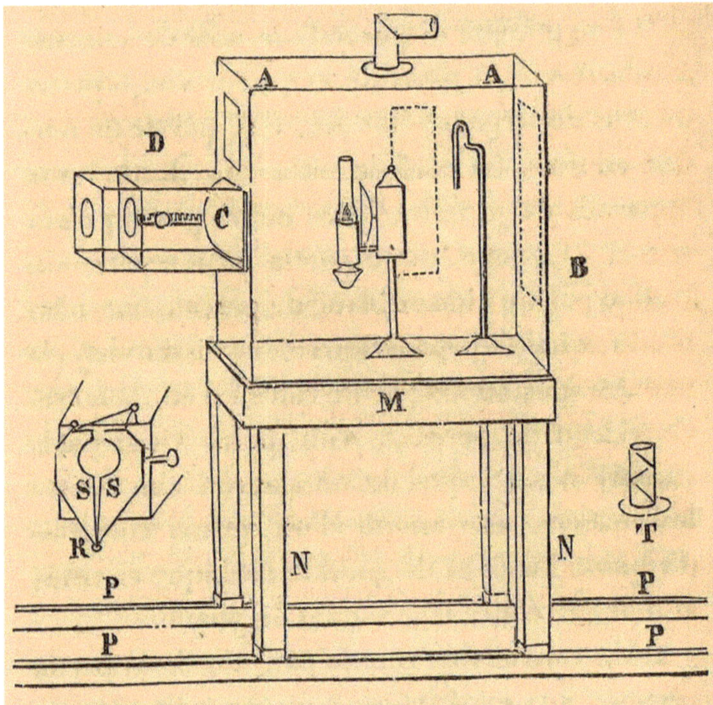

Fig. 3.38 Fantascope, Étienne-Gaspard Robertson (1831) from [110], p. 326

> The cold was not extraordinary, but the buzzing in front of the ears increased, and a strong uneasiness seemed to paralyze all mental activity. Having fortified myself by some wine, I began some experiments, with which, however, I was not satisfied. I therefore suggested to my eager companion that he should climb still higher, to which he readily consented, although he was also in a bad way. We threw out ballast again from time to time. The mercury of the barometer dropped to 12 4/100 degrees. At this altitude the cold outside the gondola was unbearable, although the thermometer was only 1 degree below freezing; we had to breathe faster, and our pulses were extremely violent. We could hardly resist the sleep that was trying to take hold of us. The blood was rushing to our heads, and Mr. Lhoest noticed that it was getting into my eyes. My head was so swollen that I could no longer put on my hat.

Robertson and Lhoest, however, survived their adventure. And so it is only on the tomb of Robertson in the cemetery Père Lachaise to his unusual life extraordinarily vivid reliefs, which are about conjuring spirits and ballooning.

This type of performance using a double magic lantern was repeated in a similar way by Henry Childe from 1822 and later reduced to a double projector by Edward Clarke with the "Biscenascope" [107]. Ludwig Döbler made this technique, known in England as "dissolving views", acceptable in the German-speaking world as "Nebelbilder" ("fog pictures").

Projection Magic

In the middle of the nineteenth century, the successful Viennese magician Ludwig Döbler thrilled the public not only with his magic tricks, but especially with his fog pictures [118].

Döbler demonstrated his fog picture apparatus in January 1847 at the Theater in der Josephstadt (Vienna). The editor of the Österreichisches Morgenblatt describes his impression:

> The present performance is certainly a new, honourable progress of his striving in the field of optics. . . . In the second section we see nebular images with moving figures. The solution of this optical task, which, as far as we know, belongs to Döbler's inventive talent alone, was at any rate satisfactory, . . . What pleased us most was the deceptive mobility of the 'Turkish juggler', the 'hoop-jumper' and the 'rope-dancer'. . . [119].

In the years following the first performance, one also finds the designation "Döbler's Nebelbilder".[46]

Döbler's fame went well beyond Vienna. He also appeared before the Russian Tsar Nicholas [118] or the King of Prussia,[47] and finally even met Goethe in Weimar. At Döbler's request, Goethe dedicated a line to him, "Not necessary to seal a diploma, you have shown us the impossible." ("Bedarfs noch ein Diplom besiegelt, unmögliches hast du uns vorgespiegelt.") [121]

At least as famous as Döbler's fog pictures was "Pepper's ghost" a few years later. The magic trick has nothing to do with pepper spirit, but is named after its inventor,[48] the London teacher Pepper. The idea behind it is simple, it is a simple mirror trick. Behind a pane, which is tilted by 45 degrees, is the stage. On this brightly lit stage the performance takes place. In another part of the stage, which cannot be seen by the spectator, the "ghost" or some person who appears unexpectedly is waiting to perform. When the ghost is illuminated, it is reflected in the tilted mirror. If the mirror is tilted appropriately, the ghost now appears to the spectator on stage as well. This effect is nowadays sometimes mistakenly called a hologram.[49]

---

[46] E. g. in various announcements of the "performance of Döbler's 'Nebelbilder'" at the National Theater Munich (1845) [120].

[47] In Töplitz in the summer of 1831 before Friedrich Wilhelm II [121].

[48] Whereby Henry Dircks should also be named as the inventor [122], who presumably provided the initial idea with his phantasmagoria "The Ghost" [123] and finally applied for a patent together with Pepper (see also [124], pp. 4–12).

[49] The "Spiegel", for example, uncritically adopted the false Reuters description of Madonna's performance at the 2006 Grammy Awards: "Madonna (L) performs from behind a holographic screen with a holographic Gorillaz performer". (from [125]).

Fig. 3.39 The thaumatrope. John Ayrton Paris (1827) from [130], p. 338

Döbel's nebular imager was able to reproduce several images in quick succession and thus already a short animation. A mechanical alternative to fading in the fog picture is the alternating display of two individual images. A disc painted with different images on the front and back gives the illusion of fading these two images when rotated about an axis. This simple explanation describes the functioning of a thaumatrope, the "miracle disc".

Of course, the simplicity of a thaumatrope is in itself almost proof of its ubiquitous presence. Seduced by this possibility, the prehistorian Azéma and the experimental paleontologist Rivère consider as a prehistoric thaumatrope bone discs found in the Stone Age with two carved images and a hole in the middle [126]. The experiment carried out by them certainly allows this interpretation, but does not provide incontrovertible proof. It could have simply been a button....

A thaumatrope can be easily created. The flip of a coin turns the images of heads and tails into a composite image, and is familiar to any schoolchild or pub-goer. The modern discovery of the thaumatrope has nevertheless been the subject of debate. In one article Claudet writes of the simplicity of the "invention" ([127], p. 425):

> I have been told that as Sir John Herschel[50] was one day making this experiment to amuse his children, in the presence of the late Dr. Paris, this gentleman was struck with the idea that if, instead of a coin, a white card was employed on each side of which one part of a design was properly arranged, the two might complete the subject during the revolution.

Paris actually put his idea into practice, as an edition of his book[51] "Philosophy in Sport" from 1827 proves [129]. There, the thaumatrope is presented as an instructive toy with which it is possible to create a combined picture from two individual images. In the first illustration the operation is shown. Tensioned on two strings, the disc, which is painted on both sides, can now move quickly around the axis formed by the string (Fig. 3.39).

---

[50]Not to be confused with his father, the no less famous astronomer Wilhelm Herschel, who had discovered not only Uranus but also infrared radiation [128].

[51]The book appeared anonymously, but is generally attributed to Paris.

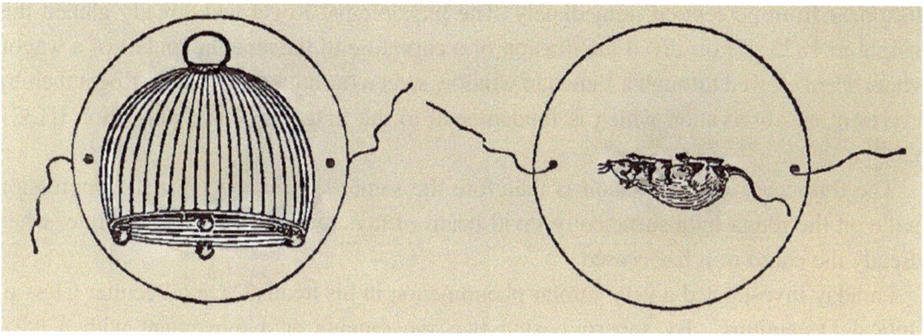

Fig. 3.40 Cage and mouse, John Ayrton Paris (1827) from [130], p. 338

Another sketch shows the basic idea of the picture arrangement. On one side of the cardboard disc an empty cage is drawn, on the other side a mouse. If the disc turns on the strings attached to the side, the mouse soon appears to be in the cage (Fig. 3.40).

Background Information

The thaumatrope is a very simple toy and the idea for it obvious. Nevertheless, almost 40 years after the conception of Paris' thaumatrope, it seemed necessary to the well-known mathematician Babbage to point out the supposed originator of this gadget.

Babbage mentions in his 1864 memoirs an incident which he believed to be the origin of the thaumatrope ([131], p. 189):

One day Herschel, sitting with me after dinner, amusing himself by spinning a pear upon the table, suddenly asked whether I could show him the two sides of a shilling at the same moment.

I took out of my pocket a shilling, and holding it up before the looking-glass, pointed out *my* method. "No," said my friend, "that won't do;" then spinning my shilling upon the table, he pointed out *his* method of seeing both sides at once. The next day I mentioned this anecdote to Dr. Fitton, who a few days later brought me a beautiful illustration of the principle.... Upon one side of this disc was painted a bird, upon the other side, an empty bird-cage. On turning the thread rapidly, the bird appeared to have to go inside the cage.

According to Babbage, the mouse in the cage would therefore only be a later interpretation of the bird in the cage and thus Herschel and Fitton would have to be named as inventors—if one is at all prepared to call such a simple thing an invention.

This change from just two single images can hardly be called an animation. But at least a cross-fading of two different pictures to one overall picture succeeds here. Although the images are not present simultaneously, but one after the other, the image does not seem to disappear from perception immediately. The physiologist Roget had already gained this insight in 1825. He observed the illusion of a curvature of the straight spokes of a wagon wheel when viewed through a Venetian window screen (a shutter with slats). Roget thereby arrives at an observation which is fundamental to the function of any animation ([132], p. 98):

The true cause of the illusion is therefore the same, . . ., namely, that the impression made on the retina by a sufficiently vivid beam of rays remains for a certain time, when already the cause of it has ceased.

Faraday investigated a very similar phenomenon in his treatise "On a Peculiar Class of Optical Deceptions". By superimposing two movements or a movement with a fixed grating, the illusion of a new movement, differently directed or deviating in speed, can be created. Faraday comes to the same conclusion as Roget ([133], p. 86)[52]:

> The cause of these appearances, when pointed out, is sufficiently obvious, and immediately indicates many other effects of a similar kind, and equally striking, which are dependent upon it. The eye has the power, as is well known, of retaining visual impressions for a sensible period of time; and in this way, recurring actions, made sufficiently near to each other, are perceptibly connected, and made to appear as a continued impression.

Kaleidoscope and Kaleidophone

Faraday then also refers in the article ([133], p. 86) to the "beautiful appearance of the kaleidophone", which will therefore be briefly explained. Wheatstone described the "Phonic Kaleidoscope" in 1827, following Brewster's "Kaleidoscope", as a device for forming figures from luminous dots. Wheatstone also recognized the apparent duration of the visual impression as the basis of the formation of the perception of a figure (from [135], p. 351).

> . . . as the property of vision which accasions the apparent duration in the same places of visible images, after the objects which excite them have changed their positions. . .

Wheatstone points out that this observation had already been made by d'Arcy in 1765. D'Arcy had presented a device by which a luminous point could be moved along a circular path. In the article "Durée de la sensation de la vue" (Duration of visual sensation) [136] an experiment of d'Arcy's is described, in which he moved a piece of glowing coal along a circular path, thus producing the appearance of a ring of fire. D'Arcy determined the duration of the impression to be 8 tercias,[53] which

(continued)

[52]See [134] p. 210 for original source

[53]A tercile or tertie is the 60th part of a second (sexagesimal second).

corresponds to 0.133 s according to modern terminology. Thus, according to d'Arcy, for the perception of a continuous movement of at least 7.5 revolutions per second would have been necessary.

Interesting is another experiment of d'Arcy's, in which he observed a scene through a revolving opaque disc. In this disc was placed an opening through which he observed the surroundings. When the disk was rotating at a sufficient speed, the observed object could again be perceived as a whole ([136], p. 21). This almost anticipates the Nipkow disk, which Paul Nipkow patented more than 100 years later (more on this later).

Plateau was also familiar with the work of Roget and d'Arcy, as a glance at his "Dissertation sur quelques propriétés des impressions produites par la lumière sur l'organe de la vue" (Dissertation on some properties of impressions produced by light on the organ of vision) from 1829 reveals [137]. His afterimage experiments with simple structures together with his knowledge of the thaumatrope[54] subsequently lead him to real animation. He describes an apparatus called "phenakisticope",[55] which, according to him, was already being sold in London at that time under the name "fantascope" (Fig. 3.41), following his idea [137].

This apparatus consisted essentially of a cardboard disc, pierced on the circumference with a certain number of small openings or slits, and having painted figures on one side. Looking at the reflection of the figures of this disk as it revolved about its center in front of a mirror, one perceived the illusion of motion of the figures. Lubschez in his "Story of the Motion Picture" [139] made a sketch of the phenakistiscope in which the slit disk is separated from the picture disk. In the drawing (Fig. 3.42), the version for a display via a mirror is shown on the left-hand side, and on the right-hand side a viewer in which a mirror is no longer needed (this is in principle Plateau's later idea, which will be presented later).

A more detailed study can be found in the Annalen der Physik und Chemie of 1834 in an article on "Stroboscopic Disks, Phenakisticope, Phantasmascope" ([140], pp. 636–649).

> The cause of these illusions is very simple; it lies in the fact that the impressions of light persist in our eye for a while, about 0.2 sec. . . .. If, therefore, an object is shown visible alternately as long as is necessary to produce an impression of light, . . . we shall believe that we see it without interruption, . . . It is also clear that we shall believe that we always see the same object, and that it is at rest, even though it is replaced at each covering by another object entirely like it and placed exactly in its place. But if, during successive coverings, the original object is successively exchanged for another, . . . which . . . differs only in that it is drawn in a somewhat

[54]Plateau was satisfied with a brief summary from the informative reference work "Manuel de physique amusante" [138].

[55]Phenakistiscope: from Greek phenax = deceiver and skopein = to look at.

Fig. 3.41 Fantascope—Serpent surgissant du centre du disque, Joseph Plateau (1833), maker: Ackermann & Co., 96 Strand," Catalogue des appareils de la Cinémathèque française, Paris, Ref N°: AP-94-212, Basile Morin (2019), Wikimedia Commons

> different position or in a somewhat different place . . . we shall believe to see one and the same object in movement. . .

Plateau, at a meeting of the Belgian Academy of Sciences on January 9, 1836, demonstrates an invention called the anorthoscope, which takes up an idea from his dissertation on the rectification of an anamorphic (distorted) image (from [141], p. 7).

> I have the honor to present to the academy a specimen of the instrument of my invention, now published under the name of anorthoscope.

An anorthoscope consists of two disks, which are mounted one behind the other, individually rotatable. The distorted image is applied to the rear, transparent or translucent disc. This disc is illuminated from behind. In front of the image disc is an opaque black cardboard disc with several slits. If the two discs are rotated in opposite directions at a defined speed, a rectified image of the distorted figures is produced (after [142], pp. 464–466). Plateau says that reading on Faraday's cogwheel experiment had inspired

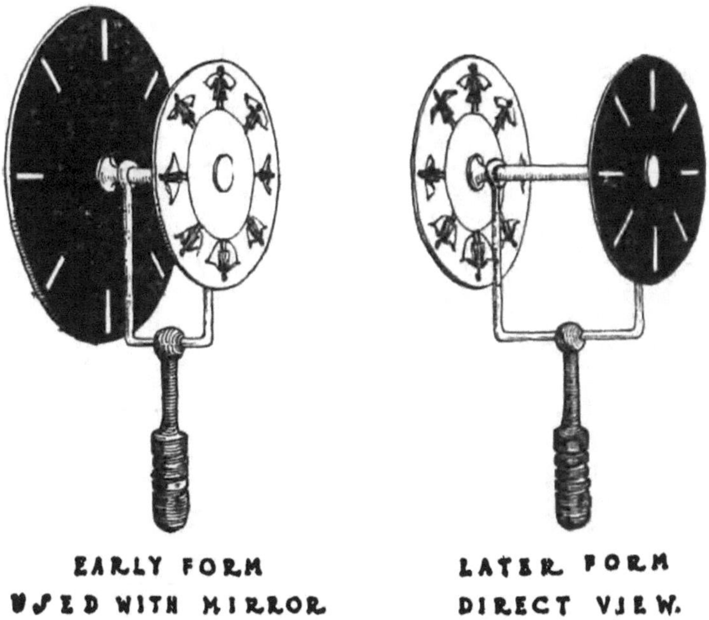

Fig. 3.42 Plateau's phenakisticope, Ben Lubschez (1920) from [139], p. 15

him to this device. In order to reproduce the superposition of the motion of two toothed wheels on millworks or two wheels with spokes that he had observed, Faraday built himself a device with two individually movable gears (Fig. 3.43). Plateau's anorthoscope is very similar to Faraday's gear construction, but Plateau represents a picture.

At this meeting, Plateau subsequently makes every effort to claim the invention of the moving image for himself and points to the "fantascope" or "phenakisticope" that he had already described. In particular, he tries to prove the earlier priority of his phenakisticope to Stampfer's stroboscopic discs.

At about the same time as Plateau (1833), Stampfer also succeeded in developing an animation with his "stroboscopic discs or optical magic discs" [143]. Stampfer developed his animation from the observations of motion behind fences (according to Roget), moving wheels (according to Faraday) and the thaumatrope of Paris, without mentioning (or knowing) Plateau. In addition to writing a good description of his, Stampfer, together with Trentsensky, succeeded in obtaining an imperial privilege for the "invention of drawing figures and coloured forms, and in general pictures of every kind, ... in such a way that, when they are passed by the eye with due rapidity by some mechanism, ... the most manifold optical illusions present themselves to the eye in connected movements and actions...".[56] This privilege helped Stampfer's discs to spread and become known under

[56]From Privilege No. 1920 in [144], pp. 406–407, granted May 7, 1833, for two years.

Fig. 3.43 Cog wheel experiment, Michael Faraday (1831) from [134] Plate 3

various names and trade names.[57] The simplicity of the technique and the resulting inexpensive production led to a multitude of variations, which, with the exception of a particularly melodious name, hardly differed from the original idea.

Stampfer, however, does not only think in terms of discs, he also proposes other arrangements of images on a cylinder or even on a strip of paper (in [143], p. 11).

> By means of several mechanical operations, the images of a period can be arranged in such a way that the last comes into contact with the first. The images can be distributed around the circumference of a disc or on the lateral surface of a cylinder, and the disc or cylinder can be rotated about its axes. In the case of a large number of pictures, they can be placed on a strip of paper or canvas, the beginning of the strip can be joined to the end, the strip can be stretched over two parallel rollers, and by means of these rollers the pictures can be set in circulation.

Horner was evidently aware of the experiments of Roget, Faraday, and Plateau in developing his device, called the Daedaleum,[58] but not of Stamper's idea of using a cylinder. In any

[57] E. g. stroboscopic discs or optical magic discs (Stampfer [143]), Wunderscheibe [145] or wheel of life, L'Ékonoscope [146], Phorolyt or Kinesiscope (Purkinje [147], pp. 62–64) circle-turner, periphanoscope, or optical magic theatre (Siebenmann [148], p. 272); pantinoscope [149]; by which, as a rule, Stampfer's, as well as Plateau's ph(a)enakisticope, were designated. As the actual inventor of the phenakistiscope Plateau is nowadays mostly mentioned.

[58] Named after Daedalus, a gifted artist and inventor in Greek mythology. He was the father of Icarus (whom he also built the wings) and builder of the labyrinth for the Minotaur (from which one could only free oneself with the famous "red thread of Ariadne"). Horner, however, refers in his naming to the walking statues of Daedalus (see, e.g., Aristotle [150], p. 37).

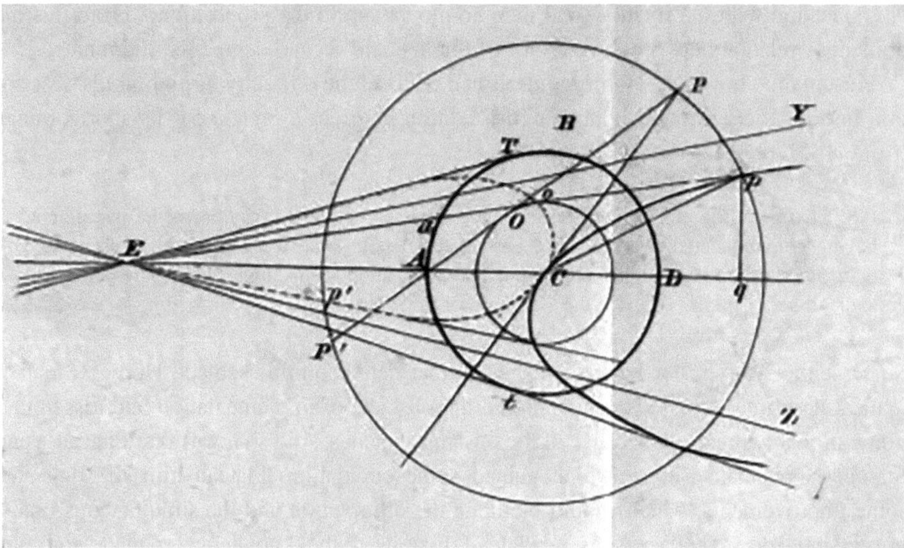

Fig. 3.44 Cross section through the Daedaleum, William Horner (1834) from [151] Plate V, Fig. 6.

case, Horner describes the improvement of a magic disc for convenient presentation to a larger group ([151][59]p. 651):

> The apparatus is merely a hollow cylinder, or a moderately high margin, with apertures at equal distances, and placed cylindrically round the edge of a revolving disk. Any drawings which are made on the interior surface in the intervals of the apertures will be visible through the opposite apertures, and if executed on the same principle of graduated action, will produce the same surprising play of relative motions as the common magic disk does when spun before a mirror. But as no necessity exists in this case for bringing the eye near the apparatus, but rather the contrary, and the machine when revolving has all the effect of transparency, the phaenomenon may be displayed with full effect to a numerous audience. I have given this instrument the name of Daedaleum, as imitating the practice which the celebrated artist of antiquity was fabled to have invented, of creating figures of men and animals endued with motion.

Since Horner was a mathematician by training, he described the Daedaleum mathematically quite extensively on the basis of a drawing (Fig. 3.44) and considered extensively the peculiarities of the curved projection of the images onto the circle.

In Fig. 3.44 is a section through the cylinder of the Daedaleum in plan view with **C** as the centre and pivot. The observer looks from the position **E** at an opening **A** of the cylinder, shown here as an inner circle with a thick line (the outer circle with a thin line is for further construction). Horner derives various mathematical functions from this simple observation

[59]From the original publication "On the properties of theDædaleum, a new instrument of optical illusion" [152].

in connection with the rotation, and then he mounts up to the proposition: "From this the immediate relation between the cissoids of the first and second degree [is] illuminated."

Presumably, however, Horner himself had realized the difficulty of putting his description into practice, in the original English edition he gives a pretext for its shortcomings (from [152], p. 41):

> P.S. I have not thought it requisite to give a more particular description of the instrument, having communicated every needful part of the detail, some weeks ago, to a respectable optician of Bristol, Mr. [John] King, jun. The publication seems, however, to have met with some impediment, probably in the sketching of the figures.

Since neither Horner nor King published further details on this subject, Horner's mathematical description then also encountered the obstacle of implementation and dissemination—probably precisely because of the missing sketches. And so it was first Plateau's and Stamper's phenakistiscope, which gained some circulation. Plateau himself suggested some improvements, which should facilitate the observation and the simultaneous image impression for several persons.[60] Plateau first used the knowledge gained from the anorthoscope and the phenakistiscope. By driving two discs at different speeds, he was able to create defined ratios between the speeds of rotation. In his publication of 1850 [153][61] he describes his construction in such detail that it should be possible to reconstruct the device. In the following, only a short explanation is given.

The black disk (slit disk) now rotates in Plateau's device four times faster than the translucent disk (image disk). 16 distorted images are applied to the image disk at 16 even angles and superimposed with 4 even columns of the slit disk. Plateau illuminates the place of image formation from behind with the directed light of a strong lamp, so that only a small, selected part of the image disk is illuminated. Thus, in a dark room, only a single moving image is seen. This basic arrangement with two discs moving at different speeds and selective illumination can possibly be called a first screen.

In his essay, Plateau also presents two further ideas on how to achieve even more impressive effects with his phenakistiscope. Both ideas come from Wheatstone, who had suggested to Plateau that the phenakistiscope be coupled with the stereoscope. To do this, simply attach two Plateau's apparatuses to either side of a Wheastone's mirror stereoscope. In order to achieve the greatest possible closeness to reality, two stereoscopic image sets (daguerreotypes) are to be created for this purpose by means of the (then) new medium of photography.

[60]Until then, to view the images, you had to hold the rotating disk close to one eye and close the other eye.

[61]The lecture was given as early as 1849; the evaluation given here is based on the German edition [154].

Let There Be Light!

For many centuries, torches, candles and oil lamps were the brightest artificial light sources. To achieve a higher brightness, one could simply increase the number of fireplaces or increase the light output by suitable light guidance with mirrors and lenses. Only in 1783 time Argand[62] improved the oil lamp with a hollow wick for an optimized oxygen supply and additional glass cylinder ([155], p. 50). In 1799, Philippe Lebon filed a patent application for a lamp to generate light and heat from gas [157], which later became known as the "thermal lamp".[63]

Because of its simplicity, economy and brightness, Argand's lamp was copied in many countries and can be traced to Germany from 1799 at the latest (see, for example, [159]). In Scherer's description, in addition to the detailed description, there is also a detailed drawing, which certainly simplified the reproduction (Fig. 3.45).

The Argand lamp or Argand burner is the original type of oil and petroleum lamp of the nineteenth century and was the preferred source of light in private rooms until the successful introduction of electric light by Edison in 1880. In public areas, theatres or street lighting, the transport of the fuel to the burner was the real challenge. The distribution of gas is much easier compared to the transport of oil, in addition, due to the lack of a wick, it also requires much less maintenance. This makes gas suitable for street lighting. Until today, about 10% of the public lighting in Berlin is realized by gas lanterns.[64]

The luminous intensity of a candle is about 1 candela.[65] A thermal or Argand lamp around 1800 emitted the light of about 10 candles (from [161], p. 44). The use of an incandescent mantle according to Auer[66] increased the brightness to 70–80 Hefner

(continued)

[62] Argand patented his lamp first in England (patent no. 1425) and later also in France. At the same time, the apothecary Quinquet and the distiller Ambroise Bonaventure Lange (L'Ange) also claimed the invention ([155], p. 50). Interesting here is the explanation in a letter of February 13, 1784, from Antoine-Alexis Cadet de Vaux (a politician and scientist in Paris) to Benjamin Franklin, one of the founding fathers of the United States. In it, the assumption is made that Quinquet and Lange had learned the details of his invention from Argand himself during their joint work with Argand on Montgolfier's hot-air balloon in the autumn of 1783 ([156], pp. 563–564).

[63] The term is a direct translation of Lebon's term "thermolampes" in the 1801 addendum to his patent ([157], p. 126) [158].

[64] Out of a total of approximately 224,000 street lights, 31,500 were still powered by gas as of March 2018 [160].

[65] Lat. for candle.

[66] However, according to current knowledge, the production of Auer mantles is not particularly environmentally friendly. At the site of the "Glühstrumpf-Fabrik Auer von Welsbach" in Vienna (glow mantle production from 1887 to 1928), contamination and radiation exposure from pitchblende, thorium and uranium were found (up to 17 μSv/h), which are still significantly above the limit values today [162].

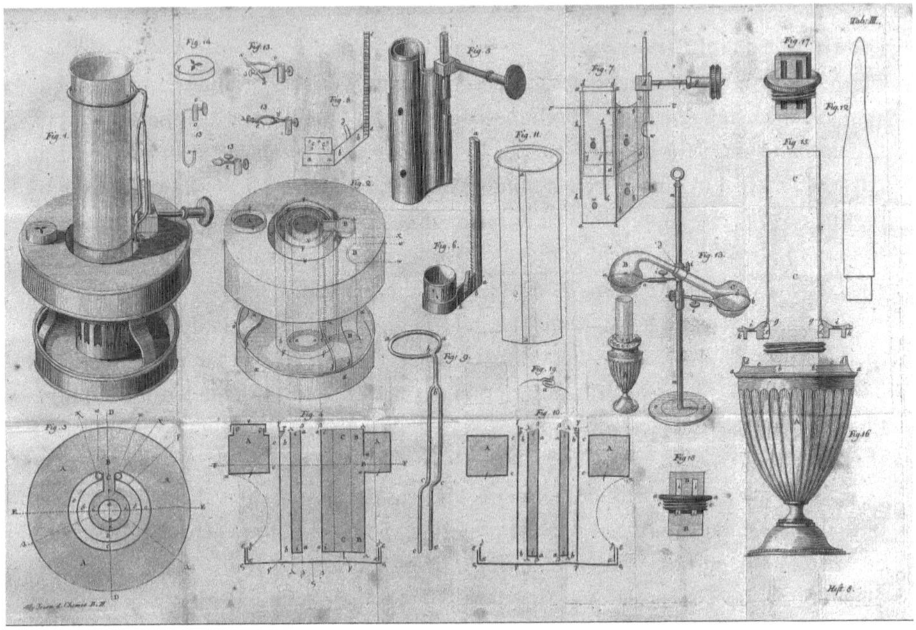

Fig. 3.45 Argand lamp, Alexander Nicolaus Scherer (1799)

candles[67] (from [163], p. 180). This corresponds approximately to the brightness of a conventional 60 W light bulb.

Even brighter was the lime light of Drummond [164], who used a combination of gas burner and parabolic mirror for long-distance signalling. Drummond's decisive improvement, however, was that the flame of the burner did not produce the light directly, but heated a material to white heat. Drummond experimented with different materials and eventually found that a glowing limestone shone 37 times[68] as brightly as an Argand burner.

The English phrase "to be in the limelight" goes directly back to Drummond's limelight.

[67] At the end of the nineteenth century, the standardized unit Hefner candle (after the standard candle of Friedrich von Hefner-Alteneck, 1845–1904) was used to define luminous intensity. 1 Hefner candle corresponds to 0.903 cd.

[68] That would be about 370 cd light intensity, which is still bundled by the parabolic mirror. According to Drummond, the light from his lamp could still be seen at a distance of 67 miles (107 km).

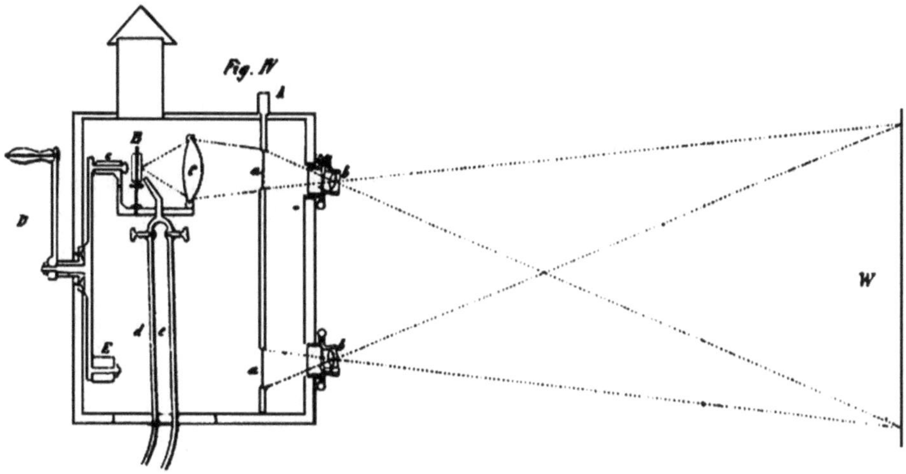

Fig. 3.46 Apparatus for displaying moving images, Franz Uchatius (1853) from [165] Fig. IV

By using these extremely bright illuminants, the wonder wheels could now also be used to project moving images, despite the heavily darkening slotted disc.

The Austrian artillery captain Uchatius had been given the task by Field Marshal Hauslab to investigate the principle of Stampfer's stroboscopic discs for an enlarged display on the wall [165].

Uchatius solved the task by designing a simple and elegant projection apparatus, which he called—somewhat less elegantly—"Apparatus for the Display of Moving Images on the Wall" [165]. Now it was possible to display 100 images in succession.[69] Uchatius first experimented with a common lamp,[70] but then failed because of the light yield of his apparatus (from [165], p. 483):

> The apparatus produced quite nice moving pictures, the size of which, however, could at most be increased to 6 inches in diameter; since at a greater distance from the wall . . ., the pictures became too poor in light because of the slits, an enlargement of the slits, however, entailed a greater indistinctness.

Since Uchatius recognized that the slits themselves lead to a darkened image on the wall, he placed a stationary, inclined objective lens **b** in front of each image of the stationary image wheel **a** (Fig. 3.46). Now, if all the images **a** were illuminated simultaneously from behind, they would appear at the same time at the same place on the wall W. Therefore, instead of the

[69] Uchatius suggests a playing time of ½ min for 100 frames, so only about 3.3 frames per second would be displayed, which does not allow for smooth motion.

[70] According to Fig. III in [165] (not illustrated here) it will probably have been an Argand lamp.

discs, Uchatius rotated the illumination (consisting of a lime cylinder **B** and the lens **C**)[71]. The illumination is slightly pivoted about the pin c, so that the illumination system always remains in an upright position. The gas supply tubes **d** and **e** are elastic and move up and down through the perforated bottom of the apparatus with each turn of the crank. E is merely the counterweight, which provides easy rotation.

As an assistant to Purkinje, Czermak had become acquainted with his phenakistiscope, known as a phorolyt or phoroscope and in 1855 it seemed desirable to him to give the representation with this device also the dimension of depth [166]. Czermak's ambition is not to represent movements of plastic figures, but in particular to produce movements in the dimension of depth. At first he used corporeal objects instead of plane images on the screen.[72] Czermak is surprised by the effect of his experiment (from [166], p. 464):

> The whole effect—although expected and foreseen—was surprising, for the illusion was complete, and hardly to be distinguished from reality—except by the silence and uniformity of the movements.

But later he also experimented with a combination of stereoscope and phoroscope. Czermak uses the terms stereophoroscope or stereophorolyt for his construction. In contrast to Plateau he does not propose the Wheatstonian mirror stereoscope, but the Brewsterian lens stereoscope. First, the stereoscopic image pairs are created and pasted side by side on the side faces of a multi-sided prism P. Horizontal slits are made around the inner prism in the faces of a circumferential cardboard belt (with the identical number of faces). Through the lenses of a stereoscope, the viewer now looks through the slits S at the inner stereoscopic image pair and receives an impression of space (see Fig. 3.47).

It is to be noted here that Czermak—without considering it worth mentioning—passes here from the usual two-dimensional arrangement of stacked discs of a phenakistiscope to the three-dimensional arrangement of a moving cylinder described by Plateau and Horner.

The cylinder shape is also used by Lincoln[73] in 1867 for a toy he calls Zoetrop[74] [169]. Lincoln does not mention or know of any preliminary work, except for the fact that an image remains on the retina for a moment. The illustration shows the simplicity of the device, which is explained by a frugal description in the accompanying text. The sketch, which is exceedingly comprehensible compared to Horner's description of the Daedaleum, becomes the actual template for the Zoetrope (Fig. 3.48).

[71] Since oxyhydrogen gas is passed through the gas supply lines and ignited, the lighting is Drummond's lime light.

[72] Czermak used pins, which he stuck into the image disk of the phenakistiscope at different depths.

[73] Meant here is William Ensign Lincoln, son of Prof. John Larkin Lincoln, Brown University (according to [167], p. 997 and in the preface to the memoirs of John Larkin Lincoln [168]).

[74] From Greek zoe = life and trope = turning.

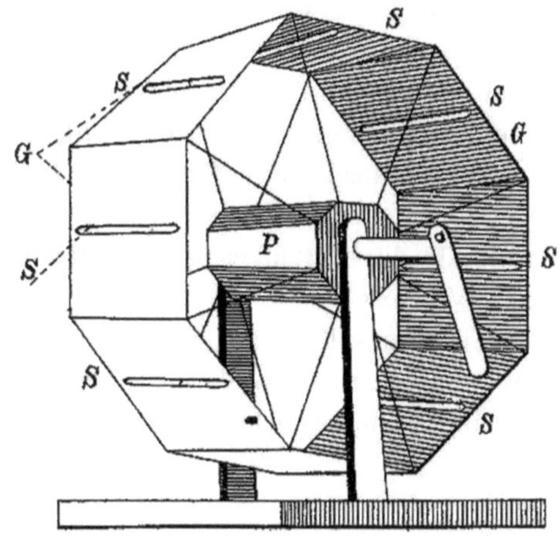

Fig. 3.47 Stereophoroscope, Johann Nepomuk Czermak (1855) from [166] Plate III Fig. 26

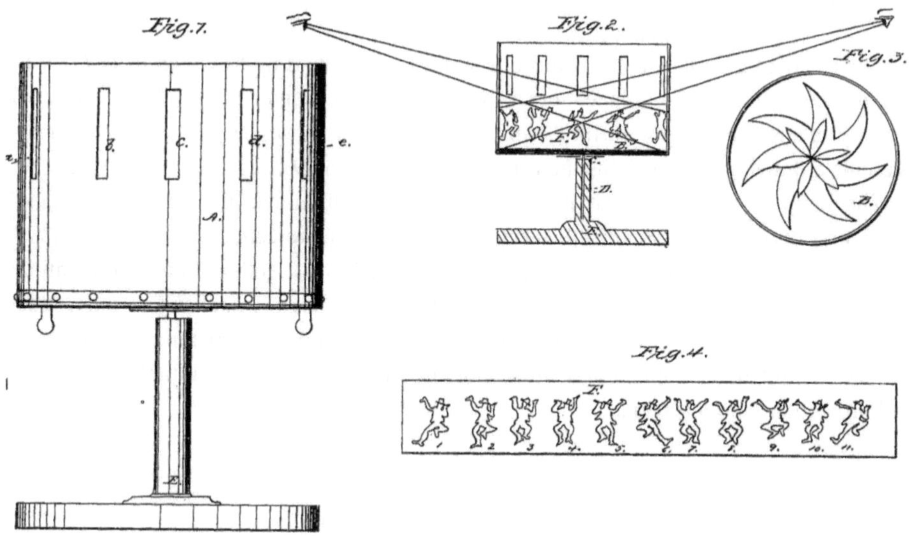

Fig. 3.48 Toy called Zoetrop, William E. Lincoln (1867) from [169], p. 1

Bradley had also applied for a patent for his toy in London in 1867,[75] which he later marketed under the name "Zoetrope".[76]

[75] Patent GB No. 629 "An improved mode of and means for producing optical illusions" (from [170]).

[76] Quite a few of these simple toys [171] and picture strips [172] survive to this day (both examples from the Penn Libraries, University of Pennsylvania), on view online at the "Magical Motion Museum".

The young Maxwell also enjoyed himself with the phenakistiscope ([173], p. 36). Several years later, the now adult Maxwell ordered a "Wheel of Life" (Zoetrope) with concave lenses instead of slits from the Glasgow instrument maker James White in order to improve the brightness of the wheel of life, as he wrote in a letter to Thomson in 1868. The focal length of the lenses in Maxwell's wheel of life is exactly the same as the diameter of the drum, resulting in images that are only half the size. Maxwell uses the device to dynamically display Rankine waves,[77] Helmholtz rings,[78] the fourth figure of the Lancer,[79] and the Volvox globator.[80] Such a zoetrope was in the possession of Mrs. Maxwell ([173], p. 485) and can still be seen in the Cavendish Laboratory, Cambridge.

Maxwell even thought through a binocular zoetrope, but feared a great difficulty in producing the images, which he described as having a 130-fold trouble ratio (from [175], p. 447):

> ... but the amount of care required to draw a presentable stereoscopic pair of pictures is about 10 times that required for presentable wheel of life with 13 pictures so that I estimate the ratio of trouble at 130/*n* where *n* is the ratio of expectation of an accurate picture in a stereoscope is the same in wheel.

Another way to give the zoetrope more brightness was found by Reynaud in 1877[81]. The praxinoscope has several plane mirrors mounted in the center of the axis of rotation, directly facing the images. In his description of the praxinoscope [179], Tissandier also presents a drawing that explains the principle well (Fig. 3.49).

The lamp in the illustration is of course used to illuminate the device. Reynaud's praxinoscope certainly gained greater fame through his projection praxinoscope [179], with which larger groups could now also see a presentation together. Like a conventional praxinoscope, this projection device consists of an inner mirror wheel and an outer image wheel (see Fig. 3.50). However, the image surfaces are now transparent and the mirror surfaces are tilted, so that a large-format moving image is created by suitable illumination and projection onto a screen.

---

[77] After William Rankine, who wrote a paper on the shape of waves on the surface of oceans [174]. Maxwell illustrates the wave motion with long plants of seaweed ([175], p. 445).

[78] In 1858, Hermann von Helmholtz published an essay on "Integrals of the hydrodynamic equations corresponding to vortex motions" [176]. In it, Helmholtz describes the hydrodynamic equations for vortex lines, vortex filaments and finally also vortex rings in water motion, to which Maxwell presumably refers here.

[79] "The Lancers" is a form of Square Dance (the Quadrille) and was very popular in England at the time of Maxwell. The fourth dance figure is called "La Trénis" after the German dance master Trenitz.

[80] Volvox Globator is a spherical, colony-forming green alga and was described by Cohn in 1856 [177]. Maxwell illustrated quite appropriately with the "Wheel of Life" the growth of a sphere until the final bursting to release the daughter spheres.

[81] British Patent No. 4244 "Praxinoscope and Praxinoscope Theatre" (after [178], p. 238).

Fig. 3.49 Praxinoscope after Reynaud, Gaston Tissandier (1879) from [179], p. 133 Fig. 2

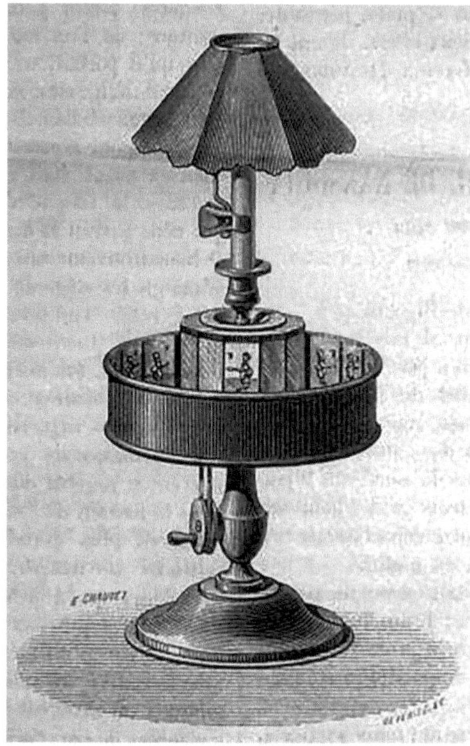

Fig. 3.50 Reynaud's projection praxinoscope, Poyet (1893) from [179], p. 61

However, Muybridge's serial photograph of a horse in motion is generally considered to be the first record of a real, moving scene [180].

The French physiologist Marey was interested in the study of movement. Marey's work on animal mechanism "La machine animale" [181] was also available in English from 1874 [182] and also contains a chapter on the movement of horses. Marey determined the pressure of the hooves on the ground in motion and found that in normal motion only two hooves touch the ground at the same time, but in gallop only one. Stanford, an entrepreneur and politician, was fascinated by the Frenchman Marey's research on animal locomotion and had hired the photographer Eadweard Muybridge to photograph his horses in motion (Animal Locomotion). In his "Descriptive Zoopraxography", Muybridge recounts a controversy between Leland Stanford and Frederick MacCrellish, the publisher of the "Daily Alta California" (where Mark Twain was also employed as a travel reporter at the time) that ultimately led to his commissioning. The subject of this dispute was the question of whether it was at all possible for a horse during a fast gallop not to touch the ground with any of its hooves, i.e. to be almost completely in the air. This is an extension of Marey's inquiry, since the latter had not yet been able to study the full gallop with his apparatus (see [182], p. 170). Muybridge was to settle this question in a photographic way and was indeed able to give a result: In Fig. 3.51, one can clearly see the "flying" horse in frames 2 and 3.

Such successive recordings with several cameras by controlled release in a defined time sequence are also used today under the name "Freeze Frame", "Time Slice" or "Frozen Moment" as a stylistic device in cinema productions. A well-known example is the "bullet time" from the film "Matrix", in which the hero winds away under the bullets in super slow motion with his coat billowing limbo-style.

Background Information

Muybridge was most certainly an unconventional figure in contemporary history. In addition to his extraordinary talent for photography and technology, he was also an exceedingly violent man. His wife's affair with an elegant and well-traveled major infuriated him so much that he set off by train and carriage, armed with a revolver, to meet his wife's lover. He still shouted to his adversary, "This is the reply to the letter you sent to my wife!"[82] and shot him straight through the heart. Amazingly, however, he was acquitted of the charge of murdering Major Harry Larkyns.[83]

(continued)

[82] [183].

[83] The defense invoked the legal fact of "justifiable homicide". The reasoning for the acquittal is incomprehensible from today's perspective: "The murdered man had been guilty of a criminal relationship with Muybridge's wife, and this was considered by the jury to be sufficient justification for the murder of Larkyns" [184].

Fig. 3.51 The Horse in Motion, Eadweard Muybridge (June 19, 1878), Library of Congress Prints and Photographs Division, Kaldari (2017), Wikimedia Commons

Fig. 3.52 Révolver photographique de Janssen, La Nature 101 (1875) from [187], p. 357

In his presentations and publications, Muybridge also liked to use the depiction of naked men and women, which clashed violently with the usual moral concepts of the time [185].

Anschütz from Berlin also took pictures with a short exposure time using a snapshot shutter, which was called instant photography ("Augenblicksphotographie") at the time [186]. Here, too, jumping horses, but also people and animals were depicted in the moment of movement. Several timed shots were called chronophotography .

The basic idea of multiple photography had already occurred to the astronomer Janssen in 1874, when he wanted to photograph the rare transit of Venus.[84] Since the planet's passage in front of the Sun lasted several hours, Janssen had a revolver made that, unlike Colt's revolver, did not emit any dangerous projectiles, but could take a series of images with long exposure times at long intervals [187] (Fig. 3.52).

Marey had something else in mind. For his research, he needed the shortest exposure time shots. So he developed an apparatus from Janssen's idea, which he called, just as martially as Janssen, the "photographic rifle" (Fusil photographique, [188]). As with a rifle, one aimed at the fast-moving object and pulled the trigger (Fig. 3.53). The apparatus took 12 pictures per second, and the exposure time for a single picture could be reduced to 1/1440 s (less than a millisecond).

[84]The passing of Venus in front of the sun (last observable in 2012, the next time will be in 2117).

Fig. 3.53 Le Fusil photographique, Étienne-Jules Marey, La Nature 464 (1882) from [188], pp. 328, 329

Figure 3.53 clearly shows the analogies to a real rifle. Today, one would certainly run into problems while recording photographs in public space in this way. However, Marey also made other chronophotographs, which he used in the laboratory for his research on human and animal movement. These are now more reminiscent of a camera [189].

With chronophotography, the possibility of capturing moving scenes had arisen, which demanded an equivalent possibility for reproduction.

Muybridge invented a zoopraxiscope for reproducing the images, which essentially uses the stroboscopic effect analogous to a zoetrope. The images were arranged on flat discs (similar to those in Stamper's "stroboscopic discs" or Plateau's "phenakistiscope"). Muybridge projected the image of the discs onto a screen for demonstration, thus achieving an effect even for a larger audience. The projection of stroboscopic discs had been successfully demonstrated by Döbler and Uchatius decades earlier. Muybridge's definitive innovation was the combination of the reproduction of large-scale moving images and the recording of real scenes. Although Muybridge used photographic plates as recording material,[85] electricity was already used during the recordings.[86]

[85]The recording was made on the then revolutionary "dry plates", a further development of the collodion wet plates that had been common until then in the versions of the German emigrant Gustav Cramer (Dry Plate Company, St. Louis) and the English emigrant John Carbutt (Keystone Dry Plate, Philadelphia). Carbutt later supplied the first celluloid film for the kinetoscope of William Dickson and Thomas Edison [190, 191].

[86]Muybridge used so-called "Electro-Photographic Exposors" as optical shutters, which are triggered one after the other by an electric motor.

Fig. 3.54 Anschütz's electric high-speed viewer, Josef Maria Eder (1891) from [192], p. 37

The reproduction of the discs in Anschütz's electric high-speed viewer works somewhat more electrically. Basically, the image field of a stroboscope disc is illuminated by a bright electric lamp,[87] which gives the electric speed viewer its name. However, the projectionist still has to turn the disc himself (Fig. 3.54).

Anschütz further developed the device into the Electrotachyscope ([192], p. 35). Now not only the name is more powerful, but the images are now mounted on a roller. This allows several series of images to be displayed side by side at the same time. The images can be viewed through window **D** (see Fig. 3.55).

Of extraordinary importance in Anschütz Electrotachyscope is the current interrupter, which briefly flashes Geissler's tube whenever an image on the roller is exactly in front of the window. Thus, at a suitable speed of rotation,[88] a sharp image is produced, which without flash (with continuous illumination) would indeed be blurred in perception.

However, the picture is small and hardly suitable for a larger auditorium. The situation was quite different in Reynaud's Théâtre optique. Reynaud used projection in his theatre from 1889[89] onwards. Reynaud already used long film strips, perforated for clean transport, so that longer screenings were possible (Fig. 3.56).

[87] Anschütz uses a Geissler tube (gas discharge tube).

[88] A frame rate of about 33 frames per second was recommended.

[89] Brevet d'invention N° 194,482 "Théâtre optique" [193].

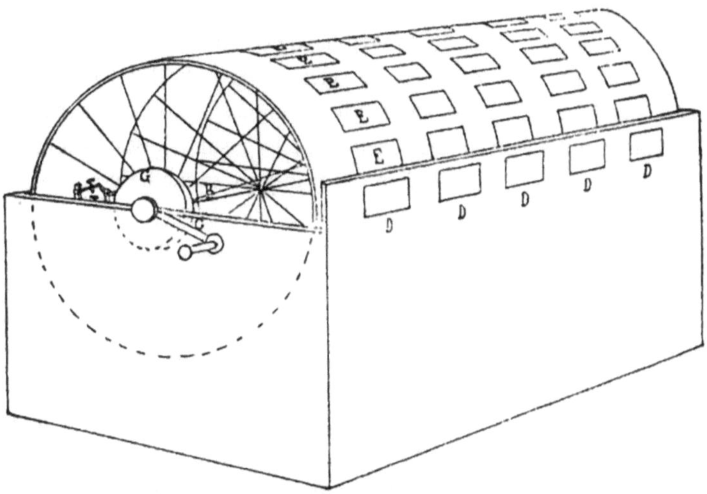

Fig. 3.55 Anschütz' Electrotachyscope, Josef Maria Eder (1891) from [192], p. 38

Fig. 3.56 Reynaud's Théâtre optique, Louis Poyet (1893) from [194], p. 128

On the occasion of a presentation in Orange, New Jersey, in February 1888, Muybridge also visited Edison's laboratory in Menlo Park and thus possibly contributed to the inspiration of Dickson and Edison, which may have ultimately led to the inventors' intensive engagement with the kinetoscope. Dickson even mentions Muybridge as one of the pioneers in his memoirs of Edison, but also that Edison had the idea for reproducing moving images as early as 1887 [195]. However, a collaboration between Muybridge and Edison did not come about, which could well have been due to Muybridge's personality.

Edison and Dickson finally presented the kinetoscope at the 1893 World's Fair, which was basically a peep box for the film strips. The periodic stop of the image strip in front of the projection lens is not solved in the kinetoscope by a suitable start-stop automatic, but much more simply by a revolving slit diaphragm in front of the light source. This created the stroboscopic effect in a mechanical way.

The Lumière brothers finally had the right idea. They used the Reynaudian concept of projection and also adopted the idea of film transport. In contrast to the concept of stroboscopic illumination, the Lumières used a periodic film transport by means of an ingenious start-stop mechanism by which the film was always transported further by a defined path inside the cinematograph (Cinématographe) [196, 197]. This had an amazing side effect: the cinematograph could be used both as a film camera and as a projector.

This meant that longer films could now be presented to a larger audience. The idea finally caught on and led to the worldwide success of cinema. Although there were constant further developments, the recording on a film was indispensable for a long time. As a result, live broadcasts were technically impossible.

3.5 Television

Seeing with Electricity

This only changed with the development of television. The discovery of the light-sensitive properties of selenium by Smith[90] [198], Adams and Day [199] led to the possibility of converting light into electricity. With the ability to detect the intensity of light, it was finally possible to create devices for seeing by electricity [200]. Since the resolution of an array of selenium receivers was rather low, the spatial distribution of the received voltages was replaced by a temporal distribution. The basic idea of image recording corresponds quasi to a single-pixel camera and can be traced back to 1884, when Nipkow patented his mechanical scanning disk ("Nipkow disk") [201].

This principle was used by Baird in his "televisor" [202] (see Fig. 3.57) and transformed by Ardenne [203] into the all-electric "flying spot scanner".

The reproduction of the pictures took place with Ardenne, as later with without exception all following commercial TV models, over a Braun tube.

The cathode ray tube was originally called the "Braun tube" after its inventor Karl Ferdinand Braun. In 1897, Braun had presented an apparatus with which the time course of variable currents could be displayed on a glow screen [204]. In addition to this original use

[90] Whereas in 1873 Smith initially only noticed that when selenium was exposed to light, its resistance changed.

Fig. 3.57 Baird's televisor in London's Science Museum (Grasnick 2017)

as an oscilloscope,[91] the cathode ray tube was used as a picture tube for the display of moving content in all television sets. The colored image raster in a Braun tube is basically formed by a dot or stripe distribution of the color phosphor (hole, stripe, or slit mask) on the display surface. This arrangement is fixed and in fact represents the ideal image raster as the projection surface of the cathode ray. Since the effect of the cathode ray—as explained above—is in principle only the image of a light spot over time, additional information (at least frame rate, line length and number) is required to assign the image signal to the display surface in two dimensions. However, this is hardly ideal, so that the display raster formed by the color phosphor is often not identical with the image raster formed by the cathode ray. Direct addressing and local assignment of individual display points is therefore difficult.

Background Information

The precise tuning of a CRT monitor to a graphics card was sometimes a task for experts. In addition to resolution and refresh rate, settings such as front or rear black shoulder,[92] line sync pulse or sync frequency were queried on the graphics card.

(continued)

[91] An oscilloscope does exactly what Braun intended: it displays the time course of variable currents (or voltages).

[92] Both together designate the horizontal scanning gap and together with the line synchronization pulse the line blanking gap, the designation front or back porch is usual.

These values in turn had to be synchronized on the tube monitor with the horizontal and vertical image position, frequency and phase.

A deviation from the ideal values resulted in unsightly image distortions as well as flickering, or stripes running through the image (if not the image itself started to run. . .).

3.6 Liquid Crystals

Carrots for the Eyes

In 1888, while investigating the hydrocarotene in the root of the carrot, Reinitzer wanted to determine the molecular weight of cholesterol with great accuracy and experimented with cholesteryl benzoate and acetate for this purpose. Reinitzer determined the melting points of the substances and discovered that the liquid, which is cloudy after melting, begins to clear at a further temperature point. In addition to these two melting points, he noticed the chromatic polarization and the optical rotation capacity of the substrates.

Friedrich Reinitzer contacted Otto Lehmann, who not only carefully examined Reinitzer's samples, but subsequently referred to the third phase between solid and liquid as "liquid crystals" [205]. Lehmann described the polarizing ability of the curious material, which was subsequently also manufactured and distributed by the Merck company from Darmstadt (and still is today).

However, even from Merck's point of view, there was no commercial application for the liquid crystals and they were found as early as 1907 [206]:

> . . .on these articles, the representation of which is costly and turnover very small, the earnings can naturally be but an insignificant one. The manufacture was therefore taken up more from the point of view of concession.

Nevertheless, liquid crystals were fascinating. The physicist Voigt had been working on the elasticity of crystals[93] since 1887 and interpreted the "crystalline medium" as a "system of molecules, which keep each other in equilibrium through their interactions" (from [208], p. 5). From this model he developed a matrix of 36[94] different coefficients with which the system can be described. Each coefficient has two indices at Voigt (e.g. D_{11}, or D_{23}) and the

[93]But also with the influence of electric fields on the optical properties of bodies (see e.g. "Magneto- and Electrooptics [207]).

[94]Actually there are 81 in Voigt (a 9 $\times$ 9 matrix), but since most of them are identical, there are only 36 different values.

symbols are arranged pleasingly in a matrix.[95] Voigt has no difficulty in extending his mathematics of crystalline elasticity to liquid crystals. In his 1910 textbook of crystal physics he writes "…so there is not the least hesitation in calling them liquid crystals or crystalline liquids." (from [209], pp. 17–18). Voigt uses the concept of tensors for the description of the molecules of crystals.

Tensors for the "Operation" of Liquid Crystals

The term tensor was introduced by the Irish mathematician Hamilton, who used the term to describe his quaternions[96] (from [210], pp. 56–57)

> …we have had occasion to consider what may be called elongation or tension, namely, another operation of the factor kind by which we pass from a given length (rather than from a given direction) to another.
>
> … the factors … may, of course, be given the general name of 'tensors' in accordance with the nomenclature used in these lectures; …

Voigt is one of the early users of tensors in his crystal physics and explains his intention (from p. 132)

> Physical functions that require the specification of a numerical value and a bipartite direction for exhaustive characterization we call tensors.… The simplest example, from which the name is also derived, is given by the uniform expansion of a body, e.g. a cylinder or a sphere.…

Tensors are a simplification in the notation of mathematical equations. In a tensor, multiple entries are packed into a matrix that can have multiple dimensions. In the simplest case, the tensor is just a single value (scalar). But a tensor can be a row or column vector or even a matrix. One replaces the matrix with the tensor in the equation and thus has a more elegant formula. This is often used in the description of three-dimensional structures (like crystals), but also for the description of four-dimensional space-time.

Einstein made lively use of tensors, for example in the theory of relativity [211] or in the description of gravitation [212].

The physicist Max Born tried to fathom the nature of these "strange substances" and is also particularly concerned with the birefringence of the liquid crystals in the electric field. Kerr had already pointed out in 1875 that the polarization of light passing through carbon

[95] This also results in the double index: D_{11} would be the value D in row 1 and column 1 in Voigt's notation, D_{23} therefore row 2 and column 3.

[96] These also have a practical use in 3D positioning in space (see also section on interaction in virtual space).

disulfide is affected when a voltage is applied to the tank [213]. Born applied this Kerr effect to liquid crystals and put forward a theory of the influence of the electric field strength on the refractive index [214]. In a second paper Born[97] considered the influence of temperature on the refractive index [215]. The Swede Oseen[98] took up Born's theory and immediately criticized its thermodynamic considerations [216]. In two sequels, Oseen now described in mathematical detail the forces acting between the molecules of the liquid crystals and influencing their orientation [217, 218].

In 1934, Marconi[99] had the idea of filling a Kerr cell[100] with liquid crystals and using it in image and video transmissions. This idea seems extremely practical, yet an actual implementation has not become known.

It was not until 1957 that an article by Brown and Shaw [219] appeared, which stimulated greater interest. Already in the following year, the Faraday Society in Leeds organized a general discussion on the "Configurations and Interactions of Macromolecules and Liquid Crystals". One of the participants was the British researcher Gray, who was later to write a remarkable book on the properties of liquid crystals [220].

Oseen's theory found its continuation in 1958 in a recent revision by Frank [221] and is now a component in the design of current liquid crystal displays as the Oseen-Frank theory.

Perhaps inspired by these recent publications, Fergason, an American physicist, attempted to use the temperature properties of liquid crystals for thermal imaging. Since Fergason's employer Westinghouse seemed to have no interest in this work, Fergason moved to the "Liquid Crystal Institute" founded by Brown at Kent State University in Ohio in 1965. The institute started the first "International Liquid Crystal Conference" in the same year. Georg Heilmeier and Richard Williams participated in this conference. Heilmeier and Williams were employees of Radio Corporation of America (RCA)[101] at that time. Both were researching the use of liquid crystals at the RCA Laboratories at the time. Williams, a chemist, knew the possibility of aligning a thin layer of the crystals by an electric field [222]. His colleague, the engineer Heilmeier, was now able to use this effect in a first liquid crystal display.

Helfrich, a former RCA employee, worked with his colleague Schadt in the Basel laboratory of the Hofmann-La Roche company. After a short time, Schadt and Helfrich were able to build functional displays from the "light control cell", which were then used commercially for the first time by the electronics company Brown, Boveri & Cie to build a digital clock—certainly obvious for a Swiss company. Brown, Boveri & Cie built the

---

[97] Here together with the Berlin physicist Stumpf.

[98] As director of the Nobel Institute the one who proposed Albert Einstein for the Nobel Prize, Max Born received it only after Oseen's death.

[99] Founder of Marconi's Wireless Telegraph Company Limited, London.

[100] In a Kerr cell (after John Kerr) the polarization state of the transmitted light is changed by applying a voltage in a birefringent liquid.

[101] RCA was one of the pioneers of television technology in the USA and the developer of many standards of picture and sound transmission.

displays in its own factory in Lenzburg, which meant that the world's first liquid crystal display factory was located in the Swiss canton of Aargau.

In the simplest case, a liquid crystal display (LCD) works as shown in Fig. 3.58. Between two glass plates (3), (5) is a layer of liquid crystal (4). The molecular arrangement in the crystal is helically twisted. This is accomplished by twisting the glass plates, to the surfaces of which the molecules adhere, 90 degrees with respect to each other. The glass plates are coated with transparent but electrically conductive material,[102] allowing them to change the orientation of the crystalline liquid. In front of a light source[103] (1) there is a polarization filter (2) which allows only the light of a certain oscillation direction to pass. In the example, only the horizontally polarized component can pass through the filter. Since the polarization follows the helical turns of the crystal structure, the polarization direction is rotated by 90 degrees when passing through the liquid crystal. This allows the light to pass through the second polarization filter (6), which is only transparent to the vertical components. The observer (7) sees the light.

This changes as soon as a voltage is applied to the conductive layer of the glass plates. The crystals align themselves in the electric field in such a way that the polarization direction of the light is now no longer rotated and the light is blocked by the polarization filter 6. When the cell is bright in the de-energized state, as in a Schadt-Helfrich cell, this is called the "Normally-White Mode".

The operation of such a cell is based on the alignment of the crystals, which is called the "nematic[104] phase". Since the crystals are twisted in the de-energized state, these cases are referred to as "twisted nematic" or TN displays.

In Fig. 3.58, the entire cell is switched, which of course gives the viewer no image impression at all. He sees either a black or a white image. Therefore, the display must be switched segment by segment. The simplest display, which was also used in the first digital clocks, is the segment display.[105] Such a display is illustrated in Fig. 3.59. On the left side the 7 segments are shown. In the middle 4 segments were switched on (red), which then form a "4". On the far right, a "2" is shown with 5 segments.

A combination of several segment displays allows the display of the time or the output of the result on a calculator. The usual Normally-White-Mode results in the digits of the display appearing dark compared to the rest of the cell. However, with a segment display, at best symbols can be displayed, but not images. It is obvious to create an image by controlling single pixels. However, this requires a large number of pixels that would

[102] The material is usually indium tin oxide (ITO), which is why this coating is often called an "ITO layer".

[103] In the case of watches and calculators, a simple reflector foil was also used to illuminate the display with the ambient light.

[104] From Greek nema = thread.

[105] Segment displays have been used since the beginning of the twentieth century (see, e.g., Wood's 8-segment display [223]), where the eighth segment is used to display the oblique stroke at the fourth position.

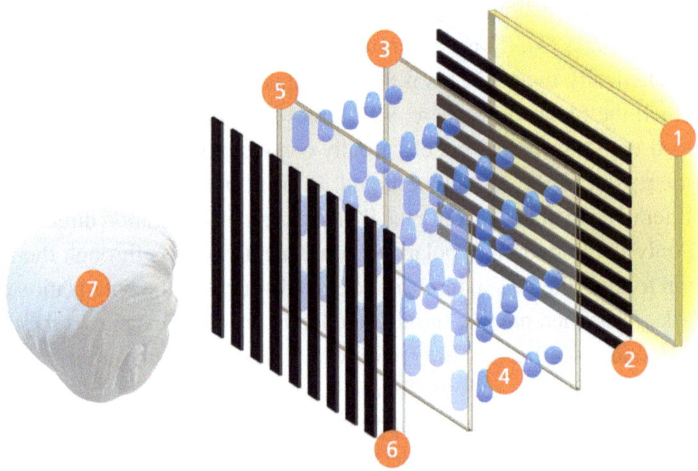

Fig. 3.58 Simplified liquid crystal cell, Grasnick (2019)

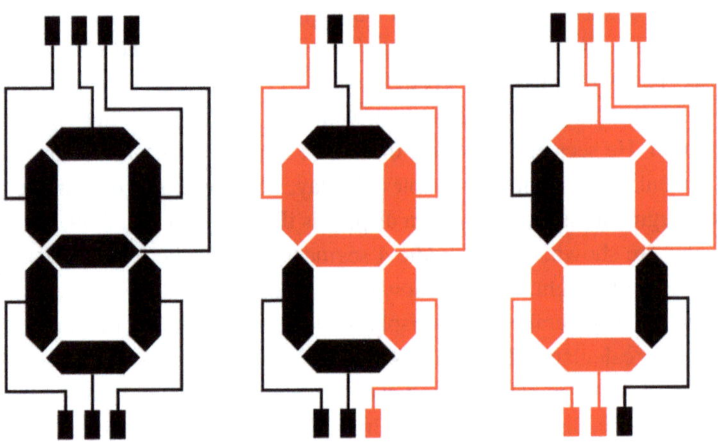

Fig. 3.59 7-segment display, Grasnick (2019)

have to be contacted individually. With a 7-segment display this is no problem; with a TV screen, according to the NTSC[106] standard of the time, about 480 lines and 640 columns are displayed, i.e. more than 300,000 pixels. Heilmeier and his RCA colleague Zanoni had therefore proposed a matrix control in which the voltage is fed to specific points of a matrix

[106]National Television Systems Committee, US-American standardization institution for television.

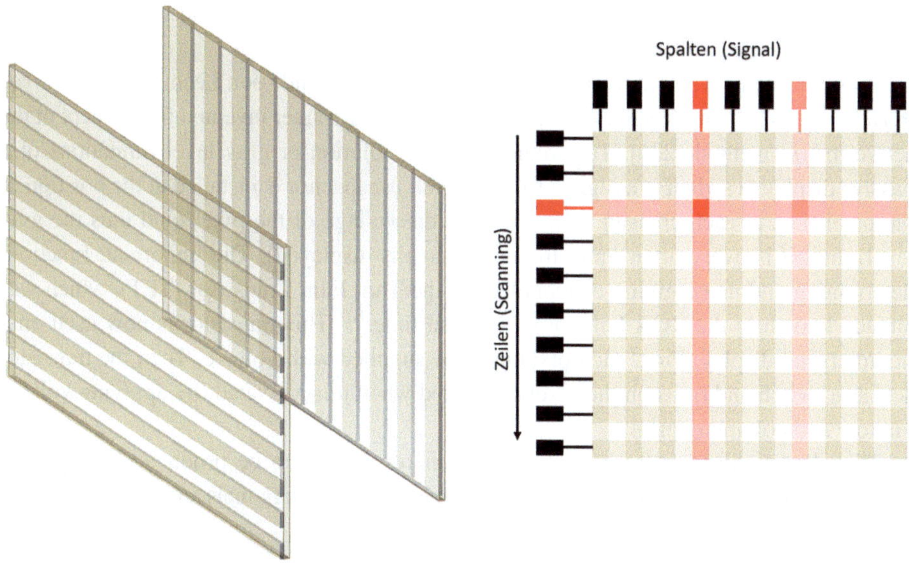

Fig. 3.60 Passive matrix approach, Grasnick (2019)

via vertical and horizontal strip electrodes [224]. Again, the traces are transparent and can be switched individually. At the crossing points of two switched traces, an electric field is created that is sufficiently strong to align the molecules in the liquid crystal. This is extremely distributive, because now instead of 300,000 individual points, only 480 rows and 640 columns would have to be controlled (i.e. just over a thousand). This kind of control is called "passive matrix" today. The problem with such displays lies in their construction. In Fig. 3.60 the horizontal and vertical tracks are shown on the left side. On the right, one column and one row are supplied with voltage. At the intersection of the current-carrying traces, the electric field is created (the small red square in the picture). It is obvious that only one single point at a time can be supplied with voltage, because if several columns and rows would be connected same time, there would exist several crossing points all times. Consequently, all crossing points to be controlled must be approached one after the other. When switching all pixels one after the other step by step, the whole pixels of the screen can form a picture together. In the practical implementation one proceeds line by line. On all columns of the screen the respective signal level is given (in Fig. 3.60 two different signal strengths are illustrated). Then the line voltage is added to the corresponding line and thus the entire image line is generated.

All lines are switched sequentially (one after the other), so that one also speaks of scanning.

Due to the voltage in the conductive paths, the desired field is not only generated at the crossing point, but also (undesirably) in a weaker form along the paths. To avoid this, a liquid crystal is chosen that only "switches" above a certain threshold value. This in turn leads to the problem that fine gradations (grey values) are more difficult to produce in the display. In practice, this results in a display with poor contrast.

These problems were motivation to search for solutions and soon Lechner, an employee in Heilmeier's Liquid Crystal Lab at RCA, had an idea. Lechner invented the concept of the active matrix method [225]. To circumvent the resolution or gray level limitations of a passive display, a switch is assigned to each pixel. A capacitor is installed at the position of the pixel (i.e., at each crossing point) to hold the voltage of the pixel (and thus the gray level) until it is overwritten (until the next image is displayed). The important, function-determining and name-giving element here is the Thin Film Transistor (TFT), which was already known in the RCA Laboratories at that time.[107] This transistor is now used on a liquid crystal display. First, each pixel must be supplied with voltage. Then the transistor acts as a switch and turns on the pixel concerned when it is activated (by applying voltage via row and column control).

This allows shorter refresh rates and higher resolutions to be produced with greater contrasts. Today, flat liquid crystal screens with large screen diagonals and high resolutions can be produced with this technology.

3.7 Luminous Crystals

The Glow of the Carborundum

For particularly large screens, light emitting diodes (LEDs) are often used as light points, as they can provide much higher brightness than LC displays[108] while at the same time enormously improving contrast. Light emitting diodes are also crystals, but unlike liquid crystals, they have a solid form.

The actual function of an LED, i.e. the glow of the light-emitting diode, is based on an effect described by the Marconi engineer Round in 1907. When a voltage was applied to a carborundum crystal (silicon carbide), it began to glow yellowish[109] [227]. Round had communicated his observation only in a short note, which subsequently received little attention. A scientific analysis of the process was only carried out by Lossev. Lossev had made carborundum crystals glow in the Nizhny Novgorod Lenin Radio Laboratory and described the effect ([228]). The process of "electroluminescence" [229] observed by the French physicist Destriau in 1935 was based on the "scintillation[110]" of zinc sulfide crystals by electricity [230]. The two effects are not completely identical. The (Round) Lossev

[107] Used by Weimer since the early 1960s [226].

[108] With liquid crystal displays, most of the light from the backlight is lost through the polarizers.

[109] In other crystals also yellow, green, orange or blue.

[110] Scintillation refers to the emission of light from certain substances by external excitation (in this case it was electricity).

effect is based on the emission of light when semiconductors are excited by DC voltage. The Destriau effect, on the other hand, is based on excitation by alternating voltage. Today, both effects are referred to under Destriau's term of electroluminescence.

Based on these effects, the RCA physicist Braunstein was able to develop simple infrared light-emitting diodes, which were converted into a real product[111] by Biard and Pittman of Texas Instruments [231]. Of course, infrared diodes are not suitable for an application with an image. For the visible range, the scientist Holonyak at General Electric developed a red LED [232]. With new semiconductor materials, it was later possible to produce green or yellow light-emitting diodes. The production of blue diodes, on the other hand, seemed difficult for a long time, since carborundum as an LED material produced only weak light. Another crystal, gallium nitride, was more promising. However, it was simply not possible to produce this semiconductor material as a single crystal in the required sizes. Without blue light, however, it is impossible to produce full-colour images. In an LC screen, each individual pixel is composed of the primary colors red, green and blue, which is why it is also called an RGB pixel. Additive mixing of the individual colors produces any desired hue, including white. Without a blue pixel, color representation would not work on an LCD any more than on an LED screen.

The scientists Akasaki and Amano at the Japanese University of Nagoya had been working on the production of gallium nitride for a long time. Only the deposition of the crystal on a sapphire substrate and the corresponding further processing[112] brought the breakthrough and the bright blue light. Independently of the two Nagoya scientists, the electronics engineer Nakamura, who was employed by the Japanese chemical company Nichia, succeeded in producing a blue LED at the same time. With this technological step, which earned the three Japanese the Nobel Prize in Physics in 2014, the possibility of creating full-color LED screens had come within reach.

These single-crystal LEDs are inorganic in nature, consisting of inanimate matter, i.e. substances that do not contain carbon. Single crystals are extremely difficult to manufacture, which is why attention had always turned towards organic materials ("animate" substances that contain carbon). Organic materials are the building material of all living things, more readily available and easy to produce. However, one needs an organic material that can emit light by electroluminescence. The Frenchman Bernanose experimented in Nancy with organic dyes such as the acridine orange[113] or quinacrine,[114] which he applied to thin films and exposed to an alternating voltage [233]. At New York University, a team led by Pope was working on electroluminescence in organic crystals

[111]Texas Instrument SNX-100 Infrared Diode.

[112]Doping: Introduction of atoms of another material (foreign atoms) to achieve certain electrical properties.

[113]A fluorescent dye used to stain samples in microscopy for the detection of nucleic acids, the carriers of genetic information.

[114]A drug for malaria prophylaxis.

[234]. Pope had elicited a light from anthracene.[115] This material was also of interest to Helfrich (even before his time at La Roche) [235] and led to an organic electroluminescent cell at Dow Chemicals [236]. The further development to a true light-emitting diode is attributed to Tang and van Slyke [237]. The key advantage of an OLED (organic LED) over an LCD is the self-illumination of the pixels. A liquid crystal display must be externally illuminated to display an image.

Each LCD image pixel must be switched light or dark via the rotation of the polarization. But this does not work perfectly in both directions. The polarizers of the LCD already block up to 75% of the light coming from the illumination. The panel's color filters, in turn, only allow a third of the right light to pass through each time, so that ultimately about 8% of the light used still reaches the viewer's eye. The lighting behind the LCD must therefore be more than 10 times as bright as the brightness of the displayed image. This is not very effective. An LED illuminates itself and in the desired color. The polarizers and color filters can be omitted and thus the brightness of the LEDs can be used completely. An LED that is turned off is actually dark, while a pixel of an LCD that is turned off still allows some light to pass through. This means that the contrast (the ratio of highest to lowest brightness value) of an LCD can already be described as good if it is above 1000:1. A real LED display, on the other hand, can easily deliver a contrast of 100,000:1.

Since the use of luminescent polymers (e.g. [238]), the fabrication of such screens can be effective. In any case, the manufacturing process can be much simpler than for an LCD: OLEDs can be printed. Another advantage arises from the elimination of polarizers. Previously, glass had to be used as a carrier for the polarizers so as not to destroy the polarization state while passing through the substrate.[116] Glass can hardly be bent without heating,[117] but this is no problem with an OLED printed on foil.

In principle, current OLED screens are usually also constructed as an active matrix, from which the designation AMOLED (active matrix organic light emitting diode) is derived. Some OLEDs are constructed like LCDs, i.e. three coloured LEDs (red, green and blue) are applied next to each other (side-by-side) and thus together form an RGB pixel. The WOLED technology (white organic light emitting diode) is interesting. Here, a white LED is used, which is then separated again into RGB via color filters.

Figure 3.61 compares the basic structure of an LC display (left) with that of an LED display. While the LC display requires illumination, line and column control, polarizers and color filters in addition to the liquid crystal, the distribution of the colored LEDs and their control is basically sufficient for an LED display.

[115] A colourless, sometimes whitish powder obtained from coal tar, also an ingredient of the aloe vera plant, used in the manufacture of dyes.

[116] As it would be, for example, when passing through a transparent plastic.

[117] Which of course depends on the thickness of the glass: Schott has produced glass with a thickness of only 20 μm, which is also supplied on rolls [239].

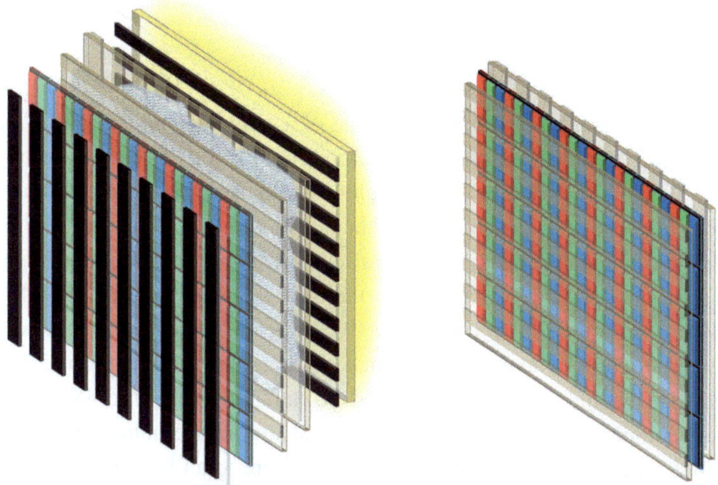

Fig. 3.61 Comparison of LC and LED display design, Grasnick (2019)

As with an LCD, there is also here the control as a passive or active matrix. The term AMOLED, for example, refers to an active matrix with an organic LED.

In current screens, the LEDs are not always used as an image matrix. Often an LCD is combined with an LED backlight. The term "LED TV" can therefore also refer to a liquid crystal display that is only illuminated by LEDs. This deliberate misnomer is certainly primarily intended to increase sales. However, it is worth taking a closer look, as a real LED screen has real advantages over an LC screen.

3.8 Radiant Plasma

Light from Charged Gas

In addition to these two main screen technologies, plasma technology was also used for a while. Just as with an LED, here too each pixel illuminates itself. The pixel itself consists of three chambers that are filled with a gas.[118] If the chamber is energized, the gas is ionized and thus becomes the eponymous plasma. The plasma emits light that is initially invisible to humans because it is ultraviolet radiation. As with a fluorescent tube, the chambers of a plasma display are coated on the inside with a luminescent material that converts the

[118]Mainly neon, plus a few percent xenon, and sometimes a little helium.

invisible UV radiation into visible light. A different phosphor is used for each color, which again allows an RGB pixel to be generated.

Plasma

The term plasma is used to describe the state of a gas in a confined spatial region in which the gas becomes ionized and electrically conductive. The name originates from a publication by the American Langmuir, who used the name "plasma" in it to describe a region of the gas containing balanced charges of ions and electrodes.[119] Langmuir had studied the behaviour of ionised gases for years at General Electric and noticed that the particle velocities of the gas deviated considerably from the Maxwellian velocity distribution that was actually to be expected. In a low-pressure mercury vapor lamp operated at room temperature, particle velocities occurred that would have corresponded to temperatures of 5000–60,000 °C. Langmuir compares the properties of plasma at low temperatures to a crystal and speaks of a type of crystal gas. Langmuir refers here to an idea of Eddington, who asked himself whether stellar gas would crystallize while remaining a perfect gas ([241], pp. 368–369).

This idea sounds even crazier than the theory of a liquid crystal. However, plasma actually has some crazy properties that distinguish it from the other states of matter (solid, liquid, gaseous) and is therefore sometimes referred to as the fourth state of matter. Plasma has the property of conducting electricity and thus also reacts to magnetic influences. Plasma does not only occur in nature in the form of lightning or auroras. It is assumed that 99% of the matter in the universe is in the plasma state (see e.g. Textbook of Plasma Physics (Lehrbuch der Plasmaphysik) [242], p. 2). The interstellar gas and all stars consist of plasma. Thus, the plasma state would be the usual state of aggregation in the universe and the states solid, liquid or gaseous known to us would be rather the bizarre exception.

A plasma emits energy in the form of radiation. If this is in the visible range, the plasma glows. In the case of a lamp or a screen, the typical plasma glow is used.

Plasma displays have been in use since the 1960s. One early application was as a computer monitor in the PLATO project at the University of Illinois. The name PLATO is certainly not coincidentally reminiscent of the ancient philosopher, but was the abbreviation for "Programmed Logic for Automated Teaching Operations", a system for computer-assisted teaching [243]. A team led by Bitzer, an electronics engineer, had developed a plasma monitor for this system in 1966 [244] and obtained a patent on it a few years later [245]. This plasma display was included in this terminal from the PLATO IV system onwards [246] and was used on several thousand systems. The system was already capable of graphics, although the graphics were still monochrome (see Fig. 3.62).

[119] "We shall use the name plasma to describe this region containing balanced charges of ions and electrons." (from [240], p. 628).

Fig. 3.62 PLATO V terminal with monochrome plasma screen from 1981, Mtnman79 (2008), Wikimedia Commons

Plasma displays were used in the first laptops,[120] but were displaced from this market relatively quickly by the increasingly cheaper LC displays. The initial advantage of plasma displays in terms of contrast and viewing angles could no longer compete with the increasingly available and full-color LCDs as early as the beginning of the 1990s. Plasma televisions with large diagonals appeared once again at the end of the 1990s. Here, plasmas

[120]E.g. the Toshiba T3200 from 1987 with 4 grey levels (orange) and a resolution of 640 × 400 pixels.

excelled over LCs with large screen diagonals. With ever larger liquid crystal and light-emitting diode displays, plasma sets lost market share and are hardly to be found on the market today.

Literature

1. Aristoteles. Problemata Physica. 4., gegenüber der 2., durchges., unverä. Aufl. Berlin: Akademie-Verlag; 1991.
2. Ibn-al-Haiṯam AA al-Hasan A. Alhacen's theory of visual perception: a critical edition, with english translation and commentary, of the first three books of Alhacen's De aspectibus, the Medieval Latin Version of Ibn al-Haytham's Kitab al-Manazir. Philadelphia, Pa: American Philosophical Society; 2001.
3. Sabra AI. Alhazen's optics in Europe: some notes on what it said and what it did not say. In: Lefèvre W, editor. Inside the camera obscura – optics and art under the spell of the projected image. Berlin: Max-Planck-Institut für Wissenschaftsgeschichte; 2007.
4. Grunmach L. Die Physikalischen Erscheinungen und Kräfte, Erkenntnis und Verwertung im praktischen Leben. Das neue Buch der Erfindungen, Gewerbe und Industrien: Rundschau auf allen Gebieten der gewerblichen Arbeit. 9. Aufl. Leipzig: Otto Spamer; 1898.
5. Bacon R. Specula mathematica. Frankfurt: Wolffgang Richter; 1614.
6. Bacon R, Lull R. Von den Geheimen Wirkungen der Kunst und Natur und Richtigkeit der Magie. Hof: Johann Gottlieb Vierling; 1776.
7. Ben Gershon L. Milchamot Ha-schem. Die Kämpfe Gottes. Religionsphilosopische und Kosmische Fragen. Leipzig: Carl B. Lorck; 1866.
8. Ben GL. Die Kämpfe Gottes. Berlin: Mayer & Müller; 1914.
9. Ben Gershon L, de Alexandria P. Instrumentum quod revelatorem secretore. Manuscript; 1342.
10. Busard HHL. Der Traktat De Sinibus, Chordis et Arcubus von Johannes von Gmunden. Wien: Springer; 1971.
11. de Lineriis J. Canon et tabulae planetarium. Messahalla; Ptolemaeus; Almansor astrologus; Ps-Hermes; Thebit ben Corat; Johannes de Lineriis; Johannes Danck: Astronomisch-astrologische Sammelhandschrift. Frankfurt a. M.: Manuscript; 1391. pp. 86r–92r.
12. de Lineriis J. Algorismus de minutiis. Johannes de Sacrobosco; Johannes Alger; Christianus de Prachaticz; Johannes de Lineriis; Astronomisch-mathematische Sammelhandschrift. Manuscript; 1409. pp. 145r–6v.
13. von Braunmühl A. Geschichte der Trigonometrie. Leibzig: B. G. Teubner; 1900.
14. Goldstein BR. The astronomy of Levi ben Gerson (1288–1344). New York: Springer; 1985.
15. Pierer JF, Choulant L, Editor. Medizinisches Realwörterbuch zum Handgebrauch practischer Aerzte und Wundärzte, Erste Abtheilung. Anatomie und Physiologie. Altenburg: Literatur-Comptoir; 1825.
16. Curtze M. Die Abhandlung des Levi ben Gerson über Trigonometrie und den Jacobsstab. Bibliotheca Mathematica, Zeitschrift für die Geschichte der Mathematik. 1889;12:97–112.
17. Curtze M. Die Dunkelkammer. Eine Untersuchung über die Vorgeschichte derselben. Himmel und Erde Illustrierte naturwissenschaftliche Monatsschrift. 1901;13:225–36.
18. Ptolemäus C. Amalgestum. Opus ingens ac nobile omnes Celorum motus continens. Felicibus Astris eat in lucem. Venedig: Petrus Lichtenstein; 1515.
19. Kopernikus N. Nicolai Copernici Torinensis De Revolutionibus Orbium coelestium, Libri VI. Nürnberg: Petreius; 1543.

20. Kepler J. Astronomia nova aitiologetos seu physica coelestis. Heidelberg: Gotthard Voegelin; 1609.
21. Kepler J. Astronomia nova. 2. Aufl. In: Caspar M, Editor. München: Beck; 1990.
22. von Fabriczy C, Brunelleschi F. Sein Leben und seine Werke. Stuttgart: Cotta'sche Buchhandlung; 1892.
23. Scott L. Filippo di ser Brunellesco. London: George bell and Sons; 1908.
24. Da Vinci L. Codex Atlanticus. 1478–1519, Manuscript, Biblioteca Ambrosiana, Mailand
25. da Vinci L. Pariser Manuskript D [Internet]. Paris: Manuscript; 1508. https://www.photo.rmn.fr/CS.aspx?VP3=SearchResult&VBID=2CO5PC6HEVP_A.
26. Ravaisson-Mollien C. Les Manuscripts de Leonard de Vinci. Paris: A. Quantin; 1883.
27. Richter JP. The literary works of Leonardo da Vinci. London: Sampson Low, Marston, Searle & Rivington; 1883.
28. Da Vinci L. The notebooks of Leonarda da Vinci. 1955. Aufl. In: MacCurdy E, Editor. Florenz, New York: George Braziller; 1955.
29. Herzfeld M. Leonardo da Vinci: Der Denker, Forscher und Poet. 2. Aufl. Jena: Eugen Diederichs; 1906.
30. Richter IA, Wells T, Kemp M. Leonardo da Vinci notebooks. New York: Oxford University Press; 2008.
31. Frisius RG. De radio astronomico & geometrico. Antwerpen: Bontius. 1545;
32. della Porta GB. Magiae Naturalis Libri Viginti. Hannover: Schleichius; 1619.
33. della Porta GB. Natural Magick in Twenty Books. In: Young T, Speed S, Editor. London: R. Gaywood; 1658.
34. Dupré S. Inside the Camera Obscura: Kepler's experiment and theory of optical imagery. Early Sci Med. 2008;13:219–44.
35. Kepler J. Ad Vitellionem Paralipomena: Astronomiae pars Optica Traditur. Frankfurt a. M: Claud Marne & Johannes Aubers; 1604.
36. Keplers KJJ. In: von Rohr M, editor. Grundlagen der geometrischen Optik: (im Anschluß an die Optik des Witelo). Leipzig: Akademische Verlagsgesellschaft; 1922.
37. Kepler J. Dioptrice seu demonstratio eorum quae visui & visibilibus propter Conspicilla ita pridem inventa accidunt. Augsburg: Davidis Franci; 1611.
38. Apelles latens post tabulam, Scheiner C. Tres Epistolae de Maculis Solaribus. Augsburg; 1612.
39. Scheiner C. Oculus hoc est fundamentum opticum. Innsbruck: Daniel Agricola; 1619.
40. Kircher A. Athanasii Kircheri … Ars Magna Lucis Et Umbrae [Internet]. Amsterdam: Joannem Janssonium à Waesberge, & Haeredes Elizaei Weyerstraet; 1671. http://digi.ub.uni-heidelberg.de/diglit/kircher1671.
41. Schott C. Magia Universalis Naturae et Artis. Frankfurt: Schönwetter; 1657.
42. Zahn J. Oculus Artificialis Teledioptricus sive Telescopium. Würzburg: Quirin Heyl; 1685.
43. Hooke R. Micrographia or some physiologicale descriptions of minute bodies made by magnifying glasses. London: John Martyn & James Allestry; 1665.
44. Hooke R. Lectures de Potentia Restitutiva, or of Spring. Explaining the power of springing bodies. London: John Martyn; 1678.
45. Hooke R. Philosophical experiments and observations. London: W Derham; 1726.
46. von Goethe JW. Schriften zur Morphologie. Stuttgart: Cotta'sche Buchhandlung Nachf; 1959.
47. von Goethe JW. Briefwechsel zwischen Schiller und Goethe – Zweiter Band von Johann Wolfgang von Goethe – Text im Projekt Gutenberg [Internet]. Accessed on 18. Feb. 2018. http://gutenberg.spiegel.de/buch/briefwechsel-zwischen-schiller-und-goethe-zweiter-band-3683/10.
48. Wollaston WH. On a new metal, found in crude platina. Philos Trans R Soc Lond. 1804;94:419–30.

49. Wollaston WH. A method of examining refractive and dispersive powers, by prismatic reflection. Philos Trans R Soc Lond. 1802;92:365–80.
50. Wollaston WH. Description of the camera lucida. In: Nicholson W, editor. Journal of natural philosophy, chemistry and the arts. London: W. Stratford; 1807.
51. de Fontana J. Johannes de Fontana: Bellicorum instrumentorum liber cum figuris. Venedig: Cod.icon. 242 (Tresorhandschrift); 1420.
52. Kranz H. Hexen, Teufel und Raketen: Technische Visionen in Johannes Fontanas ‚Liber instrumentorum'. Aachen: RWTH Aachen University; 2017 S.
53. Burckhardt J. Die Kultur der Renaissance in Italien: ein Versuch. Nachdr. Stuttgart: Reclam; 1994.
54. Alberti LB. La Pittura. Venedig: Gabriel Giolito de Ferrari; 1657.
55. Tiraboschi G. Storia della Letteratura Italiana. Rom: Luigi Perego Salvioni Stampator Vaticano; 1783.
56. Schwenter D. Deliciae Physico-Mathematicae. Oder Mathemat- und Philosopische Erquickstunden. Nürnberg: Jerimias Dümler; 1636.
57. Bettini M. Apiaria Universae Philosophiae Mathematicae. Bolognia; 1642.
58. Harsdörffer GP. Delitiae Mathematicae et Physicae. Nürnberg: Jeremia Dümler; 1651.
59. Kircher A. Athanasii Kircheri … Ars Magna Lucis Et Umbrae [Internet]. Rom: Hermann Scheus; 1646. http://digi.ub.uni-heidelberg.de/diglit/kircher1671.
60. Magia SC. Optica: Geheime doch naturmässige Gesicht- und Augen Lehr. Bamberg: Johann Arnold Cholins; 1671.
61. Martini M. Brevis Relatio de Numero, et Qualitate Christianorum apud Sinas. Köln: Johann Buseum; 1655.
62. Martini M, Blaeu J. Novus Atlas Sinensis. Amsterdam: Joan Blaeu; 1655.
63. Kohlhans JC. Tractatus opticus. Leipzig: Friederich Lankisch; 1663.
64. Harsdörffer GP. Der Geschichtspiegel: Benebens XXV Aufgaben von der Spiegelkunst. Nürnberg: Johann Anreas Endtern; 1654.
65. Vermeir K. The magic of the magic lantern (1660–1700): on analogical demonstration and the visualization of the invisible. Br J Hist Sci. 2005;38:127–59.
66. Kittler F. Die Laterna magica der Literatur. Schillers und Hoffmanns Medienstrategien. In: Behler E, Hörisch J, Oesterle G, Editor. Athenäum – Jahrbuch der Friedrich Schlegel-Gesellschaft. 1994;4.
67. Liesegang FP. Das Projektionswesen. In: Hay A, von Rohr M, editors. Handbuch der wissenschaftlichen und angewandten Photography. Wien: Springer; 1931.
68. Eder JM. History of photography. New York: Columbia University Press; 1945.
69. Dechales CFM. Cursus seu Mondus Matematicus. Anissoniana: Lyon; 1674.
70. de Monconys B. Voyage d'Angleterre, Pais-Bas, Allemagne, & Italie. Horace Boissat & Gerge Remus: Lyon; 1666.
71. Oeuvres HC. Complètes de Christiaan Huygens 1660–1661. Societe Hollandaise des Sciences, Editor. La Haye: Martinus Nijohff; 1890.
72. de Monconys B. Des Herrn de Monconys ungemeine und sehr curieuse Beschreibung Seiner … Reisen. Leipzig und Augsburg: Andreas Zeidler; 1697.
73. Huygens C. Oeuvres Complètes de Christiaan Huygens. Supplement a la Correspondence. Varia,. Societe Hollandaise des Sciences, Editor. La Haye: Martinus Nijohff; 1950.
74. Huygens C. Oeuvres Complètes de Christiaan Huygens 1662–1663. Societe Hollandaise des Sciences, Editor. La Haye: Martinus Nijohff; 1891.
75. Jacobaeus O. Museum Regium seu Catalogus Rerum tam naturalium, quam artificialium. Kopenhagen: Joachim Schmetgen; 1696.

76. Jacobaeus O. Vom Museo Danico des Oligeri Jacobaei. In: Denso JD, editor. Monatliche Beiträge zur Naturkunde. Berlin: Verlag der Real-Schule; 1752. p. 100.
77. Oeuvres HC. Complètes de Christiaan Huygens 1664–1665. Societe Hollandaise des Sciences, Editor. La Haye: Martinus Nijohff; 1893.
78. Huygens C. Traite de la lumiere. Leiden: Pierre van der Aa; 1690.
79. Huygens C. Abhandlung über das Licht. Lommel E, Editor. Leipzig: Wilhelm Engelmann; 1890.
80. Sturm JC. Collegium Experimentale, Sive Curiosum. Nürnberg: Wolfgangi Mauritii Endteri & Johannis Andreae Endteri; 1676.
81. Leibniz GW. Philosophischer Briefwechsel 1663–1685. 2., neubearb. und erw. Aufl. Schneider M, Editor. Berlin: Akad.-Verl; 2006.
82. Rossel D. Some thoughts on the bull's-eye lantern. The New Magic Lantern. 2003;9:71–5.
83. Kohlhans JC. Neu-erfundene mathematische und optische Curiositäten. Leipzig: Friederich Lankisch; 1677.
84. Zahn J. Oculus Artificialis Teledioptricus sive Telescopium. Nürnberg: Johannes Christoph Lochner; 1702.
85. Schöpf G. Historisch-statistische Beschreibung des Hochstifts Wirzburg. Hildburghausen: Johann Gottfried Hanisch's Wittwe; 1802.
86. Molyneux W. Dioptrica nova: a treatise of dioptricks. London: Benj Tooke; 1692.
87. Locke J. An essay concerning human understanding. 4th. London: Awnsham and John Churchill; 1700.
88. Project Prakash. Project Prakash [Internet]. Accessed on 1. Mar. 2018. http://www.projectprakash.org/.
89. Sinha P, Chatterjee G, Gandhi T, Kalia A. Restoring vision through "Project Prakash": the opportunities for merging science and service. PLoS Biol. 2013;11:e1001741.
90. Held R, Ostrovsky Y, de Gelder B, Gandhi T, Ganesh S, Mathur U, u. a. The newly sighted fail to match seen with felt. Nat Neurosci. 2011;14:551–3.
91. Ehrenberger BH. Novum et curiosum laternae magicae augmentum. Jena: Prelo Nisiano (Nisius); 1713.
92. Sturm JC, Ehrenberger BH. Kurtzgefasste Mathesis oder Erste Anleitung zu Mathematischen Wissenschaften. Hildburghausen: Paul Günther Pfotenhauer; 1717.
93. Balthasar T. Micrometria: hoc est, de micrometrorum tubis opticis, seu telescopiis & microscopiis applicandorum. Erlangen: Daniel Michael Schmatz; 1710.
94. Ledermüller MF. Mikroskopische Gemühts- und Augen-Ergötzung. Nürnberg: Adam Wolffgang Winterschmidt; 1763.
95. Nicolai EA. Gedancken von der Erzeugung des Kindes im Mutterleibe. Halle: Lüderwaldische Buchhandlung; 1746.
96. Ledermüller MF. Versuch einer gründlichen Vertheidigung derer Saamenthiergen nebst einer kurzen Beschreibung derer Leeuwenhoeckischen Mikroskopien und einem Entwurf zu einer vollständigen Geschichte des Sonnenmikroskops. Nürnberg: Georg Peter Monath; 1758.
97. Baker H. The microscope made easy. 2. Aufl. London: R. Dodsley; 1743.
98. Ruf O, Editor. Goethe und die Schweiz. 1. Aufl. Hannover: Wehrhahn Verlag; 2013.
99. Baker H. Das zum Gebrauch leicht gemachte Microscopium. Zürich: Heidegger und Comp; 1756.
100. von Gleichen-Rußwurm WF. Des Herrn Baron von Gleichen genannt Russworm, Abhandlung vom Sonnenmikroskop. Mürnberg: Winterschmidtische Kunsthandlung; 1781.
101. Euler L. Emendatio laternae magicae ac microscopii solaris. Novi Commentarii academiae scientiarum Petropolitanae 3. Petersburg: Academiae Scientarum; 1753. pp. 363–380.

102. Euler L. Briefe an eine deutsche Prinzessin über verschiedene Gegenstände aus der Physik un Philosophie. Aus dem Französischen übersetzt. 2. Aufl. Leipzig: Johann Friedrich Junius; 1780.
103. Schirmer S. Sketch of the performances and short account of the origin, history and explanation of the. In: Ergascopia P, Pantascopia M, c., editors. and the invisible girl. London: Large Theatre LYCEUM; 1805.
104. Tlantlaquatlapatli S, Editor H. Chronic von Berlin – oder Berlinsche Merkwürdigkeiten. Berlin: Petit und Schöne; 1790.
105. Wright WC. The portfolio comprising the flowers of literature. London: William Charlton Wright; 1825.
106. o. V. Society for the diffusion of useful knowledge. Portable diorama – dissolving views. The Penny Magazine 1843;2–3.
107. Limbird J. Dissolving views. The Mirror of Literature, Amusement, and Instruction. 1842;1:97–101.
108. Nichols and Son, Editor. The repertory of arts and manufacturer. London: Nichols and Son; 1802.
109. Robertson EG. Mémoires de Récréatifs Scientifiques et Anecdotiques. Paris: La Librairie de Wurtz; 1833.
110. Robertson EG. Mémoires de Récréatifs Scientifiques et Anecdotiques. Paris: La Librairie de Wurtz; 1831.
111. Bulletin des Lois de la Republiqe. Paris: Imprimerie de la Republique; 1799.
112. Testu, Editor. Calendrier Republicain, Décrété par la Convention Nationale, pour la IIe année de la République Françoise. Paris: Testu, Imprimeur- Libraire; 1792.
113. Neue Buchhandlung und Buchdruckerei, Editor. Kalender zum Gebrauche der vier neuen Departemente auf dem linken Rheinufer. Koblenz: Neue Buchhandlung und Buchdruckerei auf dem Paradenplatze; 1798.
114. Brewster D. Letters on natural magic. London: John Murray; 1834.
115. Robertson EG. Allgemeine Zeitung. 213. Aufl. Stuttgart; 1. August 1803;852.
116. Gilbert LW. Ueber die Luftfahrten der Bürger Garnerin und Robertson. Ann Phys. 1804;16: 257–92.
117. Vaterländische Notizen. Hamburg und Altona. II:254.
118. Österreichische Akademie der Wissenschaften. Döbler, Ludwig (1801–1864), Zauberkünstler. 2. überarbeitete Aufl., online. Österreichisches Biographisches Lexikon ab 1815. Wien: Verlag der Österreichischen Akademie der Wissenschaften; 2016.
119. Vogl JNep, Editor. K. K. priv. Theater in der Josephstadt: Hrn. Döbler's „Optische Vorstellungen" in drei Abtheilungen. Oesterreichisches Morgenblatt. 1847;12:30.
120. Sturm M. Sammlung von Theaterzetteln. Königliches Hof- und National-Theater München; 1845.
121. Pfundheller J. Skizzen aus Döbler's Reise-Erinnerungen Oesterreichisches Morgenblatt 1847;12:74.
122. o. V. from The Spectator. The Patent Ghost. The Mercury Hobart, Tasmanien, Australien; 21. Juli 1863;3.
123. Dircks H. The Ghost! as produced in the spectre drama. London: E. and F. N. Spon; 1863.
124. Pepper JH. The true history of the ghost and all about metempsychosis. London: Cassell & Company; 1890.
125. SPIEGEL ONLINE. Grammy Awards 2006: U2 Wins Big as Mariah Doesn't – Photo 3 – International [Internet]. Madonna (L) performs from behind a holographic screen with a holographic Gorillaz performer. 2006 Accessed on 22. Aug 2019. https://www.spiegel.de/fotostrecke/grammy-awards-2006-u2-wins-big-as-mariah-doesn-t-fotostrecke-12475-3.html.

126. Azéma M, Rivère F. Animation in Palaeolithic art: a pre-echo of cinema. Antiquity. 2012;86: 316–24.
127. Claudet A. A new fact relating to binocular vision. Proc Roy Soc London. 1866;15:424–9.
128. Herschel W. Experiments on the solar, and on the terrestrial rays that occasion heat; with a comparative view of the laws to which light and heat, or rather the rays which occasion them, are subject, in order to determine whether they are the same, or different. Part I. By William Herschel, LL. D. F. R. S. Philosophical Transactions of the Royal Society of London. 1800;90:293–326.
129. Paris JA. Philosophy in sport. London: Sherwood, Gilbert & Piper; 1827.
130. Paris JA. Philosophy in sport. London: Sherwood, Gilbert & Piper; 1833.
131. Babbage C. Passages from the life of a philosopher. London: Longman, Roberts & Green; 1864.
132. Roget PM. Erklärung eines optischen Betruges bei Betrachtung der Speichen eines Rades durch vertikale Öffnungen. Poggendorff JC, Editor Analen der Physik. 1825;81:93–104.
133. Faraday M. Über optische Täuschungen besonderer Art. Baumgartner A, von Ettingshausen A. Editor zeitschrift für Physik und Mathematik. 1832;10:80–101.
134. Faraday M. On a peculiar class of optical deceptions. Journal of the royal institution of Great Britain. London: John Murray; 1831.
135. Wheatstone C. Description of the kaleidophone, or phonic kaleidoscope; a new philosophical ty. Q J Sci Lit Art. 1827:344–51.
136. d'Arcy P. Sur la durée de la sensation de la vue. Histoire de l'Académie Royale des Sciences. Paris: L'Imprimerie Royale; 1768. p. 18–22.
137. Plateau J. Des illusions sur lesquelles se fonde le petit appareil appelé récemment Phénakisticope. Annales de Chimie et de Physique. Paris: Chez Crochard; 1833.
138. Julia-Fontenelle J-S-E, Editor. Manuel de physique amusante, ou, Nouvelles récréations physiques. Paris: Roret; 1829.
139. Lubschez BJ. The story of motion picture 65 B.C. to 1920 A.D. New York: Reeland Publishing; 1920.
140. Poggendorff JC, editor. Annalen der Physik und Chemie. Leipzig: Johann Ambrosius Barth; 1834.
141. Plateau J. Notice suivante sur l'anorthoscope. Bulletins l'Academie Royale des Sciences Brüssel: M Hayez; 1836. p. 7.
142. Poggendorff JC, editor. Annalen der Physik und Chemie. Leipzig: Johann Ambrosius Barth; 1836.
143. Stampfer S. Die Stroboscopischen Scheiben oder optischen Zauberscheiben. Wien: Trensensky & Vieweg; 1833.
144. Prechtl JJ. Verzeichnis der in der österreichischen Monarchie in den Jahren 1833, 1834 und 1835 auf Erfindungen Entdeckungen und Verbesserungen ertheilten Privilegien oder Patente. Jahrbücher des kaiserlichen königlichen polytechnischen Institustes in Wien. Wien: Carl Gerold; 1837.
145. Grützner P. Einige Versuche mit der Wunderscheibe. Pflüger, Archiv für die Gesammte Physiologie des Menschen und der Thiere. 1893;55:508–20.
146. Pellerin & Cie. L'ekonoscope. Lithografie Le batonniste 1868.
147. Purkinje JE. Apparat zur graphischen Darstellung von Bewegungen (Bericht über die Thätigkeit der naturwissenschaftlichen Section). Uebersicht der Arbeiten und Veränderungen der schlesischen Gesellschaft für vaterländische Kultur im Jahre 1841. Breslau: Graß, Barth und Comp.; 1842. pp. 62–64.
148. Siebenmann RS, Leuppi JR. Das Periphanoskop oder optische Zaubertheater. Schweizer Bote. No. 35. Aarau; 29. August 1833;272.
149. Journal des Demoiselles Pantinoscope 1860.

150. Aristoteles. Aristoteles über die Seele. Aus dem Griechischen übersetzt und mit Anmerkungen begleitet. Frankfurt: Johann Herrl; 1794.
151. Horner WG. Ueber die Eigenschaften des Daedaleums, eines neuen auf optischer Täuschung beruhenden Instruments. In: Poggendorff JC, editor. Annalen der Physik und Chemie. Leipzig: Johann Ambrosius Barth; 1834. p. 650–61.
152. Horner WG. On the properties of theDædaleum, a new instrument of optical illusion. Lond, Edinb, Dublin Philos Mag J Sci. 1834;4:36–41.
153. Plateau J. Troisième note sur de nouvelles applications curieuses de la persistance des impressions de la rétine. Bulletins de l'Academie Royale des Sciences 1850. pp. 30–39.
154. Plateau J. In: Poggendorff JC, editor. Dritte Notiz über neue sonderbare Anwendungen des Verweilens der Eindrücke auf die Netzhaut. Leipzig: Johann Ambrosius Barth; 1850.
155. Gehler JST. Physikalisches Wörterbuch. Leipzig: E. B. Schwickert; 1831.
156. Franklin B, Labaree LW, Willcox WB, Oberg B. The papers of Benjamin Franklin. New Haven: Yale University Press; 1959.
157. Lebon AM. 28 Spetembre 1799, Brevet d'Invention "Pour de nouveaux moyens d'employer les combustibles plus utilement, soit pour la chaleur, soit pour la lumière, et d'en recueillir les divers produits". Description des Machines et Procédés spécifiés dans les Brevets d'Invention. Paris: Madame Huzard; 1823. pp. 123–133.
158. Lebon P. Nachricht von einer ganz neuen, außerordentlichen, vom Pariser Mational-Institut geprüften, und durch ein Erfindungs-Patent authorisrten Entdeckung einer Thermo-Lampe oder eines Spar-Ofens. Paris, Stadtamhof: J. M. Daisenberger'sche Buchhandlung; 1802.
159. Scherer AN. Beschreibung einer Lampen – Vorrichtung zu chemischen Versuchen. Allgemeines Journal der Chemie Leipzig: Breitkopf und Härtel. 1799:134–45.
160. Senatsverwaltung Berlin für Umwelt, Verkehr und Klimaschutz. Gasbeleuchtung in Berlin/ Land Berlin [Internet]. Öffentliche Beleuchtung in Berlin. 2018. Accessed on 1. Apr. 2018. http://www.berlin.de/senuvk/bauen/beleuchtung/de/gaslicht/.
161. Notizen über Frankenstein's Lunar- und Solarlicht für Weingeist-. Oel- und Gasbeleuchtung. Gratz: J. A. Kienreich; 1847.
162. Umweltbundesamt Österreich. Altstandort "Glühstrumpf-Fabrik Auer von Welsbach" Gefährdungsabschätzung 2017.
163. Wilke A. Die Elektrizität, ihre Erzeugung und ihre Anwendung in Industrie und Gewerbe. Berlin: Springer; 1914.
164. Drummond T. On the means of facilitating the observation of distant stations in geodaetical operations. Philos Trans R Soc Lond. 1826;116:324–37.
165. Uchatius F. Apparat zur Darstellung beweglicher Bilder an der Wand. Sitzungsberichte der Kaiserlichen Akademie der Wissenschaften Mathematisch-Naturwissenschaftliche Classe. Wien: K. K. Hof- und Staatsdruckerei; 1853. pp. 482–485.
166. Czermak, JN. Physiologische Studien. Sitzungsberichte der Akademie der Wissenschaften mathematisch-naturwissenschaftliche Klasse 15. Wien: K. K. Hof- und Staatsdruckerei; 1855. pp. 425–521.
167. Cutter WR. New England families. New York: Lewis Historical Publishing Company; 1914.
168. Lincoln WE. In Memorian John Larkin Lincoln 1817–1891. Cambridge: University Press; 1894.
169. Lincoln WE. Toy called Zoëtrop. Patent, US 64,117 Apr 23, 1867.
170. Woodcraft B. Patents applied for and patents granted for the year 1867. London: Office of the Commissioners of Patents for Inventions; 1869.
171. Bradley M. Zoetrope, 5 component pieces, by Milton Bradley. [Internet] 1870. http://hdl.library.upenn.edu/1017/d/archives/20080204001.

172. Bradley M. Zoetrope, keep the ball rolling strip [Internet]. http://hdl.library.upenn.edu/1017/d/archives/20110304011.
173. Campbell L, Garnett W. The life of James Clerk Maxwell. London: MacMillan and Co.; 1882.
174. Rankine WJM. On the exact form of waves near the surface of deep water. Philosophical transactions of the royal society of London. London: Royal Society; 1863. p. 127–38.
175. Harman PM, Editor. The scientific letters and papers of James Clerk Maxwell 1862–1873. Cambridge: University Press; 1995.
176. v. Helmholtz H. Über Integrale der hydrodynamischen Gleichungen, welche den Wirbelbewegungen entsprechen. J Reine Angew Math 1858;55:25–55.
177. Cohn F. Über die Organisation und Entwicklung von Volvox Globator. Amtlicher Bericht über die 32 Versammlung Deutscher Naturforscher und Ärzte. Wien: K K Hof- und Staatsdruckerei; 1858. pp. 162–163.
178. Hopwood HV. Living pictures. London: Optician and Photographic Trade Review; 1899.
179. Tissandier G. Le Praxinoscope. La Nature – Revue des sciences. 1879:133–4.
180. Muybridge, E. The attitudes of animals in motion, illustrated with the Zoopraxiscope. Royal institution of Great Britain; 1882.
181. Marey E-J. La machine animale – Locomotion terrestre et aérienne. Paris: Librairie Germer Bailliere; 1873.
182. Marey E-J. Animal mechanism: A treatise on terrestrial and aerial locomotion. New York: D. Appleton and Company; 1874.
183. State Telegraph. Fatal shooting affray near Callistoga. Sacramento Daily Union. 48/7345. Sacramento; 19. Oct. 1875.
184. Russian River Flag. The late murder case. Daily Alta California. 27/9083. San Francisco; 14. Februar 1875.
185. Herbert S. Projecting the living image. Eadweard Muybridge: The Kingston Museum Bequest. East Sussex: The Projection Box; 2004. pp. 107–153.
186. Voß G. Die Augenblicksphotographie. Pecht F, Editor. Die Kunst für alle: Malerei, Plastik, Graphik, Architektur. 1889;4:148–151.
187. Flammarion C. Le passage de Vénus. La Nature – Revue des sciences. 1875;356–358.
188. Marey E-J. Le Fusil photographique. La Nature – Revue des sciences. 1882;326–330.
189. Marey E-J. Movement. London: William Heinemann; 1895.
190. Ramsaye T. A million and one nights: The history of the motion picture. New York: Simon & Schuster; 1926.
191. McCosker MJ. Philadelphia and the genesis of the motion picture. Pennsylvania Magazine of History and Biography 1941;65.
192. Eder JM. Jahrbuch für Photographie und Reproductionstechnik. Halle a. S.: Wilhelm Knapp; 1891.
193. Reynaud E. Théâtre optique. Patent FR 194 482 Dec. 1, 1888.
194. Tissandier G. Le Théâtre optique de M. Reynaud. La Nature – Revue des sciences. 1892:127–8.
195. Dickson WKL, Dickson A. The life & invention of Thomas Alva Edison. London: Chatto & Windus
196. Lumiere A, Lumiere L. Le Cinématographe. Patent, FR 245 032 Feb 13, 1895.
197. Lumiere A, Lumiere L. Kinetographic camera. Patent, US 579,882 March 30, 1897.
198. Smith W. Effect of light on selenium during the passage of an electric current. Nature. 1873;7: 303.
199. Adams WG, Day RE. The action of light on selenium. Philos Trans R Soc Lond. 1877;167:313–49.
200. Perry J, Ayrton WE. Seeing by electricity. Nature. 1880;21:589.
201. Nipkow P. Elektrisches Teleskop. Patent, DE 30105 Jan 5, 1884.

202. Baird JL. Television and like system. Patent, US1735946A Oct. 15, 1926.
203. von Ardenne M. Leuchtfleck-Abtaster (Flying spot scanner). Tabellen zur angewandten Physik: Elektronenphysik, Übermikroskopie, Ionenphysik. Berlin: Deutscher Verlag der Wissenschaften; 1962. p. 275.
204. Braun F. Ueber ein Verfahren zur Demonstration und zum Studium des zeitlichen Verlaufes variabler Ströme. Annalen der Physik und Chemie. 1897;296:552–9.
205. Lehmann O. Flüssige Kristalle. Leipzig: Wilhelm Engelmann; 1904.
206. Merck KGaA. Merck Geschäftsbericht 2003: Fortschritt für Lebensqualität. Darmstadt; 2004.
207. Voigt W. Magneto- und Elektrooptik. Leipzig: B. G. Teubner; 1908.
208. Voigt W. Theoretische Studien über die Elasticitätsverhälnisse der Krystalle. Abhandlungen der Physikalischen Classe der Königlichen Gesellschaft der Wissenschaften in Göttingen. Göttingen: Dieterichsche Buchhandlung; 1887. p. 6.
209. Voigt W. Lehrbuch der Kristallphysik (mit Ausschluss der Kristalloptik). Leipzig: B. G. Teubner; 1910.
210. Hamilton WR. Lectures on quaternions: containing a systematic statement of a new mathematical method. Dublin: Hodges and Smith; 1853.
211. Einstein A. Hamiltonsches Prinzip und allgemeine Relativitätstheorie. Sitzungsberichte der Königlich Preussischen Akademie der Wissenschaften, Sitzung der Physikalisch-Mathematischen Klasse. Berlin: Verlag der Königlichen Akademie der Wissenschaften; 1916. pp. 1111–1116.
212. Einstein A. Näherungsweise Intergration der Feldgleichungen der Gravitation. Sitzungsberichte der Königlich Preussischen Akademie der Wissenschaften, Sitzung der Physikalisch-Mathematischen Klasse. Berlin: Verlag der Königlichen Akademie der Wissenschaften; 1916. pp. 688–696.
213. Kerr J. A new relation between electricity and light: Dielectrified media birefringent. Lond, Edinb, Dublin Philos Mag J Sci. 1875;50:337–348.
214. Born M. Über anisotrope Flüssigkeiten. Versuch einer Theorie der flüssigen Kristalle und des elektrischen KERR-Effekts in Flüssigkeiten. Sitzungsberichte der Königlich Preussischen Akademie der Wissenschaften, Sitzung der Physikalisch-Mathematischen Klasse. Berlin: Verlag der Königlichen Akademie der Wissenschaften; 1916.
215. Born M, Stumpf F. Über anisotrope Flüssigkeiten. Zweite Mitteilung. Die Temperaturabhängigkeit der Brechungsindizes senkrecht zur optischen Achse. Sitzungsberichte der Königlich Preussischen Akademie der Wissenschaften, Sitzung der Physikalisch-Mathematischen Klasse. Berlin: Verlag der Königlichen Akademie der Wissenschaften; 1916.
216. Oseen CW. Versuch einer kinetischen Theorie der kristallinischen Flüssigkeiten. Kungliga Svenska Vetenskapsakademiens Handlingar. Stockholm: Almqvist & Wiksells Boktryckeri; 1921. pp. 1–39.
217. Oseen CW. Versuch einer kinetischen Theorie der kristallinischen Flüssigkeiten – Zweite Abhandlung. Kungliga Svenska Vetenskapsakademiens Handlingar. Stockholm: Almqvist & Wiksells Boktryckeri; 1922. pp. 1–48.
218. Oseen CW. Versuch einer kinetischen Theorie der kristallinischen Flüssigkeiten – III. Abhandlung. Kungliga Svenska Vetenskapsakademiens Handlingar. Stockholm: Almqvist & Wiksells Boktryckeri; 1923. pp. 1–40.
219. Brown GH, Shaw WG. The mesomorphic state – liquid crystals. Chem Rev. 1957;57:1049–157.
220. Gray WG. Molecular structure and properties of liquid crystals. London: Academic; 1962.
221. Frank FC. I. Liquid crystals. On the theory of liquid crystals. Discuss Faraday Soc 1958;25:19.
222. Williams R. Liquid crystals in an electric field. Nature. 1963;199:273–4.
223. Wood FW. Illuminated announcement and display signal. Patent US 1908;974:943.
224. Heilmeier GH, Zanoni LA. Electro optical device. Patent US 3,499,112 1967.

225. Lechner B, Marlowe F, Nester E, Tults J. Liquid crystal matrix displays. 1969 IEEE International Solid-State Circuits Conference Digest of Technical Papers [Internet]. Philadelphia, PA, USA: IEEE; 1969. Accessed on 12. May 2019. pp. 52–53. http://ieeexplore.ieee.org/document/1154684/.
226. Weimer P, The TFT. A new thin-film transistor Proc IRE 1962;50:1462–1469.
227. Round HJ. A note on carborundum. In: Sze SM, Editor. Semiconductor devices: pioneering papers. World Scientific; 1991. pp. 879–879.
228. Lossev OV. Luminous carborundum detector and detection effect and oscillations with crystals. Lond, Edinb, Dublin Philos Mag J Sci. 1928;6:1024–44.
229. Destriau G. Électroluminescence -Luminiscencence dans les champs électriques et phénomènes èlectroniques dans les semi-conducteurs. In: Comptes rendus Academie des sciences. Paris: Gauthier-Villars; 1939. p. 36–7.
230. Destriau G. Recherches sur les scintillations des sulfures de zinc aux rayons α. J Chim Phys. 1936;33:587–625.
231. Biard JR, Pittman GE. Semiconductor radiant diode. Patent US 3,293,513 1962.
232. Holonyak N, Bevacqua SF. Coherent (visible) light emission from Ga [. . .] junctions. Appl Phys Lett. 1962;1:82–3.
233. Bernanose A. Electroluminescence of organic compounds. Br J Appl Phys. 1955;6:S54–5.
234. Pope M, Kallmann HP, Magnante P. Electroluminescence in organic crystals. J Chem Phys. 1963;38:2042–3.
235. Helfrich W, Schneider WG. Recombination radiation in anthracene crystals. Phys Rev Lett. 1965;14:229–31.
236. Gurnee EF, Fernandez RT. Organic electroluminescent phosphors. Patent US 3,172,862 1960.
237. Tang CW, Van Slyke SA. Organic electroluminescent diodes. Appl Phys Lett. 1987;51:913–5.
238. Burroughes JH, Bradley DDC, Brown AR, Marks RN, Mackay K, Friend RH, u. a. Light-emitting diodes based on conjugated polymers. Nature 1990;347:539–541.
239. Schott. Die Zukunft liegt biegsam in Ihrer Hand. [Internet]. Geschichten über die Faszination innovativer Spezialgläser – SCHOTT Innovation. 2017 Accessed on 12. May 2019. https://www.schott.com/innovation/de/ultraduennesglas/.
240. Langmuir I. Oscillations in ionized gases. Proc Natl Acad Sci. 1928;14:627–37.
241. Eddington AS. The deviation of stellar material from a perfect gas. Mon Not R Astron Soc. 1928;88:352–69.
242. Cap F. Lehrbuch der Plasmaphysik und Magnetohydrodynamik. Wien: Springer; 1994.
243. Bitzer DL, Johnson RL. PLATO: A computer-based system used in the engineering of education. Proc IEEE. 1971;59:960–8.
244. Bitzer DL, Slottow HG. The plasma display panel: a digitally addressable display with inherent memory. Proceedings of the November 7–10, 1966, fall joint computer conference on XX – AFIPS '66 (Fall). San Francisco, California: ACM Press; 1966. p. 541.
245. Bitzer DL, Slottow HG, Wilson RG. Gaseous display and memory apparatus. Patent US 3,559,190 1966.
246. Stifle J. CERL Report X-20: The PLATO IV Architecture. May 1972. Urbana: Computer-based Education Research Laboratory, University of Illinois; 1971.

Virtual Illusions: The Generation of Spatial Images

4

Summary

Humans have the natural aptitude for spatial vision. The visual apparatus is connected to the brain, which is capable of correcting image errors and interpreting the recorded image. In principle, this basic understanding of space can also be acquired with one eye from a fixed position. By changing position, each new position also creates a new perspective. Small changes in position are enough to gain a real impression of space. This works just as well if you do not move your head but look at the real scene with both eyes. The brain is easily able to generate the perception of space from these slightly different perspectives. In the stereoscope, the two slightly offset images are offered to the eye, creating the impression of a real scene. For this, however, one has to look into a peep-box. It is more comfortable to put on 3D glasses and enjoy an equivalent impression on a big screen. The only unpleasant thing that remains is the glasses. But there is a remedy for this too. For more than a century, there have been methods for 3D presentation without 3D glasses. What started with integral photography at the beginning of the twentieth century later became autostereoscopy and is now on its way into the twenty-first century under the catchword light field.

4.1 Natural Spatial Vision

The True Perception

Since the beginning of evolution, humans have seen with two eyes and thus spatially. When seeing, the eyes only provide the optical part of the image. The final perception of what is seen requires downstream data processing, which generates a visual impression from the physical information. The human visual system is based on the connection of the

A. Grasnick, *Basics of Virtual Reality*,
https://doi.org/10.1007/978-3-662-64201-6_4

optical apparatus with the mental power of the brain and its memory of experience. The imaging quality of the eyes is extremely limited. The physicist Helmholtz puts it in a nutshell in a frequently reproduced quotation [1].[1] The quality of pure imaging by the eyes is poor. And not only that. The transmission of reality through the eye is not only fundamentally flawed, but also varies from person to person, and even varies within a person's pair of eyes. The physical dissimilarity of the eyes usually leads to the dominance of one eye in most people (see, e.g., [2–4]). This dominance is by no means fixed, but may well vary in strength [5] and may also alternate in a person over time (see, e.g., [6, 7]). The dominant "leading eye" has an influence on spatial vision, as has been demonstrated by a valence test with the ophthalmic testing device "Polatest"[2] [8]. The balance of stereo vision also influences the perception of direction. Valence refers to the "Valuability" of the eyes during binocular vision, where stereo equivalence means the equivalence of both eyes and stereo prevalence describes the occurrence of preference of a particular eye [9].

The guiding eye does have an influence on daily life, which is particularly noticeable in directional walking. Even if the person walking quite consciously tries to keep a certain direction, the path will shift in the direction of the "higher-order" eye. Such a test had been performed by the physician Rosenbach in 1903 [2].[3] A similar test was performed again at the Max Planck Institute for Biological Cybernetics (Max-Planck-Institut für biologische Kybernetik) more than 100 years later. Here, too, a tendency to walk in circles in unknown terrain could be demonstrated [10], which is at least an indication of the actual effect. We may therefore assume that the eyes of the observer have an influence on the perception of real environments.

Not only the valence of the eyes plays a role. During the focussing of near objects, not only accommodation and convergence but also the pupil diameter are coupled with each other. Scheiner had already noticed this effect about 400 years ago. Scheiner observed the constriction of the pupil when looking at near objects ([11], p. 31). Helmholtz, who explains Scheiner's experiment again in detail ([12], from p. 95), and Bach ([13], from p. 59) also followed this assumption, although Bach ascribes the initial function to convergence here. The term "pupillary near reflex" was used since the 1950s by the American ophthalmologist Marg [14]. This term is still somewhat less precise than the term "near triad" used since the 1970s [15, 16]. This name, which is common in the specialist literature, is also increasingly used in the German-speaking world in the German translation "Naheinstellungstrias".

---

[1] "Now it is not too much to say that to an optician who wished to sell me an instrument which had the latter defects, I should think myself perfectly justified in using the harshest expressions about the negligence of his work, and in returning his instrument to him with protest." ([1], p. 21).

[2] Vision tester for binocular tests (binocular vision) from the Zeiss company.

[3] "Further, I would note that right-sided monocular fixation in binocular vision may perhaps serve to explain the fact that the vast majority of people, when walking, ..., have a tendency to deviate to the right from the straight line (direction line)."

If the coupling of accommodation, convergence and pupil diameter is involuntary and compelling, one could use this effect to infer from the given knowledge of one parameter to the others. However, this will not work completely because, for example, pupil size may also depend on other factors. Even the stress of concentration while reading changes pupil size [17]. But more known is the constriction or dilation of the pupil caused by emotions [18]. Basically, it can be said that pleasant stimuli or impressions enlarge the pupil and unpleasant things constrict the pupil. Taking these factors into account, the measurement of pupil size and convergence can allow conclusions to be drawn about accommodation.

Perception is not based on pure imaging and its primitive evaluation, but on the further processing of the image data in the brain of the observer. The human visual image processing is conditioned to correct image errors. The retinal images are enriched with the experiential data of the observer and thus support the ability to interpret. The optical resolution of the eye is greatest in the retinal fossa (fovea centralis) and can reach a value of up to 200,000 receptors per mm^2. In peripheral vision, this can drop to less than 10,000 per mm^2 [19]. The resolution of the eye is not evenly distributed compared to a camera sensor. In addition, the density of cones and rods in the eye will not only be different in different individuals, but will vary between eyes even in the same person. Inclusions, scars, or deposits on the cornea, in the lens, or in the vitreous body of the eye will further degrade the image. Optical imaging with a primitive, defective lens also leads to distortion and color errors at the edge of the image. Surprisingly, however, this often does not lead to a deterioration of perception.

The perceived image enhancement is partly explained by the fact that the eye keeps projecting new image elements onto the site of sharpest vision through continuous fixation and search movements. This effect is not dissimilar to the concept of "super-resolution" in the context of "computational imaging". Moreover, when an object is fixated, both eyes of the observer converge to each other such that an object point of the fixated object falls on corresponding retinal locations preferably in (or near) the fovea centralis. Additional saccades[4] provide further evaluable image information that is used by cognitive data processing and converted into perception.

In addition, information that promotes spatial perception is also obtained by seeing with one eye. First and foremost this includes perspective, which has already been discussed as central perspective or color perspective or aerial perspective ("blueing") and blurring with increasing distance. Nevertheless, an example should be given here once again that proves the knowledge of this effect in the Renaissance. The version of the Mona Lisa from da Vinci's workshop,[5] which is now in the Prado, is a good illustration of the sensible use of aerial perspective (Fig. 4.1).

[4]Involuntary micromovements of the eye.

[5]Painted by one of his students, perhaps Melzi.

Fig. 4.1 La Gioconda del Prado, Leonardo da Vinci workshop (zw.1503 u. 1519), Museo Nacional del Prado, Room 56B, Madrid, Escarlati (2012), Wikimedia Commons

The color of the objects in the background of the picture goes with increasing (apparent) distance from the viewer more and more into the blue and becomes more and more blurred. This corresponds to the actual perception in reality.

Texture perspective is nothing more than a form of central perspective, in which the reduction of the distance between known structures can be used to infer their distance from the observer. Such textures converging in distance can already be found in Alberti (see Fig. 2.41), whose achievement is especially the description of the construction of such a perspective. This works not only for textures, but also for known objects, where by their relative size, their spatial arrangement can be inferred. The relative height of an object describes its position in the image and can help to estimate a distance. First, when viewing a

Fig. 4.2 Apparent size, relative position, and occlusion using the Stanford Bunny as an example, Grasnick (2019)

landscape image, it can be assumed that objects that are located lower in the image are also located closer to the viewer. This is illustrated in Fig. 4.2 with a simple example.

It is obvious that with the reasonably known size of a rabbit, the perception of a smaller animal must be attributed either to a younger specimen or to a further distance. In addition to apparent size, the relative position of the bunny in the image provides an additional clue. It is decisive at which relative height the object is located to the horizon line. At the same absolute object height, an object should be relatively closer to the horizon if it is further away. A hare sitting higher up is therefore presumably further away than one crouching at the lower edge of the image. The way occlusion works is also obvious. An object that obscures another must necessarily be in front of it.

Occlusions can sometimes be revealed in the case of nearby objects that appear close together in the perceived image, but are actually offset in spatial depth, by moving the head or assuming a more favorable position to the object of observation. This produces the effect of an apparent movement of objects that are actually stationary. This apparent change is parallax or, when produced by motion, motion parallax. The value of the apparent motion of an object is an indication of its distance from the observer. A distant object will not move much in position, while a close object will have apparent distinct motion.

Fig. 4.3 Circular rings with shadows cast, Grasnick (2019)

Another indication for understanding space are the shadows that objects cast. The position and length of the shadows allow conclusions about the shape and size of the objects.

This will be illustrated by a small sketch. In Fig. 4.3 on the left side there is a grey ring. If one were to infer the 3D shape of the drawn object from this image alone, one would assume that it is a flat disk. The same shape also allows other interpretations depending on the shadow. For example, the object in the middle could be a shell-like structure, while the one on the right is more like a truncated cone.

However, this presupposes the mental positioning of a light source at the top left. If one shifts this in the mind to the lower right, the perceived depth is also reversed.

In real life, you do not need to think long. The casting of the shadow is sufficiently known from experience. Experience is an important tool for knowledge anyway. An object whose size is known can be easily estimated in distance. From a cast shadow, the spatial shape of the illuminated object can be inferred.

The importance of monocular depth criteria for spatial perception can hardly be overestimated. Especially at large distances, the difference between the retinal images of both eyes is too small to be evaluable. Therefore, the monocularly obtained (or just binocularly identical) information must be sufficient to obtain the necessary knowledge for the respective situation. The visual system is trained to this perception in such a way that the spatial arrangement can be inferred even in a two-dimensional one, and is even recognized unconsciously.

In natural spatial vision, it is automatically assumed that 3D perception can only occur when the object is viewed with both eyes simultaneously. It was explained that other effects such as perspective, motion parallax or occlusions also contribute to the understanding of space, whereby insights into the structure of 3D space can also be gained with one eye. The resulting impression is two-dimensional in nature, only conclusions about the spatial arrangement are drawn, compared with previous experiences and finally blended with the 2D impression. In this process the brain takes over the downstream image processing and tries to understand the scene. Only the comparison of the evaluation with the experience values of the sighted person enables a classification into the known world of knowledge and thus the interpretation of the scene. All this happens completely unconsciously and involuntarily. This cognitively enriched impression can certainly appear spatial to the observer. However, a closer look reveals that the depth gradation of objects in space can also be correctly interpreted monocularly, whereas the perception is only apparently a spatial one. Although the scene can be interpreted in spatial coordinates, there is no real spatial effect. If one has a direct comparison (e.g. by viewing a scene alternately with one and both eyes), one immediately notices the missing spatial impression with monocular viewing.

Missing-the-Bottle-Test

Sit on a chair in front of a table and close both eyes. Meanwhile, your assistant places a bottle on the table in front of you. It does not matter if the bottle is filled or with what, it is only important not to know the location of the bottle. After you are asked to open one eye (and only one!), try to grab the bottle from the side. You will probably find it difficult or even fail.

If you repeat the experiment, opening both eyes, you should be able to grip the bottle easily.

Missing-the-Bottle-Test-increased Difficulty

A very similar effect can occur in the event of poisoning. To bring about such a situation, you can slightly modify test 1. It is important that a bottle with a high-proof alcoholic content is available. The contents of the bottle must be consumed gradually until one is so drunk that control over one's own eye movements is significantly reduced and double vision occurs. In the subsequent test, it does not matter whether the test is performed with one or both eyes—the test is very likely to fail, since the person under test is functionally one-eyed anyway as a result of the drunkenness.

4.2 Binocular Spatial Perception

Seeing with Two Eyes

Binocular spatial perception, or quite simply seeing with two eyes, is not a recent discovery. Already in antiquity, the two-eyed vision is thematized.

This raises the question of how it is possible that, despite different images, only one object is ever seen. Empedocles has a rather sympathetic explanation for this. He saw in the goddess of love Aphrodite the creator of the eyes, which she had essentially assembled from flames and with a slight admixture of earth, and had mounted and united the eyes with a kind of "love nails". And thus it comes to the result all known to us (from [20], p. 254):

> One will be the sight of both eyes.

Aristotle's students are not satisfied with this. In the post-Aristotelian writings "Problemata physica", which were published as textbooks by members of the Peripatos, the philosophical school of Aristotle, there is also an explanation of double vision in book XXXI “Concerning the eyes” (from [21], p. 265).

> Why do things appear double to those who squint? Because the movement of each eye does not reach the same point. Therefore, our soul interprets the double vision as the vision of two.

This is well observed. The cross-eyed person sees double because the visual axes do not intersect at one point and thus sees two objects instead of one.

Surely this is not the whole truth. Even with single vision with two eyes, one sees different images. Thus, Euclid describes the larger visual field when seeing with two eyes (from [22], p. 18).

> Theor. 26. If a sphere is seen with both eyes, and its diameter is equal to the straight line between the two eyes, half of it is seen.

There are two important observations in this. One is that what is seen by each eye is different, since one sees only half of it. Furthermore, the consequence of what has been said is that with both eyes one sees both halves, combined the whole again.

A basically geometrical analysis of seeing with two eyes is also followed by Ptolemy and verified with various experiments. According to Ptolemy, the visual cones of the two eyes merge to form a common visual cone when both eyes align themselves with the same object. The visual axes leading from each eye to the object, the common viewing point, are enriched by a virtual axis. This axis effectively describes the cyclopean viewing point between the two eyes. On this fictitious axis appear the things that are fixed (see [23], p. 286).

> Those [things], namely, which lie on the axes themselves seem distinctly to assume a single place, which is the common axis.

It is certainly not wrong to include a small sketch to describe the common visual axis of Ptolemy. In Fig. 4.4, the observer looks with both eyes at the object (blue axes). The common axis is formed by the red line.

The realization that binocular vision produces different images for each eye and yet sees a common, single object has regularly preoccupied those who had turned to optics anyway. One simply cannot help but notice this mystery in daily life as well. Of course, Alhazen had also dealt with binocular vision and, just like Ptolemy, had established a common axis of the eyes when seeing. This central axis is also crucial in Alhazen’s avoidance of double vision. In Risner’s drawing of Alhazen’s remarks (Fig. 4.5) some lines are drawn showing cones of vision of the eyes, the axes to the considered points and the respective common axis.

The enclosed text illuminates the drawing and lets us see that Alhazen also has the cones of vision depart from the eyes. At the distance where the visual axes of both eyes meet (i.e.,

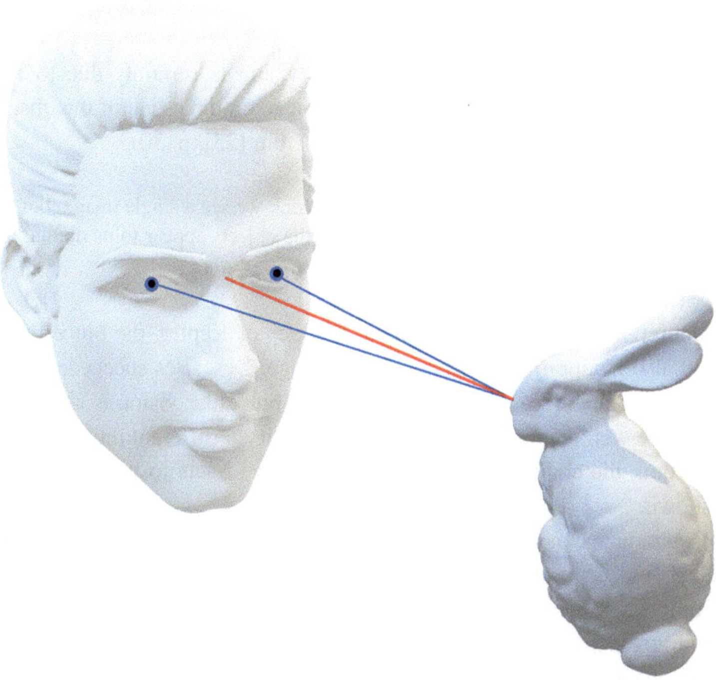

Fig. 4.4 Ptolemy's common visual axis, Grasnick (2019)

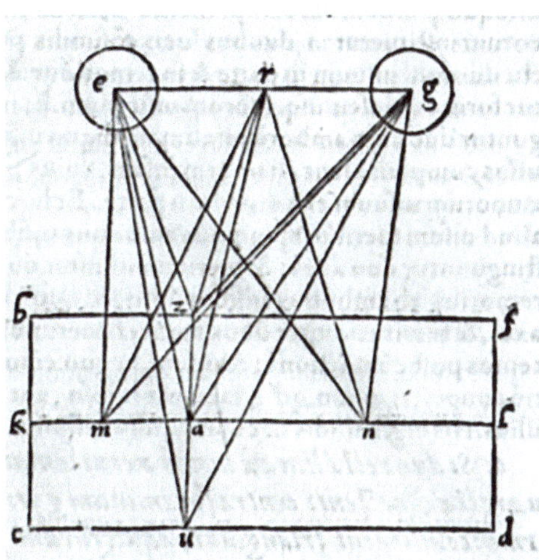

Fig. 4.5 Binocular vision. (According to Alhazen, Friedrich Risner (1572), from [24], p. 77)

at the fixation point), one can trace out an area that is parallel to the eyes. In this area, single vision would be possible.

The introduction of the concept of image in vision is certainly due to Kepler, who actually described the points of an object as an image on the retina of the observer.

Kepler created a theorem on binocular vision (from [25], p. 30),

> If both retinas are stimulated in the same way, we think we see a single image; but if the two retinas are stimulated or painted in unequal ways, the objects appear to us double instead of single.

With this definition and the detailed optical theories of Kepler, the binocular process of seeing could also be approached as an optical image. Descartes succeeded excellently in this in his "Dioptrique" [26]. His understanding of binocular vision is shown in Fig. 4.6. Here one can see well the mapping of the object points at different retinal positions of the eyes. In addition, the connection of the eyes via nerve tracts to the brain of the sighted person is shown.

The optical radiance of a Kepler or Descartes were immense at this time. It was not long before optical devices based on their preliminary work were produced.

The "Compleat System of Opticks" [27] presented by Smith in 1738 contains some thoughts on binocular vision. In the German translation of 1755, an actual spatial effect is found in the chapter "Vom doppelten Fernrohre" (A Binocular Telescope) (from [28], p. 344):

> The focuses of the two telescopes there are two equal rings as usual, which terminate the pictures of the objects there formed; and of consequence the visible area of the objects themselves. These equal rings by reason of the equal eye-glasses, appear equal and equally remote when seen separately by each eye, the other being shut; but when seen united by both eyes, they appear much larger and remoter too; and the objects seen through them do also appear larger, though circumscribed by the united rings in the same places as when seen separately

Smith describes here an effect that occurs by horizontal displacement of two similar objects and their observation with parallel view. The double telescope in this case acts like a stereoscope. He makes another observation with pair of compasses which he holds in front of his eyes (from [28], p. 345):

> But by compressing the legs with your hand, the two inner points will come nearer to each other; and when they unite, (having stopt the compression,) the two inner legs will also entirely coincide and bisect the angle under the outward ones, and will appear more vivid, thicker and longer than they do, so as to reach from your hand to the remotest object in view, even in the horizon it self, if the points be exactly coincident.

Smith experiments with other similar objects (e.g. tobacco pipes and straws) and comes up with the following explanation:

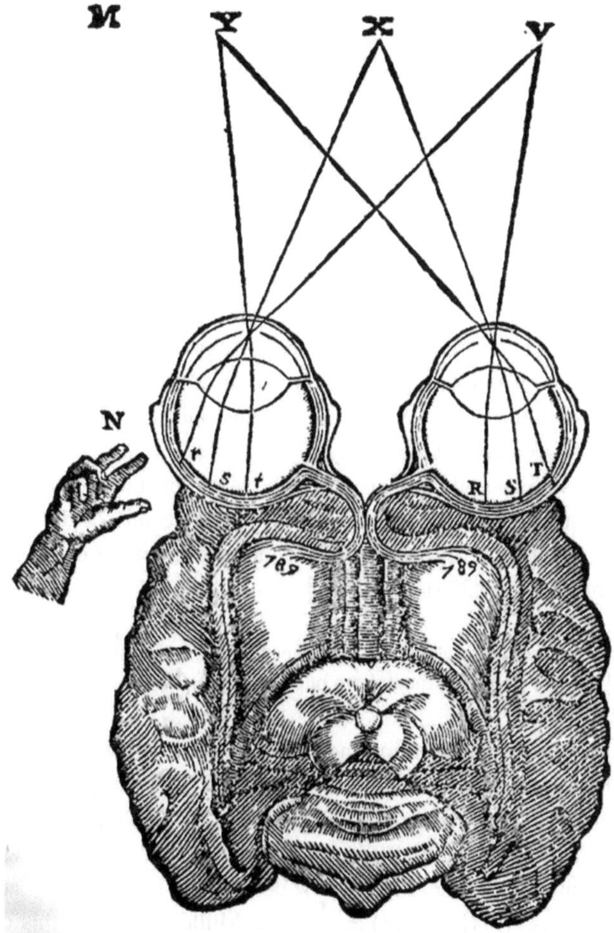

Fig. 4.6 Binocular vision. (According to Descartes, Renè Descartes (1657) from [26], p. 49)

> I apprehend the reason of the phenomenon to be this: by shutting the eyes *d,*e alternately, it will be evident that the points *d, a, f,* are in one straight line to the, and the points *e, b, f* in another; and so the reason of the apparent union of the points *a, b* at the object *f,* is because their pictures are upon the same points of the retina's as those of the object itself. For the same reason, while the legs of the compasses are gradually compressed, if your eye be kept fixt upon the imaginary point of the united legs, this point will appear to approach gradually…

This explanation includes an aspect that is necessary for spatial vision. The pixels of an object point fall on different parts of the retina of the respective eye. Although these points are different, they originate from the same object point and are therefore involuntarily connected. This is why we speak of corresponding image points. The distance between two corresponding image points as a deviation from the fixation axis[6] is called disparity.[7] The term disparity is used specifically to describe this situation in spatial perception. In addition to the monocular depth criteria mentioned at the beginning, disparity is the central cause of spatial vision.

In his drawing (Fig. 4.5) Risner had already quite correctly depicted the facts explained by Alhazen. Admittedly, the sketch is not self-explanatory. Descartes drawing is clearer. But here the object points all lie in one plane. Disparity occurs when the object points are at different distances.

Here again the question arises, which has been current since antiquity, when one sees an object single and when one sees it double. The object space in which objects must be located in order to still be perceived as a single object with both eyes is called the "horopter".[8] The name is not new, but was already used in the early seventeenth century in the books of Aguilonius [30]. With it Aguilonius had described a mental surface on which objects can be perceived binocularly.

In the centuries since Aguilonius proposed the horopter, the idea has been constantly refined. At first it quickly became clear that a line or plane surface did not correspond to actual experience. The physician Baer explained the horopter in his own way when looking at an object C (from [31], p. 223).

> If the horopter is the surface in which we simply see objects, it is, strictly speaking, not a plane, but a spherical surface passing through both pupils and the point C.

The surface of the horopter is therefore not flat, but has a spherical shape. This roughly corresponds to the observation which the Dessau teacher Vieth[9] practically verified with pencil and compass point in 1817 [33]. The physiologist Müller had similar thoughts, which he investigated scientifically. Presumably unaware of Vieth's observation, whom he at least did not mention by name, Müller now determined the horopter "according to the example of Aguilonius" ([34], p. 178) as a circle. Müller described this circle, which is known today as the "Vieth-Müller circle" (from [34], p. 240):

> All objects, then, which lie in the circular plane of the horopter are seen single and distinct, all objects outside it are seen double and indistinct.

---

[6] The distance of each point is measured from the position of the imaged fixation point (the axis of an eye).

[7] Also "transverse disparity" with purely lateral displacement.

[8] Horopter: from Greek gr. horos = border + opter = scout, limit of vision, plane of projection, the epitome of the points of external space that image on corresponding parts of both retinas and are therefore easily seen [29].

[9] Vieth is certainly better known as a gymnastics father and as the author of an encyclopedia of physical exercises [32].

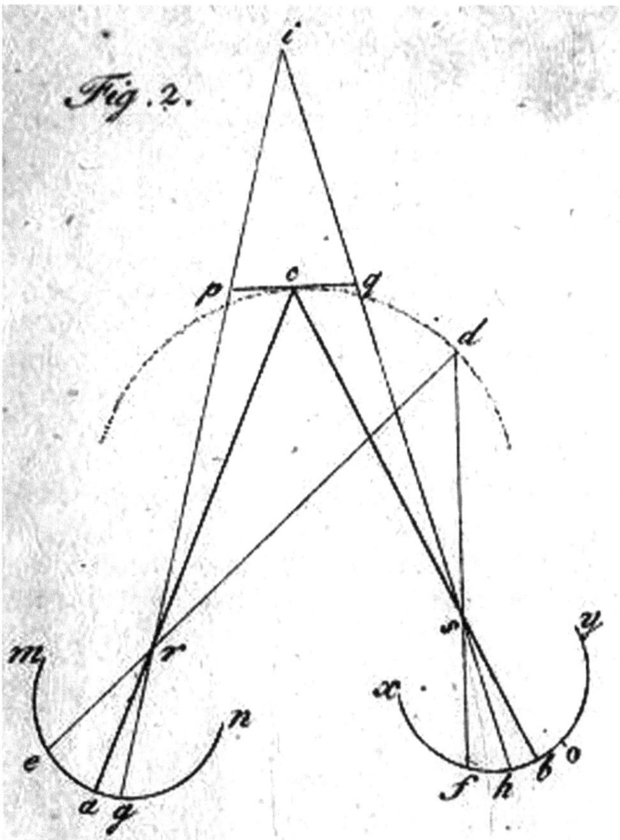

Fig. 4.7 Theoretical horopter. (Johannes Müller (1826) from [34], plate IV, Fig. 2)

This defines the theoretical horopter and is shown in Fig. 4.7. Again, the eyes are directed to a common point (c), to which both visual axes are directed (represented by the distances ac and bc). The points lying on the circular segment of the horopter are seen single—the rest are not.

Of course, the actual horopter does not follow Müller's theoretical specifications. The physiologist Hering criticized the treatises of Helmholtz about the horopter,[10] because he could not confirm its calculations already with his own seeing [35]. Now with Hering the horopter decomposes into a mathematical and empirical part, which proves the true horopter also by experiment [36]. The concept of an empirical horopter with Hering's division between "real thing" and "seen thing" is also adopted by Hillebrand, a former

[10]Criticized here from Helmholtz's treatise are especially "The Distribution of the Corresponding Points in Both Fields of View", the "Form of the Horopter", and the "Significance of the Horopter in Vision".

collaborator of Hering, some years later [37]. Hillebrand, a philosopher, describes the process of seeing quite vividly and finds different horopter forms as a result of seeing at different distances. All in all, the findings of Hering and Hillebrand allowed the recognition of the deviation of the theoretical from the practical horopter and found their way into the technical literature as "Hering-Hillebrand horopter deviation".

The Danish physiologist Panum had intensively studied "seeing with two eyes". He recognized that even when looking at things that are not on the horopter, double images do not always necessarily occur [38]. Thus, if the horopter is consistently referred to as the area in which the objects within it are always seen single, the original simple shape enlarges into a space called the Panum area. In this area, the images of the left and right eyes can fuse into a common three-dimensional visual impression. The value of the quality of fusion and thus of stereo vision, stereopsis, can be taken to be the maximum disparity of the individual images produced in a test by objects staggered in space. Today, this can be tested very easily with a 3D monitor or VR goggles by offering the test subject test images with defined disparity and determining the disparity at which the test subject begins to see double images.

But even outside the Panum area, depth vision could occur if the perception of double images can be avoided by deliberate blurring of the image of objects outside the fixation point.

4.3 Spatial Disturbance

Perception Problems with 3D Glasses

Spatial perception is thus primarily achieved through the retinal images and their evaluation. These different images on the retina can be generated not only by real environments, but also quite specifically by the presentation of a stereo image pair.

In reality, however, spatial perception can also be disturbed. The binocular spatial effect of a stereoscopic 3D presentation cannot be perceived by all people. Visual disorders (e.g. strabismus or eye tremor) complicate binocular perception or provide double images. Functional monocularity[11] or practical monocularity[12] can even lead to "stereo blindness" [39] and thus complete loss of binocular spatial vision. The emphasis here is deliberately on the amendment "binocular". Through monocular depth perception and especially through motion parallax, even stereo-blind people can find their way in real space without problems. However, a binocular illusion cannot help these persons to experience space.

Irrespective of various diseases of the eye, which will not be discussed further here, perception also depends on other factors. First of all, the distance between the eyes must be

[11] Inability to binocular spatial vision.

[12] Lack of vision in one eye.

mentioned here. Although the distance between the eyes is within a certain range[13] for most people, it does have an influence on depth perception. A child with a small interpupillary distance will experience less retinal image disparity than an adult with a huge head and considerable interpupillary distance. Of course, this does not necessarily mean that depth is perceived differently. As is well known, the visual system overlays the images with the experiences and thus creates a "correct" perception. However, the interpupillary distance has a considerable influence on artificial stereo vision with 3D or VR glasses. If the representation is not adjusted to the eye distance of the viewer, the viewer will see an image that does not match his usual perception. In 3D cinema, the disparity between the images of the stereoscopic image pair is already determined during the recording by the base distance between the stereo camera (the distance of both lenses from each other). Nothing can be changed during projection, so all visitors see the same thing. The film is made for an average visitor with "normal" eye distance. The consequence of this is that the big guy with considerable interpupillary distance will now be offered an image pair that does not correspond to his interpupillary distance (his stereo base), but to that of the stereo camera. Since this base distance will be significantly smaller, the viewer will perceive an image that is considerably reduced in depth. The opposite case occurs with the child. The child now sees an image that is significantly enlarged in depth and becomes frightened. It has already been said that perception is also trained to a large extent and is also automatically conditioned over a certain period of time.

The Influence of Image Size on 3D Vision

In addition, of course, the disparity of the two frames of the 3D movie depends on the size of the projection on the screen. The larger the cinema screen, the greater the disparity. Let us now assume that the child mentioned above had earned a commendation and was allowed to go with his parents to the cinema with the largest screen in Germany, the Leonberg "Traumpalast" with a screen of 38 × 22 m. If the operator wanted to play a standard 3D film there, the reward would quickly become a punishment. The disparity of the 3D images would be so strong that the child would not have the slightest chance to merge the images. After a few admonitions by the big-headed parents to please stop acting like this, the child would eventually despair of the constant double images and perhaps even throw up.

The fellow with the large eye distance might want to watch the same movie on his latest 3D TV. Even with a very large TV (say 65″), the horizontal extent of the TV would still only be about 1.4 m, or 1/25 of the giant theater. Our viewer's 3D impression would be disappointing. The images would be flat, the depth barely perceptible. So he would lament and perhaps consider reversing his purchase.

[13]Typically, interpupillary distance since Wheatstone is 63.5 mm [40] with a spread between 47 and 78 mm [41].

This example already shows that it is important to know the output medium for a 3D content.

The distance between the objects in the individual images of a stereo pair may only ever take the value of the interpupillary distance. This corresponds to the parallel position of the eyes when looking at long distances. An outward rotation of the eyes, the "binocular excess", is possible to a certain extent for a short time, but it is unnatural, exhausting and unpleasant.

In addition, there is no motion parallax when seeing with 3D glasses. In real vision, the slightest movements of the head or eyes create new perspectives again and again. The number of these perspective views is not limited in nature, but only given or limited by the scene. This motion parallax provides strong stimuli that promote spatial perception. The use of 3D glasses prevents motion parallax. The stereoscopic image pair now contains only the two perspectives for the left and right eyes. When the head moves, a viewer unknowingly expects the corresponding change in visual impression. For the most part, this is hardly a hindrance when viewing a few 3D images or a stereoscopic short film. A full-length 3D movie may reveal this error after prolonged viewing. As expected, the perceived images and the resulting spatial impression should change with self-motion. However, contrary to everyday experience, perspective does not change, which is inconsistent with real-world experience. The conflict between expected and delivered result can lead to discomfort or nausea with increasing observation time. The more clearly the difference between expectation and result can be perceived and the longer the deviation lasts, the more likely a physical reaction to what is seen is.

VR glasses worn on the head offer advantages here. The corresponding image pair is calculated for each new position or viewing direction. However, this can result in a new problem. The calculation of the images from the 3D model and their representation on the displays of the VR glasses takes time. This means that the display only takes place a certain time after the movement has been executed. Even shutter glasses need some time to execute a pass (i.e. to display a stereo image pair). The shutter glasses hide or allow the view of the screen. The images and the shutter are synchronized so that the image content matches the open shutter (left image for left eye). The sequence of images must be fast enough not to be perceived as flickering. An image sequence consisting of individual images played back in rapid succession can no longer be resolved into its individual images above a certain frame rate. As a result, the images merge together and a continuous change in position from image to image can be perceived as motion. Mach, in an analysis on the sense of time of the ear, had attempted to measure the time interval at which two signals are still distinguishable [42]. He measured the "smallest perceptible time" as exactly 47 ms and described it as "vanishing to the eye". According to Mach, two images would have to be played back at a frame rate of 21 frames per second[14] to be perceived as a single image. During the

[14]Equivalent to 1/0.047 second per frame.

development of the film viewer "Kinetoscope ", Edison mentions an extremely current value in a note with the specification of 25 frames per second (from [43], p. 2):

> It is probable that 25 per second will be sufficient to give the illusion

The value 25 frames per second (40 ms frame interval) is the usual frame rate for a full frame in the television signal. The psychologists Wittmann and Pöppel wanted to know more. They stated that the smallest temporal interval between two perceptions would have a length of only 30 ms [44]. Pöppel also gives a possible explanation for the merging of signals. The transport of optical signals in the visual system already requires more than 20 ms from the stimulus to its perception [45]. Thus, an observer is not able to react to a stimulus within this period of time. Thus, everything that affects the observer within a time window smaller than 20 ms of optical stimuli is perceived as simultaneous.

In everyday perception, the observer is rarely exposed to only a single stimulus. Especially in real movement, different stimuli affect the body at the same time. The perceiver constantly changes his position or at least moves his head. These changes are noticed by the sense of balance and self-perception and are accompanied by a simultaneous, involuntary change of perspective.

That is different when playing interactively with VR glasses.[15] There, too, the motion effects are perceived together with the images on the displays. But, as already mentioned, the calculation of the images and their representation on the display of the glasses takes time. This results in a temporal offset. The new perspective is perceived later than the movement that led to this new perspective. If this time offset (delay) lasts longer than the said 20 ms, it can be perceived by the player (consciously or unconsciously). It makes sense to compare this to natural experience. Such a time delay between actio (the movement) and reactio (the perception) is actually only ever observed when the function of the senses has been disturbed in some way.[16] The reaction of the body when using VR glasses is therefore similar to the reaction to poisoning. The person reacts quite unintentionally with malaise and nausea. However, this simulator sickness[17] only develops its pathological effect if the latency cannot be pushed below 20 ms, as predicted by Pöppel [48].[18] It is probably the case that the perception of temporal latency also has an increasingly marked effect on the subject with a longer duration of disturbance. It is therefore no exaggeration to speak of a temporal divergence.

Another problem that occurs when seeing with 3D glasses is the conflict between accommodation and convergence. What is meant by this is the following. Natural vision

[15] The problem of "simulator sickness" is also known to the manufacturers of such glasses. In a blog, Oculus recommends a strategy to "mitigate" the latencies [46], which also includes the 20 ms.

[16] Which is the case, for example, poisoning or excessive alcohol consumption.

[17] Or simulation sickness, see, e.g., [47].

[18] DiZio and Lackner investigated 19 ms.

with both eyes uses the inward rotation of the eyes to a common fixation point (convergence). Quite involuntarily, the focus is directed to this point at the same time as the convergence. This forces the adjustment of the focal length of the eye lenses by means of contraction of the eye muscles to change the radii of curvature as well as the central thickness. As a result, the fixed point can again be sharply imaged on the retinae. Accommodation and convergence are jointly related to pupil diameter for near triad. This improves the depth of field in addition to the quality of image[19] on the retina when focusing on near objects. In stereoscopic vision, this original link no longer works. For example, when a 3D object appears to arise far in front of or behind the 3D display in stereoscopic vision, both eyes involuntarily converge on this illusory aerial image. The close-up linkage now also forces accommodation to converge on this very point. But this fails to have its usual effect, since the screen is not at the point of convergence. Thus, the eye cannot produce a sharp image. Perception, in order to avoid double images, has a compulsive urge to fuse. This compulsive urge to fuse now also forces a suitable convergence position of the eyes. Thus, the natural coupling of the visual process is disturbed. This has an unfamiliar effect on inexperienced 3D consumers, and viewing is then often perceived as strenuous.

4.4 Stereoscopes

A Peep-Box for Two Eyes

The perception of a spatial effect through the depiction of different perspectives requires the assignment of an image to the eye intended for it. However, if the eye sees an image that is not intended for this eye, a false 3D impression is created. Since a false impression should not be created under any circumstances, suitable measures must be taken for image separation and eye assignment.

A surprising finding from the history of representational optics is the fact that it was not until the nineteenth century that an instrument was invented with which three-dimensional perception could be specifically provoked by the representation of perspective images. Already in antiquity, philosophers such as Empedocles, Euclid or Aristotle had dealt with two-eyed vision. Witelo, Alhazen and Kepler, after all the most renowned opticians of their time, had even produced drawings illustrating the different positions of the retinal images and thus the basis of spatial vision. Even the universal genius da Vinci was apparently unable to recognize this connection, although this discovery would have been well within the realm of possibility due to his knowledge of perspective and imaging.[20] It was not until

[19] The fixation point is imaged at the point of sharpest vision.

[20] Apart from the image pair of the Mona Lisa: The portrait of the Mona Lisa exists in duplicate (one painting in the Prado in Madrid, another in the Louvre in Paris); almost identical, but with small differences. In 2013, Carbon and Hesslinger showed the partial spatial effect of the image pair in a

Fig. 4.8 Dr. May looks through Wheatstone's stereoscope at King's College. (David Tett (2016), courtesy of David Tett)

1838 that the English physicist Wheatstone also described the phenomenon of binocular vision as a function of imaging two different perspectives [40]. Wheatstone referred in his investigation to da Vinci's "Trattato della pittura" [51] and ventured the thesis that da Vinci might have noticed the divergence of retinal images in the viewer's eyes if he had used a cube instead of his preferred sphere. Wheatstone developed an instrument to test his binocular theory of vision, to which he gave the name "stereoscope".[21] This name is still used today for devices that produce a 3D effect by "pretending fake images," that is, by assigning different perspective images to the observer's eyes. Wheatstone actually used mirrors to reflects the perspective images into the eyes. In Fig. 4.8, Wheatstone's mirror stereoscope is shown in function. The observer looks through the small mirrors at the pair of images attached to his left and right.

With Wheatstone device, even large-format drawings or photographs can be spatially represented in a realistic way. The 3D effect becomes more obvious when artificial objects are combined with the natural 3D scene. A good example of this is the photograph of the physicist Govi, which the stereophotographer Dubosqe had placed in an artistic frame (Fig. 4.9). The frame artificially copied to the image makes the shot one of the first "mixed

stereoscopic superimposition [49], but did not provide evidence for the stereoscopic intention, but only for a (partially) perspective-correct difference (see also [50], p. 6).

[21] "The frequent reference I shall have occasion to make to this instrument, I will render it convenient to give it a specific name, I therefore propose that it be called a Stereoscope, to indicate its property of representing solid figures." (from [40], p. 374).

Fig. 4.9 Professor Gilberto Govi, Wheatstone Collection at King's College, London, K/PP107/8/47, 8/48, stereophotograph by Carles Dubosque, taken between 1862 and 1886. (Courtesy of King's College London Archives)

reality" images, a mixed scene of real and artificial reality. In the figure, an anaglyph image is shown next to the original images. When viewed in 3D, Prof. Govi appears to be sitting in the frame, which now acts like a window into virtual space.

Dubosque was not the first photographer to produce stereoscopic photographs for the Wheatstone stereoscope. Already in the year after the invention of the stereoscope, Wheatstone himself experimented with the newly developed photographic processes talbotype[22] and daguerreotype[23] (see [57], p. 7). In Germany, at about the same time, Moser developed an understanding for the spatial images of Wheatstone's stereoscope and used the "representation of stereoscopic figures by photographic means" (in [58], p. 384).

Kohlhans had already designed his "dark boxes" in 1677 not only for observing the surroundings as a camera obscura, but also as an object or image viewer. This "panscopium" was in principle the continuation of Griendel's device. What Griendel understood by a panscopium is communicated by the optician in his "Specificatio" at position 25 (from [59], p. 97) (Fig. 4.10):

> Panscopium, my new invention, is at the same time a Telescope, through which one can see with two eyes, also a Microscopium, Polemoscopium, Helioscopium, Lucerna Optica: a wonder of all kinds of mirror work, and rare optical presentations.

A panscopium could also be fitted with lenses for the purpose of magnification. Wheatstone's reflecting stereoscope, on the other hand, was not even equipped with eyepiece lenses at the time of its creation (Fig. 4.11). Viewing the images without an

[22]Also "calotype" [52] or "photogenetic images" [53], developed by Talbot, who is probably better known today for the self-imaging effect or Talbot effect (see [54], pp. 403–405).

[23]Developed by Daguerre [55] based on Niépces heliography [56].

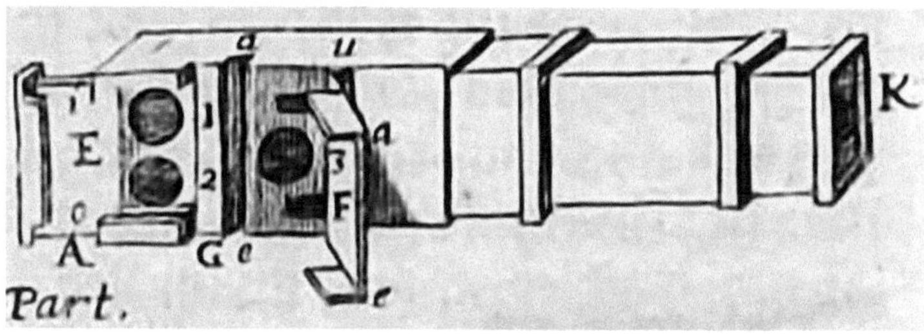

Fig. 4.10 Panscopium. (Kohlhans (1677) from [60] plate VIII copper stucco)

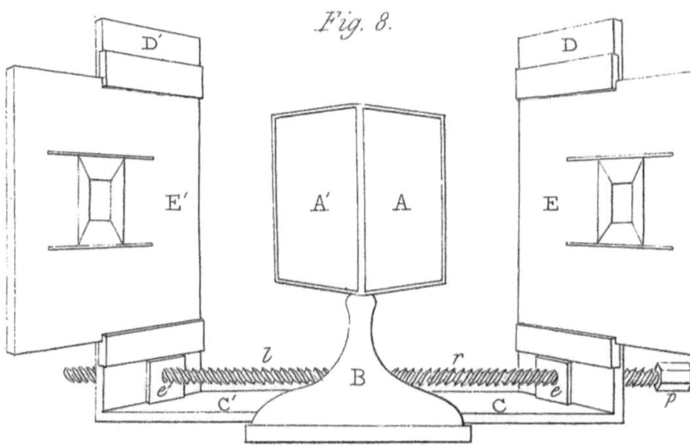

Fig. 4.11 Wheastone's stereoscope. (Charles Wheatstone (1838) from [40], Fig. 8)

eyepiece was at least sometimes difficult with this setup due to the lack of magnification of the images in relation to the viewing distance, which was probably also noticed by Wheatstone himself.

In a later publication [57] Wheatstone now proposes the use of lenses for viewing the stereo image also for a reflex stereoscope, referring to the already introduced use of lenses in refraction stereoscopes by Brewster ([57], p. 9). In Fig. 4.8 it can be seen that such lenses were indeed used. Brewster had used prismatic ocular lenses in the form of a split lens in his stereoscope before [61] (see also Fig. 4.13).

The prismatic lens shape is not chosen at random, but helps the viewer to focus the stereoscopic image pair. The physiological close-up triad forces the viewer to turn his eyes inwards towards the focal point. However, the image pair of the two perspective images is not in the same place, but in Brewster's stereoscope in placed on the image stage left and right next to each other. A very nice example of this was given by Brewster himself (see Fig. 4.12).

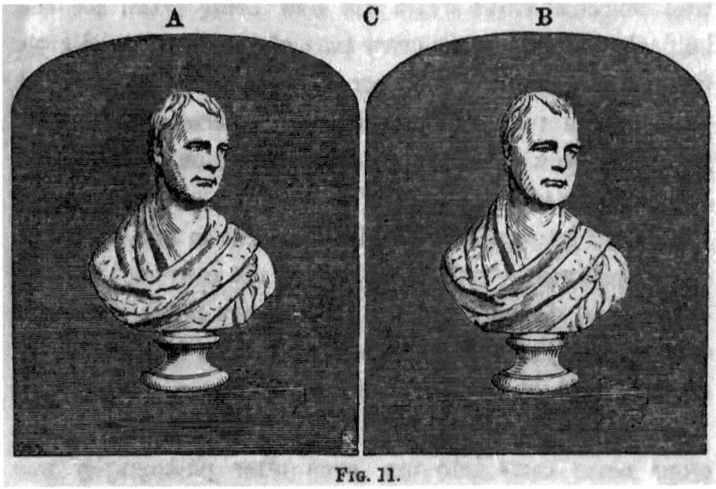

Fig. 4.12 Portrait of a Gentleman, stereoscopic image pair. (David Brewster (1856) from [62], p. 64, Fig. 11)

Due to the prismatic lens effect, the viewer's gaze is effortlessly redirected to the image center of the respective perspective image. The virtual 3D image is created at the position of the original viewing direction (see Fig. 4.13).

Due to the magnifying effect of the lenses, the Brewster stereoscope can be very small and compact. This design has hardly been improved to this day and in a sense represents the archetype of today's VR glasses. In a direct comparison of the historical Brewster stereoscope (Fig. 4.14 left) with a now widely used Cardboard stereoscope (Fig. 4.14 right).

In principle, only two things are necessary: Eyepiece lenses and a stereo image. The obvious difference between a Brewster stereoscope and a current stereoscope lies in the way the image is displayed. Whereas Brewster used drawings and later photographs for this purpose, today displays are used for this purpose and the automatically illuminated image content can thus be changed elegantly. The decisive difference, however, is what one does not see. The use of the display of a mobile phone is accompanied by the presence of many sensors. A current smartphone not only detects its own position or the cardinal direction in which it is held, but also acceleration and vibration. With this, you can now set up the stereo image based on the position. You thus get a position-adaptive stereoscope.[24]

The eyepieces of the stereoscope and must be pressed tightly against the eyes of the viewer to view the 3D image, then as now. The magnifying effect of the lenses results in a virtual image. Early on, people were interested in extending this limited view so that the stereoscope could be used at a greater viewing distance and for a larger audience. Again it

[24]Below are some representatives of such position-adaptive stereoscopes (with or without phone): e.g. Oculus Rift, HTC Vive, Sony Playstation VR, Samsung Gear VR, Zeiss VR One, Google Cardboard, ...

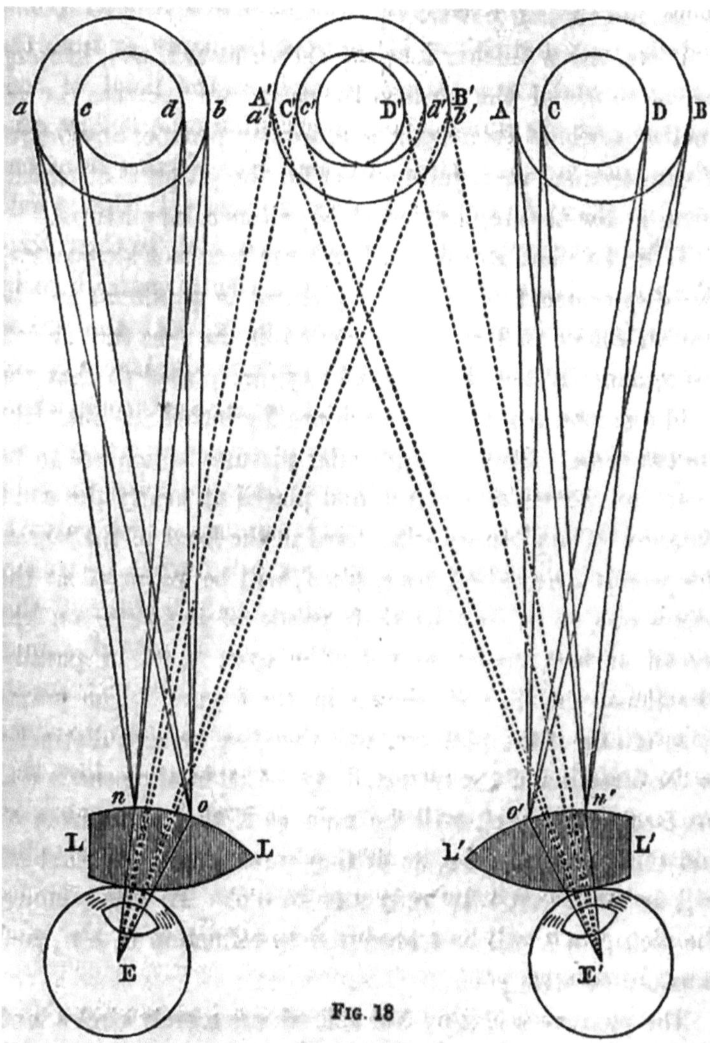

Fig. 4.13 Prismatic lenses in Brewster's stereoscope. (David Brewster (1856) from [62], p. 74, Fig. 18)

was Maxwell who had an idea for this. With his "real image stereoscope", Maxwell first imaged the stereo image as a real image via two converging lenses. In a second imaging step, a second larger[25] field lens is placed in the image plane. An observer can thus stand about 50–60 cm in front of the lens and still perceive a real three-dimensional image without any aids, which would describe an autostereoscopic display (see [63], p. 148).

[25] For Maxwell, that's 3 inches.

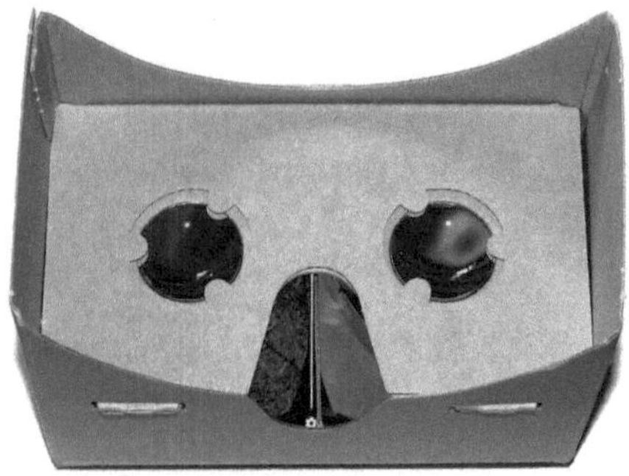

Fig. 4.14 Stereoscope comparison. (Brewster (1856) from [62], p. 67, Fig. 14, Grasnick (2019))

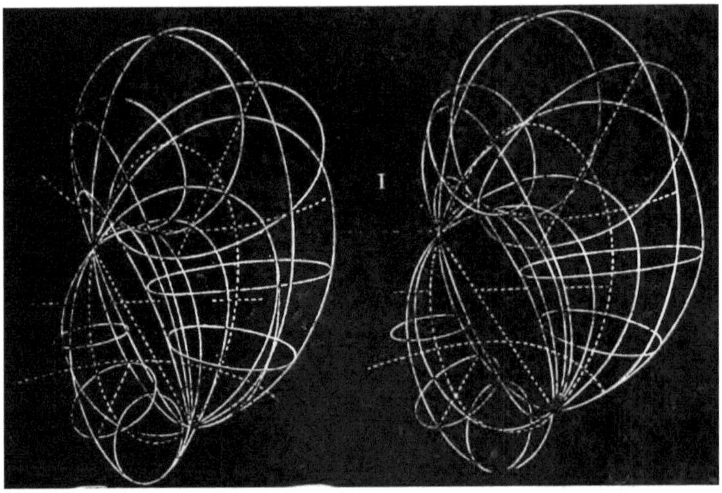

Fig. 4.15 Pair of horns cyclides. (James Clerk Maxwell (1890) from [63], p. 159, plate XI, Fig. 1)

Maxwell used this free-eyed stereoscope to display his cyclides, the ring-shaped surfaces and formations in three-dimensional space. And this makes perfect sense. Figure 4.15 shows a drawing of one such cycloid. Viewed as a flat image, the structure is barely comprehensible and seems to represent only a jumbled collection of curves. In the stereoscope, the image reveals its spatial structure. This now also explains the name "pair of horns" chosen by Maxwell.

A stereoscope does not always have to stand on a table or be worn on the head. Brewster had already sketched stereoscopes to be mounted on the wall or to be held in the hand (Fig. 4.16).

What seems to be most interesting here are the glasses. Stereoscopic glasses can be held in the hand and are inexpensive to produce. This makes them a suitable candidate for inclusion in magazines or books. The representation of the stereo image, introduced by

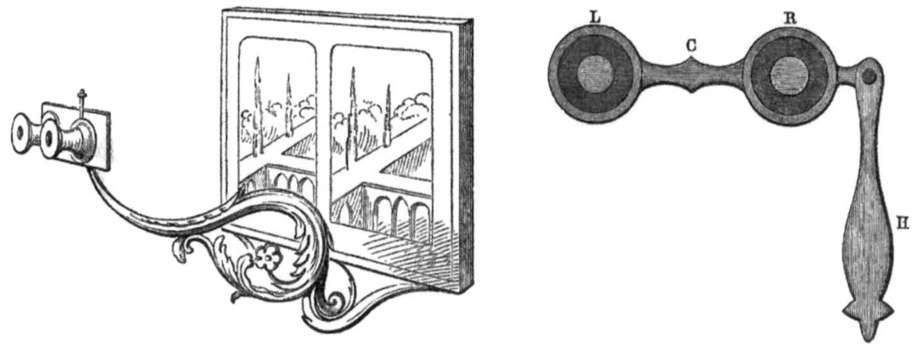

Fig. 4.16 Wall and hand stereoscopes. (David Brewster (1856) from [62], p. 69, Fig. 15 u. p. 70, Fig. 17)

Brewster, in which the two images are arranged side by side, is obvious but not obligatory. In principle, there is nothing to be said against placing the images on top of each other (over-and-under or top-down). Three scientists of the University of Hohenheim had thought about such an image design and presented it in a book that is really worth seeing [64]. There, macro images of animals, plants and minerals are presented on several pages, which can be viewed three-dimensionally with the KMQ viewing device. The surnames of the developers Christoph Koschnitzke, Rainer Mehnert and Peter Quick gave the method the name "KMQ". The book also shows an advantage of this viewing device. If one wanted to display the individual images in widescreen format in a book or magazine on a page next to each other, one would need a very wide, unusual page format. With an over-and-under arrangement, two full A5 pages would easily fit on one DINA4 page. However, the promise of a "world first" printed on the title is quite a strong statement.

The American officer Bernier had already set up a "Stereoscopic Photographic Department" at Wright-Patterson Air Force Base in Dayton, Ohio, shortly after the Second World War and introduced such a format there. This process was not only used by the military, but also by Warner Bros. Entertainment and Paramount Pictures for the 3D cinema of the 50s and 60s. The aspect ratio of the film in Bernier's system was already, as in the much later KMQ system, an alternating left and right image in full size (one above the other). This is also the origin of the name "Alternate-Frame Technique" chosen by Bernier (see [65], p. 602).[26]

To convert a stereoscope for a horizontally arranged stereo image pair into one for a vertical image pair, it is sufficient to merely rotate the prisms of the viewer by 90°. The simplicity of the prismatic stereo lorgnette[27] is also a problem. Although the aid can be held quite easily in front of the eyes or put on as a pair of glasses, a very specific distance from

[26] Whereby the observation in Bernier's system presupposed polarization glasses, which will be discussed later.

[27] A lorgnette is a pair of spectacles without temples that are held in front of the eyes by a stem.

the image must be maintained in order to perceive the 3D image correctly. This does not always work out well, so in the 1990s a system was introduced in the USA using a mirror deflector that was slightly more tolerant of distance mismatches simply because of the size of the image field [66]. But also the "View Magic" viewer did not have the quite big success.

4.5 3D Glasses

Image Separation to Put on

It is not always necessary to create the stereoscopic effect by forcing the viewer to look in a certain direction. With certain methods, both images can also be generated at the same location and, using suitable glasses, can be split into the two individual images again when viewed. In doing so, the eyepieces of the stereoscope and the image stage can be spatially separated to a large extent. The image separation into left and right image is accomplished by different 3D filters or optics.

In the early days of stereoscopy, Rollmann, a German physics teacher, noticed and noted the monochromaticity of a mixed image while studying stereoscopic fusion without a stereoscope. Rollmann had fused a red and a green image with his naked eyes and noticed something else besides the spatial effect. The spatial image appeared grey to him [67]. From this observation Rollmann developed a color stereoscope, in which the individual images are extracted from the overall image of the double drawing by their respective colors [68]. The separation of the two-color drawing (Rollmann names blue and yellow as the colors) was carried out by Rollmann using a red and a blue glass.

Around the same time as Rollmann, the French physics teacher d'Almeida also experimented with stereoscopy. d'Almeida, however, built a real device for this purpose. Using two projectors, d'Almeida projected two individual stereo images onto the identical position of a screen [69]. A color filter was placed in front of each projection lens. A red filter was mounted in front of one lens and a green filter in front of the other. The viewers used the same color filters to cover the corresponding eye with red or green respectively. This can easily be used to equip a pair of glasses as well. The last two paragraphs of d'Almeida's text are interesting. He first proposes to also use full-color images that the viewer looks at through a rotating disk with alternating apertures. A synchronous second disc covers and opens the lenses with the appropriate portion of the stereo image pair. This describes nothing less than a mechanical shutter arrangement. Finally, he informs us that he is working on a device to combine his idea of the with a phenakistiscope. With this, at least short, but above all moving 3D scenes were then possible.

However, it was Du Hauron who introduced the name "anaglyph[28]", which is still in use today [70]. Du Hauron certainly made a contribution not only with the naming, but

[28] From Greek ana = on + glypho = to represent, thus approximately to represent one another.

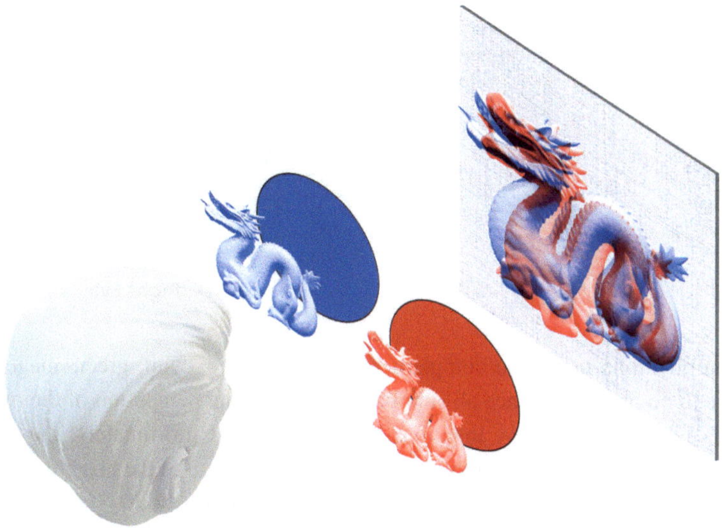

Fig. 4.17 Illustration of anaglyphs, Grasnick (2019)

especially with the practical implementation of his color photography and stereoscopy, which supported the spread of this technique.

Due to its simplicity, the anaglyph technique is still used today. The necessary color filters can be produced cheaply, the stereoscopic images are easy to print and can be displayed on any monitor. One only needs the inexpensive 3D glasses, their lenses with different, ideally complementary color filters. The illustration in Fig. 4.17 shows a red-blue filter pair. The respective filters always allow only that part of the image to pass that is intended for the eye in question.

An anaglyph to be viewed on the LC monitor would ideally be composed of the primary colors of the subpixels. For the three primary colors red, green and blue (RGB), the number of possible variants (permutations) corresponds to the factorial of three.[29]

Since a color or a color component cannot be used equally for the right and left image channel, the six possible combinations for anaglyph glasses on the RGB-LCD result as follows:

- Red–green
- Red–blue
- Red–cyan
- Green–blue
- Green–magenta
- Blue–yellow

[29] The product of all numbers less than and equal to the number to be permuted ($3! = 6$).

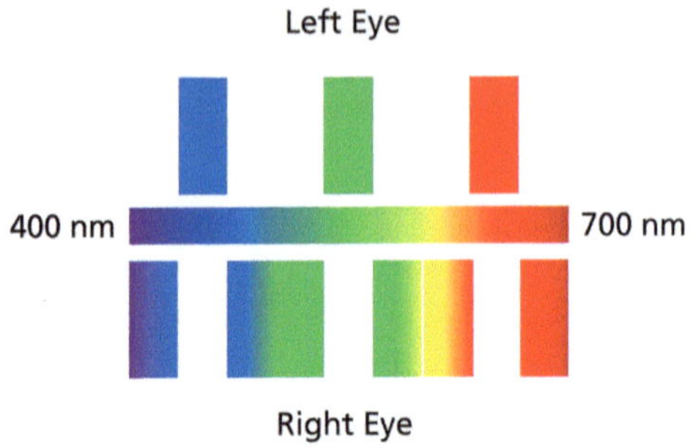

Fig. 4.18 Infitec anaglyph. (Grasnick (2019) adapted from [71])

The combinations using the complementary colors (red–cyan, green–magenta, blue–yellow) are to be preferred, since all subpixels are used. The perception of the mixed color then results in the "vivid" gray mentioned by Rollmann. Unfortunately, the color filters nevertheless lead to a distortion of the colors and to a reduction of the brightness, especially when trying to display colors (which Rollmann already pointed out).

In d'Almeida's work there is also a reference to the solution of the color problem. D'Almeida explains that the two colored glasses must not have the same parts of the spectrum in common [69]. This is certainly a bit of a quibble, but in principle it describes what the Dolby 3D filter does in 3D cinema. Dolby uses technology from the German Daimler subsidiary Infitec. The developers there also use color filters, but each glass is always simultaneously permeable to different narrow areas (bands). A picture should illustrate this. In Fig. 4.18, the entire visible range from 400 to 700 nm wavelength is shown in the middle. Above this are the narrow-band RGB filters for the left eye, and below them the bandpass filters for the right eye.

If it is done ideally, both filters complement each other again to the full spectrum and the maximum possible brightness can pass the filters.

These filters are now much better than a cheap red-green glasses, but also much more expensive. The advantage of these glasses compared to the polarization technique is the possible elimination of the need for a special screen, because a special screen must be used to maintain the polarization state after the reflection of the projected images.

And that's the segue to polarization.

In addition to color and intensity, light can have another property: polarization.

Polarization occurs in nature in the quite incidental, for example, the reflection of light on a water surface or the scattering of light in a slightly cloudy sky. Some creatures can use polarization for navigation,[30] humans need aids for this. Whether the sunstones of the Vikings were sufficient for this purpose remains controversial [73]. However, it is undisputed that the Vikings used birefringent materials in their sunstones. One particularly clear and birefringent material is Iceland crystal.[31] Because of its unusual optical properties, this

[30] As Frisch noted, bees also navigate by polarization, among other things [72].

[31] Original "Silfurberg", also "double spar", today "calcite".

crystal has certainly been the subject of serious investigation. Bartholin described his experiments in 1670 [74], which Huygens was able to explain a few years later by the wave nature of light [75]. Malus furthermore discovered polarization by reflection [76] and also invented the term "polarization" [77]. Malus' real merit may be the mathematical formulation of the intensity of polarized light after passing through a linear polarizer, which is known today as Malus' law.[32] This simple formula is still perfectly adequate today to define the position of polarizers for a 3D application. However, the first attempts to use polarization for stereoscopic projections were made only at the end of the nineteenth century. The English optician Anderton took advantage of the possibilities of his time and used available technology for the polarizers of the projectors and analyzers of the glasses [78]. Especially in the case of spectacles, all the variants proposed by Anderton, such as Nicol prisms, tourmaline crystals or even a stack of plane-parallel plates, are heavy and costly. From Herapath's artificial tourmaline [79], which at that time could already be produced in larger plates [80], polarizers could certainly have been fabricated. But Anderton seems to have been unaware of this. So it took some time until the mineralogist Bernauer simplified the production process. Bernauer had the herapathite crystals crystallized between two glass plates [81]. Carl Zeiss in Jena used Bernauer's invention to produce polarizers under the trade names Herotar and Bernotar [82], which were also used for 3D film[33] from 1937 [85]. Better known than Bernauer's polarizing filters, however, are those made by Polaroid. The founder of the company, Land, worked at the same time as Bernauer on the development of large-format polarizers. Unlike Bernauer, Land aligned the herapathite needles directly in the substrate by applying electrical or mechanical force and was thus able to dispense with glass plates [86]. These polarizers were used for 3D film from 1936 [87] and provided an early alternative to the anaglyphs used until then. Norling, who was even nominated for an Oscar with his 3D anaglyph short film "Audioscopiks[34]", also recognized the possibility for 3D color film representation with polarizers and presented a 12-minute film for the 1939 World's Fair based on polarization with Land's Polaroids [89]. Polaroid polarizers made it possible to produce inexpensive glasses and were therefore increasingly used in 3D cinema. The viewing of the film was as simple as with the anaglyph film. The two individual images are projected onto the screen and viewed with polarised glasses (Fig. 4.19).

Besides the already mentioned need for a special screen, two projectors were also required. Bernier solved the problem for his alternating frame method by rotating the

---

[32] The intensity passing a linear polarizer is the product of the original intensity multiplied by the squared cosine of the angle of rotation. The prerequisite is that the original intensity is already linearly polarized.

[33] Advertising film of the "Volksfürsorge" life insurance company: "Zum Greifen nah" (Close enough to touch) by Curt A. Engel and Karl Schröder, premiered at the Ufa Palast in Berlin in 1937 [83], re-released at the Berlinale in 1980 [84].

[34] According to Norling's own description, a 3D sound fim ([88], p. 285), a phonofilm based on the plastigram (see also [89], p. 612).

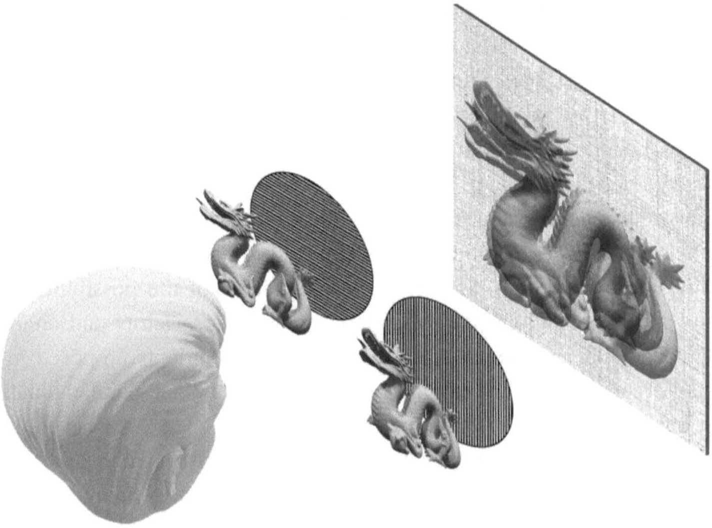

Fig. 4.19 Viewing with polarizing filter, Grasnick (2019)

polarizing filter in front of the projector with the respective image displayed (see [65], p. 603). So one position for the left image, suitably rotated for the right. The rotation must now fit the linear polarizing filters in the spectacles of the viewer, which are rotated 90 degrees against each other. However, these glasses only work well as long as the viewer keeps his head absolutely straight. If he tilts his head to one side, Malus' law strikes inexorably. The polarizing filters no longer block the unwanted image completely, but allow part of the "wrong" image to enter the eye. The channel separation decreases, the 3D image becomes worse [90]. Land had already fully recognized this situation in 1937 and proposed a solution[35] [91]. However, almost 70 years passed until the company RealD implemented Land's idea and made it known to a wide audience [92]. RealD took over the small company StereoGraphics of the American stereo filmmaker Lipton and thus also adopted his idea of active rotation of the polarization state. With the advent of liquid crystals, a technology was available that no longer required mechanical rotation of the polarizing filters (as Bernier still did), but allowed for an elegant electro-optical method. Lipton et al. proposed to rotate the polarization angle via a timed shutter [93]. This modulator, referred to as "ZScreen" by StereoGraphics [94], is essentially nothing more than an LCD without polarizers. Polarized light is rotated through the crystals by a certain angle depending on the applied voltage. Thus, the angle of polarization can be changed depending on the image being displayed at any given time. Since the analyzers in the RealD glasses consist of circular polarizers, the light from the ZScreen must also be circularly

[35] He proposed to use so-called retardation plates in the thickness of ¼ of the wavelength (λ/4-plates) to generate circular polarization.

polarized. This also works there with a retardation plate (see [93]). This kind of polarization rotation is suitable for projection as well as for displays. For a projector, the modulator is simply placed in front of the lens and synchronized with the film. For an application as a screen attachment, two further requirements arise. The obvious requirement is that the modulator must be at least the size of the display. Less obvious is the second requirement. The image build-up and image change must be fast in order not to be perceived as flickering. As a reminder: In the early days of LCD technology, the image build-up was slow and therefore the refresh rate was low. In order to no longer perceive the image change and to avoid image flickering, the image must be displayed with a refresh rate of at least 50 frames per second.[36] However, this only applies to a single image. In the case of a 3D display, both images must be displayed alternately and this increases the requirement to double the frequency. Now it would even be 100 Hz that have to be displayed. For a long time, this was impossible with liquid crystal displays.

To solve this problem, Faris had proposed a "micro-polarizer array" [95]. All the individual elements of such an array, which may be true polarizers or retarder (waveplates), are matched to the size and shape of the underlying pixels. The elements are always arranged alternately in rows and columns or in a checkerboard pattern. Microstructured polarization elements are popular for 3D display on the LC screen because the implementation is quite simple. As already explained, the light is completely linearly polarized when passing through an LCD. This means that only waveplates[37] or strips are needed to alternate the polarization line by line. The viewer wears passive polarization glasses with the associated twisted polarizing filters—and the 3D TV is ready. Here, too, an additopnal $\lambda/4$-delay plate is often used in order to have circularly polarized light again. For this, of course, the glasses must then also be equipped with circular analyzers. You can find such structured retarders in monitors and TVs that adorn themselves with the title "passive 3D", since in the monitor and glasses actually only the polarizing filters are responsible for the 3D effect without any electrical control. JVC[38] used the name "Xpol Circular" [96] for its filters, LG[39] the name "Film Patterned Retarder" [97]. By using the structured polarizers, it is now possible to watch a 3D movie almost in cinema quality on a 3D TV (Fig. 4.20).

However, placing additional elements in front of the pixels of a screen does not only bring advantages. Initially, the production of large-format structured polarization elements was expensive. This negated the advantage of inexpensive glasses. In the meantime, the introduction of printed structures [98] has significantly reduced the price. The necessary halving of the usable lines per field can certainly also be tolerated with today's screen

[36] 1/s = Hz (Hertz).

[37] In this case, they are $\lambda/2$-plates, which do not produce circular polarization, but only rotate its direction.

[38] Victor Company of Japan, Yokohama.

[39] LG Electronics, South Korea.

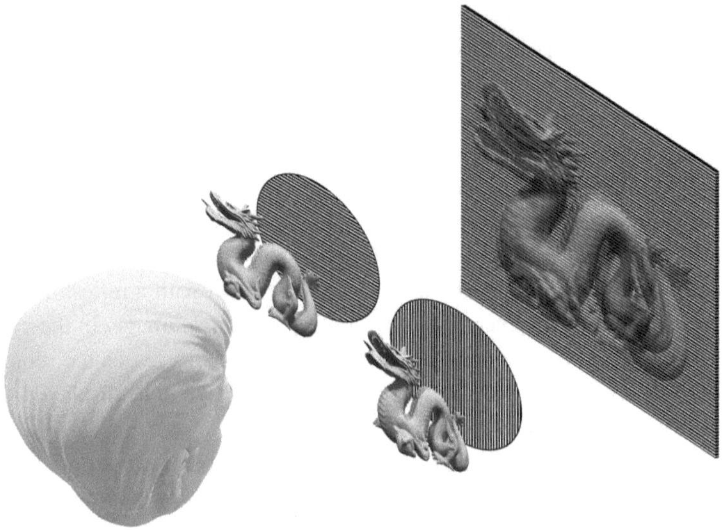

Fig. 4.20 Viewing structured polarizers, Grasnick (2019)

resolution of 4 k[40] or more. After all, even half the resolution is enough to meet the current Full-HD[41] television picture standard. However, a problem that stems from the geometric structure of the display remains. In front of the color filters of the LCD are physically necessary additional layers (cover glass, polarization filters), which have a measurable thickness. These may be thinner than 1 mm in total [99], but still create a gap between the imaging LC layer and the micro-polarizer array. Especially from short viewing distances or when viewed from above or below, the overlap between the alternating image lines and polarizers is no longer ideal and leads to perceptible image degradation [100]. JVC also admits this: "View 3D images at the recommended viewing distance. ... The 3D effects may differ when the screen is viewed from different angles" [101].

One can avoid this parallax error by doing without the additional polarizers. The assignment of the images can then also no longer take place via polarization. One possibility, which d'Almeida had already proposed in 1858 [69], is the temporally alternating masking of the images. It is not known whether d'Almeida developed the experiments he announced to a system. This is known only from the 1920s. At that time, the engineer Hammond used the rotating shutters known from film recording technology for a 3D projection [102]. This was a good idea at first. But now, in the cinema, a shutter synchronized to the projector had to be attached to each seat in the form of a pair of glasses. This was a considerable expense, but it was actually undertaken once The Selwyn Theatre

[40] 4 k = 4000, resolution 3840 vertical × 2160 horizontal pixels.

[41] Resolution 1920 vertical × 1080 horizontal pixels.

in New York was equipped with Hammond's "Teleview" technology[42] and a 3D film was shown in it.[43]

It seems understandable that the elaborate technology could not establish itself and Hammond's shutter did not spread. It was again the LC technology that provided an alternative to the mechanical shutters. In principle, the shutter works like Lipton's ZScreen. A polarized image is rotated by the modulator in one or the other polarization direction. If you put a second polarizing filter directly in front of the LC, you have the basic structure of an LCD again. A shutter can be a simple Schadt-Helfrich cell, where the liquid crystal is located between two polarizing filters and transparent electrodes. When used as a shutter, the LC does not even have to be patterned. Since you only need two states across the entire glass (namely On/Off), a single pixel is sufficient. One lens of the shutter effectively represents the shutter of a camera in front of the viewer's eyes. Lipton used his experience from the development of electro-optical shutters and packed the technology of the ZScreen into 3D glasses. The StereoGraphics shutter glasses "CrystalEyes" [105] were a great success and are still offered today.[44] Such liquid crystal glasses were also produced by other companies.[45]

It is again the simplicity of the system that allows the application in the cinema[46] and the use on the home 3D TV.[47] Again, an illustration is provided to illustrate the function. The time sequence of the function (the opening and closing of the shutters) is now shown in two moments one after the other (in Fig. 4.21 above and below).

It may now make sense to briefly compare the two authoritative processes of our time. Without a doubt, both systems are suitable for showing a 3D film in the cinema or the 3D TV experience at home. The difference lies first of all in the presentation of the image. For the polarized glasses, the stereoscopic image pair is broken down into strips and these are alternately reassembled on the screen. It is now clear that each image can only have half the original screen resolution. For shutter glasses, a whole image is always displayed, but it is only seen with one eye. So in terms of resolution, the shutter is superior to the polarizing filter. But it is the case that for the 3D effect the optical "shutters" of the glasses have to be opened and closed alternately. In the same time that only one image is displayed for the polarizing glasses, the shutter needs two in succession. As already mentioned, display and shutter have to work with double frequency. This is where the polarized glasses win.

In principle, it can also be described as a type of image representation of the two individual images of the stereo image in the display (so-called multiplexing). With the polarizing filter there is a spatial multiplexing, with the shutter it is a temporal multiplexing.

---

[42] Puchalski determines the number of seats as of 1918 to be 964 [103].

[43] Teleview Movie "M.A.R.S." [104].

[44] Now as RealD CrystalEyes.

[45] E.g. in Germany by ELSA as "REVELATOR" [106].

[46] E.g. XPAND shutter glasses [107].

[47] E.g. Samsung 3D active shutter glasses [108].

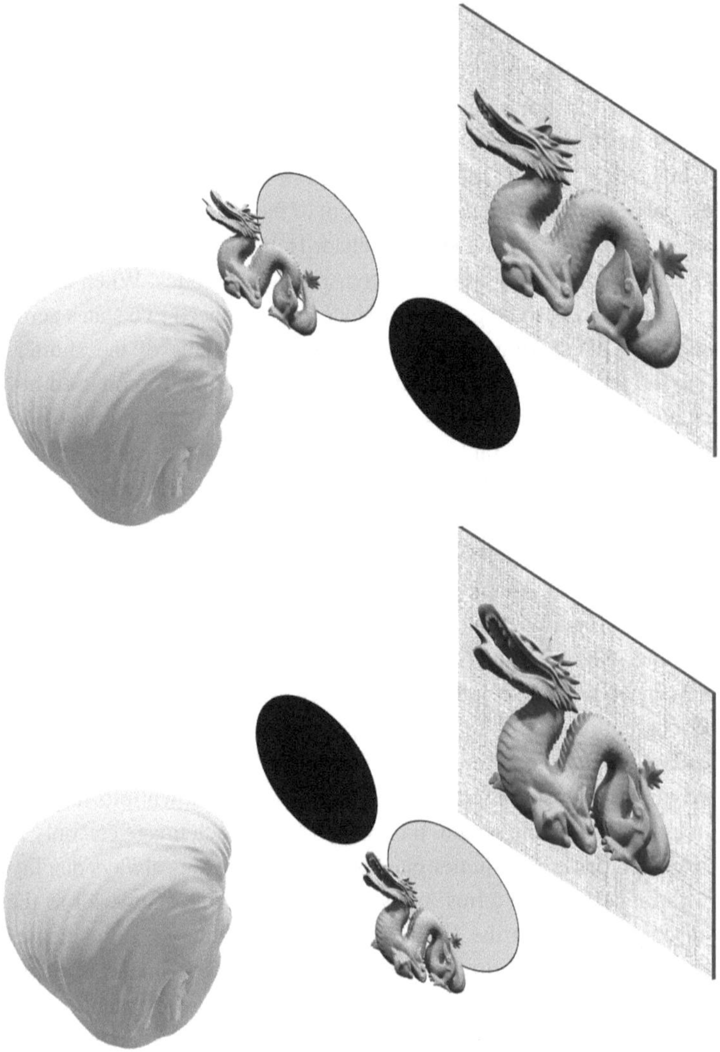

Fig. 4.21 Viewing with shutter, Grasnick (2019)

Temporal multiplexing is also responsible for the Pulfrich effect. But the effect, named after the physicist Pulfrich, was first noticed by the astronomer Wolf. And it came about this way. While surveying fixed stars, the director of the Heidelberg Observatory (Wolf) made the observation that the stars sometimes seemed to "float" [109]. Since Wolf used the Zeiss stereo comparator to compare two temporally offset images, Pulfrich, a Zeiss employee, felt called to investigate the effect. He reproduced the observation and discovered the "Pulfrich effect" (now named after him), which he called "time parallax" [110].

The spatial perception by the Pulfrich effect results from the difference in brightness of the two retinal images, which cause a temporally delayed perception of the darkened image. Pulfrich describes the effect as follows: "The time difference … arising from the difference in brightness of the two eyes … has been transformed into a space difference …"[48] ([110], p. 559). Again, temporal multiplexing is present.

The fact that Pulfrich's discovery is mentioned here is mainly due to the "Pulfrich glasses". The key inventor was the sound engineer Beard, who used this effect to view films on a television screen by converting "lateral motion"[49] into a stereoscopic image [111]. In this process, one eye of the viewer is obscured so that a moving object is perceived later with the obscured eye. Depending on the movement of the object or the camera, a true 3D effect is produced. Beard's technology was first commercialized in the US by the company Nuoptix [112]. Since the 3D effect required neither special recording technology nor special playback systems, but only cheap cardboard glasses, the technology spread quickly. The simple Pulfrich glasses were distributed by the millions and used especially for advertising and documentation. This was mainly due to the fact that the continuous movement or rotation in the film is quite tiring in the long run. One would certainly not want to have to watch a feature-length Pulfrich film. Nevertheless, the process was also used in German television at the end of the 1990s under the name "Telcast" [113], but again only for a very limited time.

4.6 Naked Eye

Stereoscopy Without 3D Glasses

First of all, the question arises once again whether the use of 3D glasses is always required to perceive a 3D effect. Of course, humans can also obtain spatial information from two-dimensional images within certain limits. Basically, however, it must be said that the condition for 3D vision is essentially the fusion of two suitable images. In his description of the stereoscope, Brewster had shown a way to fuse two individual images even without glasses. For this, he chose the example of a patterned wallpaper at which the viewer stares ([62], from p. 97). Meanwhile, the viewer somehow manages to cross the visual axes of the eyes in front of the wallpaper, i.e. to fixate an imaginary point in the air. Figure 4.22 illustrates the process. The observer looks at the wallpaper (which is somewhat translucent to show the points of impingement of the visual axes) and fixes crossed two ornaments. At the top of the picture, the two sections that the viewer now sees are shown as examples.

---

[48] Whereby he sees in it "in a sense" the confirmation of a saying of the Grail knight Gurnemanz (from Wagner's Parsifal): "You see, my son, time becomes space here" (ibid.).

[49] Translation or rotation.

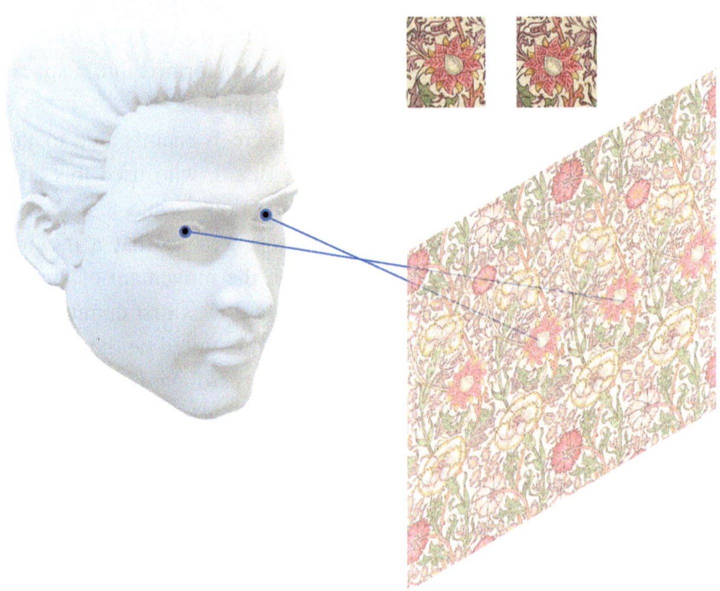

Fig. 4.22 Cross view of Brewster's wallpaper, Grasnick (2019)

As a result, the viewer will have the impression that the pattern of the wallpaper is between the wall and himself and also changes with his movement.

In this context, Brewster tells the story of Professor Christison, who was plagued by a wall decorated with stars. The stars danced in front of the wall and the professor, who was just recovering from a fever, was worried about his health. Indeed, it must be noted that small objects (in Christison's case, stars) on an otherwise unpatterned wallpaper lend themselves admirably to cross-eyed fusion.

Background Information

The observer of the wallpaper pattern experiences something that today in stereoscopy would be called negative parallax. In this case, the images of the left and right images are shifted relative to each other in such a way that the observer must converge in front of the screen in order to merge the two images. If the images of the stereo image pair were placed side by side (as shown in Fig. 4.22 above), the objects would be displaced from each other.

The terms negative and positive parallax (see e.g. [114]), are based on the assumption of a reference point from which the deviation is measured. If one

(continued)

assumes the first column as the horizontal reference point of an image, then all horizontal positions of the objects in the image result as x-values based on the column number. For example, if we look at the edge of the left flower in Fig. 4.22 above, we find that the position of the flower (as viewed from the left edge) is further to the right than that of the right flower. Thus, the object has a higher column number. The parallax is now calculated as the difference in the x-coordinates (or column number) of the right and left positions[50] [115]. Here the left value is greater than the right and, voila, the result is negative. So, negative parallax is to be expected and an object dancing in front of the screen.

That a stereo image can also be viewed with parallel view requires no further explanation. It is immediately obvious that when viewing the left image with the left eye and the right image with the right eye, a 3D impression must be created. It turns out, however, that it is not quite that simple after all. Wheatstone had built himself a mirror stereoscope for this purpose, Brewster forced the visual axes onto the right path via prismatic lenses, and Rollmann erected a vertical partition for this purpose [67]. The rationale for this is simple. Due to the natural fusion constraint in vision, which is responsible for the fact that one usually does not see double images in everyday life, the eyes automatically turn towards the fixation point. In order to keep the eyes parallel, the point of observation would therefore always have to be at a very great distance. But this is not always the case. That is why Dove, in his 1851 proposal to use double vision as a stereoscope, had already spoken of the need to practice this gaze (from [116], p. 187):

> He who has practiced double vision can place the two stereoscopic images side by side, transform them by double vision ... into four, bring the two middle ones to the ceiling, and then obtain the relief in the middle between his two projections.

Dove considers parallel view exhausting and even discourages its use.

The binocular computer-generated patterns fabricated by the Hungarian scientist Julesz from 1960 onwards [117] can be viewed with parallel view as well as with a stereoscope. The amazing thing about Julesz's random-dot stereograms, however, is the effect they have when viewed over a longer period of time (Fig. 4.23). Out of the jumble of seemingly random dots, a square rises above the background after prolonged viewing.

The way Julesz's stereograms work is relatively simple. First, both patterns are completely identical. Then, in the area that is to appear in another layer, the respective black dots are colored white.

But the effect still requires a stereo pair. Following Julesz's stereoscopic process, the neuroscientist Christopher Tyler, together with the programmer Maureen Clarke, found a

[50] Right minus left.

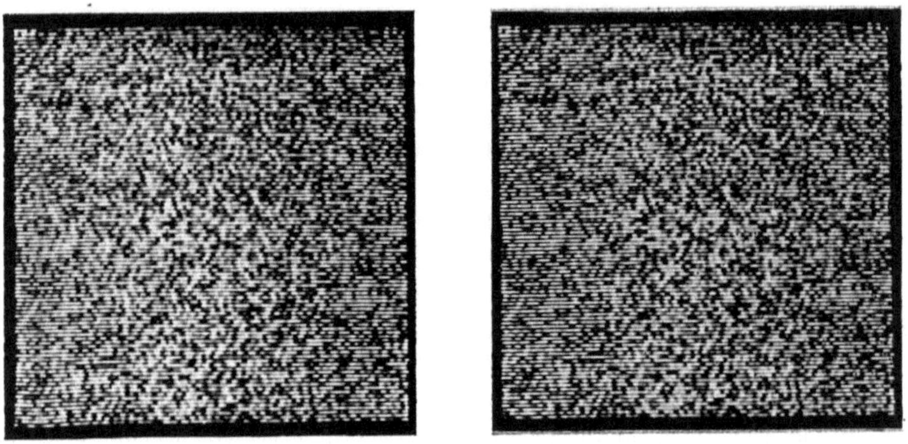

Fig. 4.23 Random-dot stereogram. (Béla Julesz (1960) from [117], p. 1129, Fig. 4)

solution to achieve the 3D effect with a picture as well. In principle, he combined the idea of Brewster's wallpaper with the basic idea of the random-dot stereogram [118]. Through a small offset or by recoloring individual pixels in the pattern, a stereoscopic disparity is created, which the viewer can perceive as a spatial image with cross view. The term autostereogram, which Tyler and Clarke used, is still the valid term[51] today, although it is also often called "Single Image Random-Dot Stereogram" (SIRDS) in the scientific environment.

This technique of creating a spatial image from a seemingly random pattern was refined into real art by the computer scientist Baccei [119]. The exhausting cross view mutated into the somewhat less stressful parallel view[52] when using magic pictures, but this time it did not seem to bother him. Baccei's book "The Magic Eye" sold millions of copies.

There is another way to turn a single image into a 3D image. In this case, only the colors of the image are used and any kind of stereo image representation is abandoned. Since Newton in the seventeenth century separated light into its colors through a prism by means of dispersion [120], the possibility of assigning a defined angle to a certain color after it has passed through the prism is known. It can be said that if the properties of the prism are known, the color of the light can be determined from the angle of dispersion or, conversely, the angle can be determined from the color. These properties were exploited by Brewster in his chromatic stereoscope ([121], from p. 126). The effect of color decomposition may be obtained by prismatic half-lenses or true prisms. The difference here is that the visual axes are no longer bent as in the lens stereoscope, but can converge to a common point.

[51] Also for "The magic eye."

[52] Because of the the "relaxed" infinity position of the eyes.

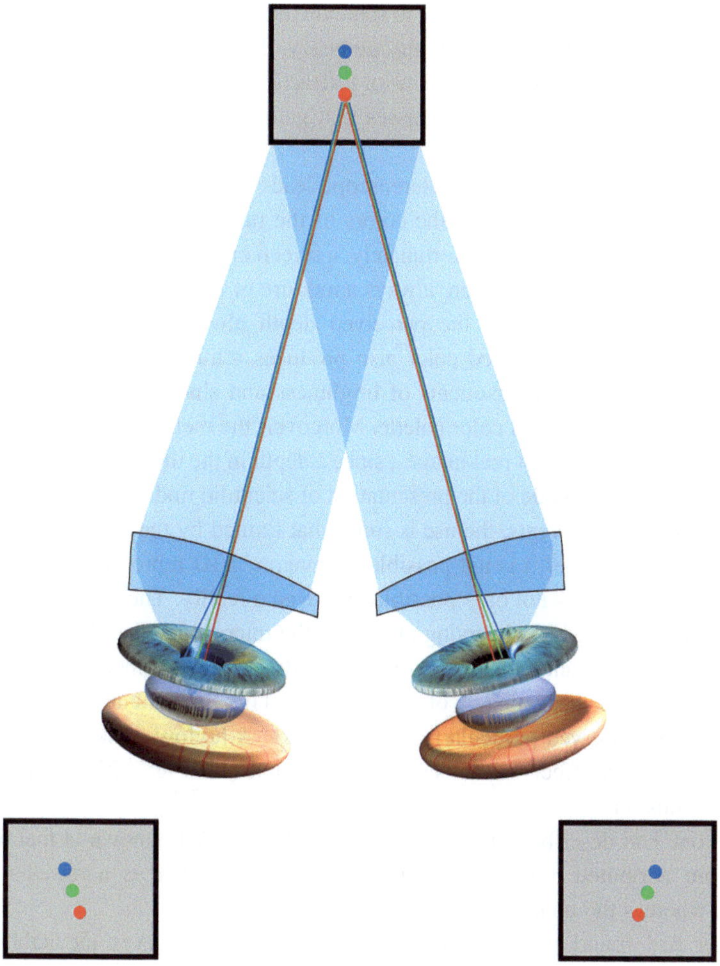

Fig. 4.24 Chromatic stereoscope according to Brewster, Grasnick (2019)

The mode of action is quickly explained. When passing through the prisms of the chromatic stereoscope, the short-wave, blue portion of the light is deflected more than the long-wave, red portions. A red dot would be deflected a little, a green one a little more, and the blue dot at all to the maximum. Since the prisms are rotated 180° to each other, the deflection angle is also exactly opposite. This creates a stereo image in the viewer's pair of eyes, in which the position of the dots depends only on the colors of the image. The stereoscopic color depth effect is illustrated in Fig. 4.24. The figure above shows the three color dots viewed through the prismatic half-lenses as suggested by Brewster. Depending on the color, the respective point is more or less deflected. At the bottom of the picture are the two retinal images, which now show a disparity.

Again, the stereoscope does not have to stand on the table or hang on the wall, but can be put on as glasses. Since large glass prisms are heavy, this was only usefully achieved with Steenblik's suggestion to use microprisms or diffractive elements [122, 123]. This made it possible to produce inexpensive eyeglasses that gained some notoriety under the name "ChromaDepth" [124].

The functioning of the chromatic stereoscope leads to a problem in connection with a screen. In principle, you could use the colors of the rainbow to create a continuous 3D structure with many gradations. Unfortunately, a screen only provides the three basic RGB colors. All other displayed colors are always a mixture of the primary colors. As has been shown, the deflection and thus the perceived depth always corresponds to a certain wavelength. As a result, a mixed color also produces a mixed depth. One can achieve improvements by specific adjustments of brightness and shading [125], but still cannot solve the problem of a limited color palette. Moreover, the method is hardly suitable for a stereoscopic representation of real images, since a depth in the image is always represented with a specific color. Outside of the presentation of scientific findings (e.g. [126]) or in the field of advertising and events, the use is somewhat limited by the color restriction.

One might ask whether it is not possible to imagine a 3D representation without such limitations. In fact, this is not only possible, but even a reality. The botanist Gmelin from Tübingen had noticed when observing objects in the microscope that the depth effect can be perceived inversely and concave (curved inwards) surfaces sometimes appear convex (curved outwards). This inversion of depth, which Gmelin attributes to the casting of shadows, can be observed particularly well in masks. A hollow mask that is rotated can again have a convex effect when viewed from behind (see Fig. 4.25). Hence the name hollow-face illusion.

Wheatstone had described the cameo-intaglio effect, already known at that time ([40], p. 383), and attributed it to monocular observation. A cameo is a raised relief on a gemstone, whereas the intaglio (cameo) is an incised relief.

Brewster had found that viewing a flat image with both eyes inevitably leads to revealing the planarity of the stereoscopic representation. However, if one eye is closed, the spatial impression of the image seems to increase ("Monocular Relief" see [62], p. 46).

The photographer Claudet, who had already worked with the thaumatrope, also made stereo images for Wheatstone's and Brewster's stereoscopes ([127], p. 124). Through his involvement with stereoscopy, he was familiar with the relief-like image impression produced by a spatial image. He constructed an apparatus, the stereomonoscope, in which a spatial impression could also be produced with an image [128]. Here, the stereo image is imaged with two lenses on a ground-glass screen and can still be seen stereoscopically [129].

The Jena optician von Rohr was hardly enthusiastic about this, and so he succinctly referred to the optical image, which even the ground-glass screen does not change ([130], pp. 103–104). Rohr, however, had been inspired by the physiologist Gullstrand to conceive a binocular viewer for Brewster's monocular relief [131]. Thereupon Rohr developed the "Verant" to present the single image to each eye again at the same angle of vision as they

Fig. 4.25 Hollow-face illusion, Grasnick (2019)

had been when the image was taken. In addition, Gullstrand's condition should be observed, which states that the eye pivot must coincide with the crossing points of the principal rays ([132], p. 33). This allows the eye to roam over the image at will without changing perspective. It is certainly helpful to introduce such a verant once, since the device is no longer used today. Therefore, now a picture from the instruction manual of that time (Fig. 4.26).

Rohr was not only a connoisseur of binocular instruments, but also a keen inventor. The knowledge of the "natural" representation helped him to the idea of the synopter, with which both eyes could see the same image [134]. Von Rohr thought that when looking at a photograph with the synopter, an impression of space could also be created, which would be nullified when seeing with both eyes without the synopter.

Monocular depth vision was later studied again by the physiologist Ames [135]. Ames now thought, among other things, that a stronger spatial impression would be gained if the fixed action point could be moved away from the image plane.[53] He also suggested the use of lenses for this purpose. Ames had conducted some more extensive experiments to test monocular and binocular perception. He had designed the Ames room named after him in such a way that when observed with one eye through a peephole, the illusion of a certain position made the room appear not only spatial but perfect in its dimensions [137]. The illusion is then revealed when people enter the room. In the photograph (Fig. 4.27), the same Ames room is photographed at two different times. The proportions do not seem quite normal, but if you look at both pictures you can easily see the trick. The room is in fact distorted, but creates the monocular illusion of correct architecture via the pattern of the floor and the edges of the walls. With this setup, Ames had given an example that seeing with one eye can also lead to an impression of space. Here, however, to a false impression.

[53] Ame's hypothesis was tested and confirmed in 2016 at the School of Psychology & Neuroscience, University of St Andrews [136].

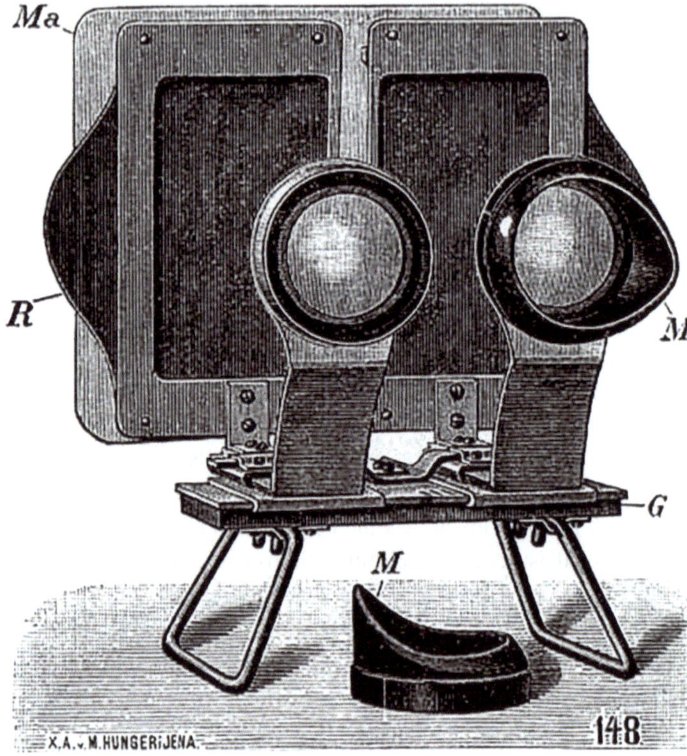

Fig. 4.26 Zeiss double verant. (Carl Zeiss Jena (1902) from [133])

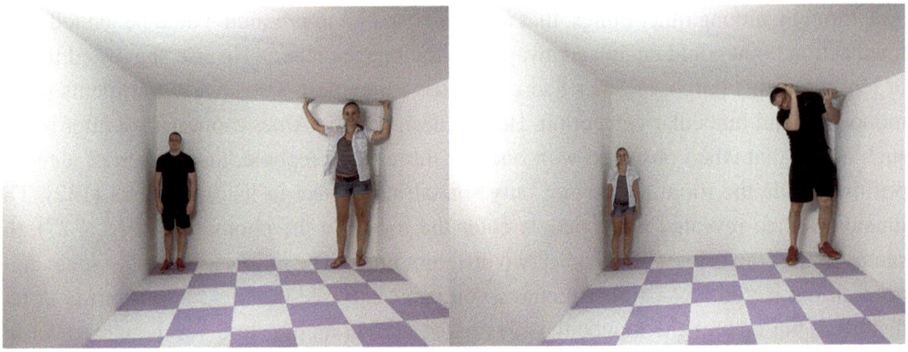

Fig. 4.27 The Wessons in Ames room, Grasnick (2019)

Ames' idea of using glass lenses to enhance the effect of a flat image was based on the fundamental conviction that a realistic impression could be achieved by means of mathematics and optics. The mathematician John Dee wrote a preface to the English translation of Euclid's elements [138], in which he describes, among other things, the many possible applications of mathematics. Besides such obvious applications in geometry, architecture, navigation, or perspective, there are also exceedingly unusual uses such as astrology, anthropography,[54] thaumaturgy,[55] or zography.[56] Zography in Dee's sense describes the art of creating a virtual visual perception that is of extreme realism. A technical implementation of this idea was described by the mathematician Kohlhans in his description of optical curiosities in 1677. In the chapter "Von finstern Kästlein" ("From dark boxes")—by which, of course, a camera obscura is meant—he reports on the effect of a lens placed in front of the peephole (see [60], p. 295):

> But if one puts another *vitrum convexum*[57] . . . into the seeing-hole, one sees the *objecta* . . . as they appear from the outside *libero occulo,*[58] in width, breadth, roundness and distance.

By this is meant that when the objects inside the dark box are viewed through the lens, an effect is obtained which corresponds to natural vision. This type of display became popular in the following years and led to the use of a simple optical device for viewing images, the zograscope.[59] A zograscope is nothing more than a large lens through which to view an image. For practical use, the lens of the zograscope is set in a frame and mounted on a tripod (see Fig. 4.28). Behind the lens there is often (also in this case) a mirror mounted, which is tilted 45 degrees with respect to the table top (in the picture already slightly tilted to the lens). Thus, one can simply place the images on the table for viewing.

It seems unbelievable, but the impression of space arises involuntarily, without effort and immediately during the zograscopic observation of any image. The basic principle is magnification through the lens. The image to be viewed is near the focal point, as in a magnifying glass. This allows the eye to accommodate relaxed at long distances, the eyes do not converge. As a result, the normally existing binocular disparity, i.e. the difference between the images of the two eyes, is eliminated. The view of the observer becomes cyclopic.[60] Additionally, since no motion parallax is expected at large viewing distances (as suggested by the lens), all physiological indicators that might suggest the image is flat are now eliminated. In addition, the edges of the image are no longer in the viewer's field of view due to the magnification of the lens. The perception of image depth is now based solely on physiological effects derived from the design of the image and experience (see also "Natural Spatial Vision"). The effect is confirmed by most observers (see e.g. [140]). If the effect is now largely based on the monocular depth factors, the spatial effect can be

[54] Description of the external human appearance.

[55] Miracle work.

[56] From (Greek) Zoe = life and -grafie = drawing.

[57] Convex glass.

[58] With a naked eye.

[59] From Greek. Zoe = life + -grafie = drawing + -skop = to look at, quasi a "life-picture-viewer".

[60] One-eyed, according to Greek mythology. The Cyclopes are the one-eyed sons of Gaia and Uranos, in Homer those with whom Odysseus had some trouble [139].

Fig. 4.28 Zograscope (late eighteenth century). (Susan B. Price (2006) Wikimedia Commons, edited)

maximized if all factors are actually included. This has been used for centuries in the technique of trompe-l'œil,[61] which attempts to give the impression of a real object by means of photorealistic representation and especially by the correct use of light, shadow and perspective. With a simple trick a further increase can be achieved. By using real or virtual passepartouts (see e.g. [141]), the spatial arrangement becomes clear and the depth effect is enhanced (see e.g. Fig. 4.29).

Since the function of spatial vision always works particularly well when both eyes see exactly the same thing, 3D vision is also stimulated when the scene is also viewed monocularly without any aid, for example by completely covering or closing one eye. The single-eyed Pulfrich effect during motion was observed in a patient in 1982 and described as an abnormality [142]. In fact, however, 3D vision with one eye is completely normal and occurs even in completely healthy individuals. Movement introduces temporal parallax into the image, which produces an enhanced spatial effect.

[61] French tromper = to deceive + l'œil = the eye.

Fig. 4.29 Virtual passepartout, Grasnick (2019)

The Monocular Pulfrich Effect—a 3D Display for Everyone

Pulfrich himself mentioned that he had been blind in his left eye due to an injury suffered in his youth and would never have seen the "Pulfrich" stereo effect with his own eyes (from [110], p. 555). Nevertheless, Pulfrich could also have seen stereo with the remaining eye.

And it goes like this.

First, find a good quality video that has a lot of movement or rotation. Videos with fast camera movements and flyovers[62] are well suited for this. The content is of little importance. It is only important that there is a lot of movement and that the objects in the scene are arranged differently in space.

If you now look at this scene with both eyes, you already get a certain understanding of the scene structure through the movement alone. If you now cover one eye, the scene actually appears in 3D to most viewers after a short time. This effect is so realistic that it can hardly be achieved with other methods.

Which eye is closed is not decisive, nor is the direction of the movement. Contrary to the classic Pulfrich effect with glasses, the single-eye Pulfrich effect is extremely tolerant. It

[62] An internet search will quickly turn up some examples if you type "3D" and the phrase "one eye closed" into a video search.

does not matter in which direction the scene is moving. Only the movement itself is important.

4.7 Autostereoscopy

About Barriers and Lenses

Stereoscopic viewing is based on the assumption that a maximum of two images are required to create a spatial effect. In principle, this is also sufficient if the images are presented in a suitable manner. The presentation is always suitable if one eye only sees the image that is intended for that eye. So that the left eye only sees the left image, a wide variety of stereoscopes, 3D glasses or visual techniques can be used to enforce this assignment.

But what happens if this assignment is not possible? If, for example, a 3D display is to be made in a large format and in a public space, where the viewers appear randomly and unprepared in front of the 3D display? In this case and in all other cases where the visitors cannot be equipped with glasses, one tries to create the 3D effect with an autostereoscopic display.[63] In this case, the image content is displayed in space via optics in such a way that a viewer can position himself in front of the display and does not need any aids such as 3D glasses or a stereoscope.[64]

The simplest type of such optics is based on a transparent film to which a non-transparent (opaque) striped grid is attached. The London publisher Symons proposed such a grid to be placed over a picture in 1896.[65] The individual elements of the picture were made up of different lines,[66] which meant that when superimposed on the "grid fence" of the transparent film a certain pattern always had to emerge.[67] Based on this idea, the London publisher Bliss, Sands & Co. published the "Motograph Moving Picture Book"[68] in 1898, in which a number of suitable pictures were illustrated and the film supplied.

Figure 4.30 shows the overlay of image and raster. On the left side you see the printed image without the raster, on the right side the same image with the raster overlaid. You can now clearly see the dark moiré stripes whose position, direction and size change when you move or rotate the screen in front of the image. This creates the impression of movement.

[63] See, for example, the "Handbook of Autostereoscopy" ("Handbuch der Autostereoskopie") [143].

[64] Wheatstone's or Maxwell's stereoscopes also managed without aids, but the viewing freedom in front of the stereoscope was extremely limited.

[65] "Improved movable devices for producing variable optical illusions" [144, 145].

[66] Partly horizontal, partly vertical, often wavy.

[67] What is known today as the Moiré effect.

[68] The cover picture is by the famous French painter Toulouse-Lautrec.

Fig. 4.30 The Magic Disc. (F. J. Vernay, Yorick (1898) from [146])

With such a grid, however, not only the illusion of movement can be created. By suitable preparation of a stereo image pair, a spatial impression can be created from the grid and the image without a stereoscope. An idea for this came from the French photographer Berthier in 1896, who in his publication "Images stéréoscopiques de grand format" [147] presented not only an outline of stereoscopy but also the proposal of superimposing image and raster. As in the case of an alternating image, Berthier composes the raster image from the image strips of two perspective views. The separation, however, now takes place via the strip raster in front of it. The process is illustrated by the images used by Berthier. In the upper image of Fig. 4.31, the strip grid is superimposed on the composite image. Unlike a "motograph", a space is needed between the image and the grid. The composite image is shown at the bottom left and the two images that an observer would see with the left and right eye are shown next to it on the right.

The 3D effect will be difficult to observe here, as the wide black lines will disturb the overall picture impression.

The American Ives did this unmistakably better. Frederic Ives developed the cromolinoscope camera for his cromoscope[69] [148]. In doing so, Ives invented a simplification for capturing stereoscopic images with a single lens. This worked quite easily when a strip grid was mounted between the film stage and the lens. The first presentation of the

[69] Also cromoscope or chromoscope, a stereoscope for displaying full-color photographs.

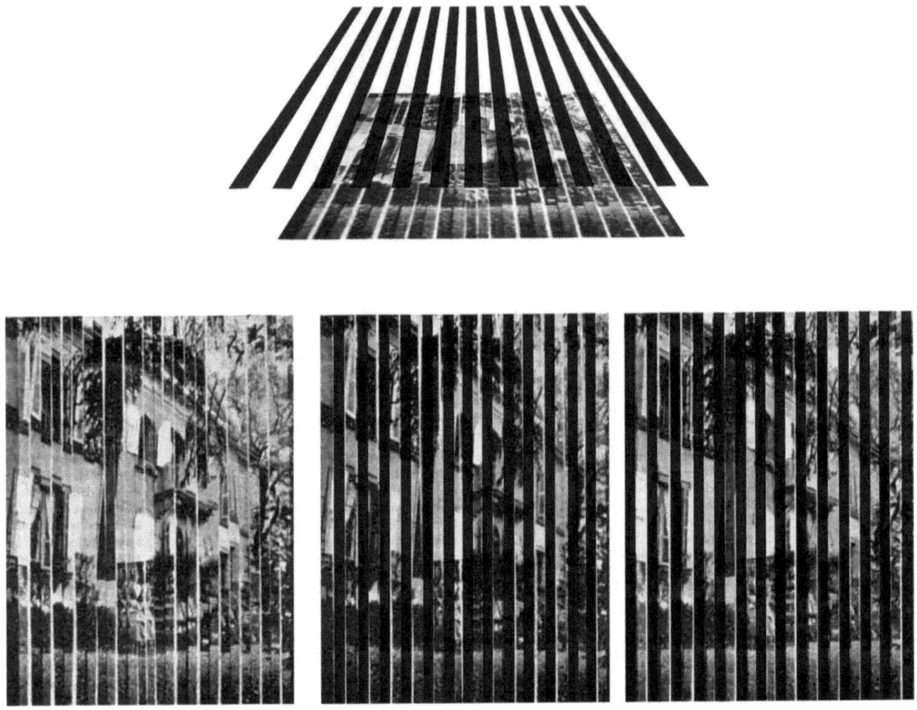

Fig. 4.31 Stéréoscopies lignées. Auguste Berthier (1896) from [147], p. 330, edited (Grasnick 2019)

idea [149] describes the use of the strip grid also for autostereoscopic reproduction.[70] Finally, Ives gives his invention the name "parallax stereogram".[71] Shortly afterwards, in a patent publication [150], he describes in particular the recording of the stereoscopic image. Here the photographic film is arranged directly under the fringe grid. The viewing of a parallax stereogram is done via a viewing screen,[72] which corresponds to the strip grid of alternating opaque and transparent lines. Even at that time, this allowed the stereogram to be viewed from a defined viewing position[73] without the use of any aids.

Ives describes here the basis of an autostereoscopic monitor without using this term. The adjective "autostereoscopic" (autostéréoscopique, from [151]) is introduced only a few years later by Estanave. Estanave, who was familiar with the concepts of Berthier and Ives (as can be seen from his patent, see [152], p. 1), also extended their ideas to a stereoscopic

70"...a stereogram which should require no stereoscope..." (from [149], p. 51).

71"...I propose to call them parallax stereograms." (from [149], p. 52).

72"transparent line-viewing-screen" (from [150], p. 1).

73"...a stereogram ... seen from a suitable view-point without the aid of any optical device..." (from [150], p. 1).

double raster projection [153]. Here, the stereoscopic image pair is projected through a first raster onto a diffusing screen ("glace dépolie", frosted glass). This splits the individual images into stripes and simultaneously meshes them together. A viewer looks through a second grid onto the combed striped image and gets a true stereoscopic impression. Estanave used the barriers not only for autostereoscopic display, but also for presenting classical alternating images [151]. A number of autostereoscopic works by both Ives and Estanave have survived [154]. Among them are[74] alternating images [155] that are of high quality even by today's standards. In the images from the early twentieth century, the barrier structure was already so fine that it was barely visible to the naked eye. The problem with a barrier is always the loss of light that inevitably results from the obscuring lines. However, this disadvantage can be circumvented by imaging through lenses.

A small note by the French physicist Lippmann showed the possibility of spectacle-free viewing through many individual lenses. In 1908, Lippmann had outlined a method with integral photography [156] in which the 3D effect is achieved without glasses simply by looking at the image. Lippmann, who was to receive the Nobel Prize for his color photography[75] in the same year, had equipped his highly sensitive photo layer with an optical surface for this purpose. Lippmann's drawing (Fig. 4.32) shows the function of his idea. The object point ***A*** is focused by the convex front side of the small rod lenses onto the also convex back side of the lenses. On this side was also the photosensitive layer. Each pixel ***a*** is imaged under the rod lens at a slightly different position. After development of the film, it could be viewed again in 3D through the same lenticular.

It has been handed down by a contemporary witness[76] that Lippmann used small spheres instead of rod lenses [157]. Although the spheres had a diameter of only about 0.25 mm,[77] a complete image was recorded under each sphere. This was mainly due to the high resolution of the film. A present-day photographic film can have a resolution of more than 100 lines per mm.[78] A holographic film has a resolution of several thousand lines.[79] Lippmann's photo emulsion was even finer. Lippmann's color film had a resolution of about 10,000 lines per millimeter[80] [160]. Thus, with a sphere size of only 0.25 mm, 2500 lines can still be imaged under each glass sphere, which would be roughly equivalent to a

[74]E.g. in the collections of the "Société Française de Photographie".

[75]Heliochromia, from Greek helios = sun + chroma = color.

[76]Presumably Jaques Boyer, referred to in the source only as "the French Correspondent of the Scientific American".

[77]"less than a hundredth of an inch".

[78]E.g. Fujichrome "Velvia 50" with approx. 16 lines per mm [158].

[79]E.g. Thorlabs PHG63 "Green Holography Plate" with about 3000 lines per mm [159].

[80]The high resolution in Lippmann's film was necessary because Lippmann's heliochromy was based on the interference oflight (hence sometimes called interferential photography).

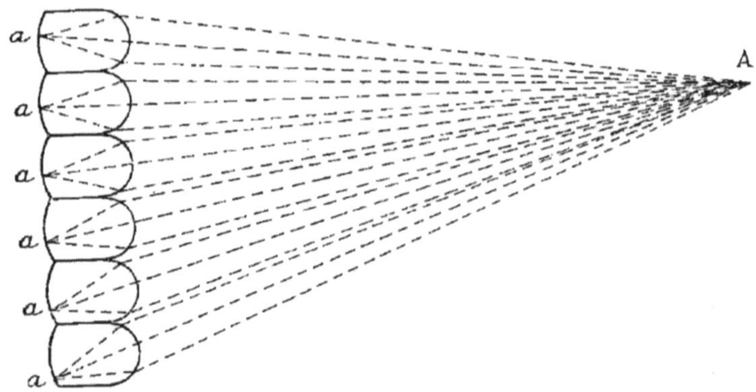

Fig. 4.32 Lenticular grid for integral photography. (Gabriel Lippmann (1908) from [156], p. 823, Fig. 1)

full HD resolution image. If the size of a Lippmann photo plate[81] is added to this, it can be quickly determined that an incredible 86,400[82] images fit onto one heliochromic photography.

Looking at these images again through the same lenticular, a 3D space appears to open up to the viewer when looking at the plate. However, the effect appears inverted in depth, front becomes back and back becomes front. As with a hollow-face illusion (see also Fig. 4.25), one might still be able to gain a 3D impression, but it must collapse when one moves in front of the image. A solution would be to turn the image around to look at it from behind. Then it would be laterally inverted, but at least depth correct. In this case, the lens structure could act as position-dependent illumination if the image were sufficiently transparent. Although the usual mode of operation of a heliochrome color image is based on the reflection of incident light off the silver particles of the photolayer, in this case the transmissivity of Lippmann's plates seems to have been present. To see the 3D image, the photograph was backlit with a "good light" [157], perhaps simply turning it over. In an early analysis of Lippmann's technique, the Russian Sokolov[83] had suggested exactly this procedure [162]. Some years later, Ives[84] had also recognized this problem [163] and presented a technical solution for it, in which the individual sub-images now appear again laterally correct by means of a further optical imaging [164]. Lippmann himself is not known to have been intensively involved in pseudoscopy. However, one could assume that Lippmann's hint for double rectification[85] (photographic and geometrical) was aimed in this direction.

[81]Which usually had a size of 6 × 9 cm (see, e.g., [161], p. 66).

[82]240 × 360.

[83]Of whom not much more is known today than his name and the fact that he was a professor at Moscow State University.

[84]Herbert E. Ives.

[85]"un double redressement, photographique et géométrique." [156], p. 823.

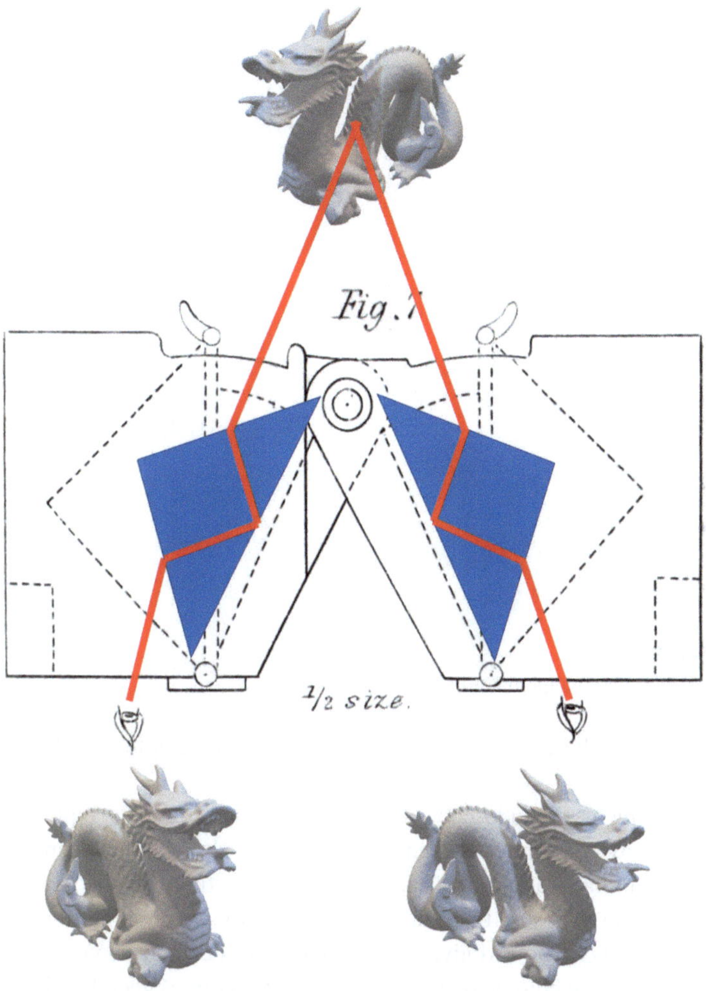

Fig. 4.33 Wheatstone's pseudoscope, Charles Wheatstone (1852), Grasnick (2019)

The basic problem of inverted depth due to the interchange of stereo images was already known to Wheatstone, who developed a device he called a pseudoscope ([57], p. 11) to deliberately create the false impression. When looking through the pseudoscope, this effect can be created artificially. Wheatstone had sketched the device in an essay, to which only a little addition of the ray path has been made here (see Fig. 4.33). When looking at an object (following the red lines), the image for each eye is mirrored on the inner surface of the prism (blue). The respective images are shown below the eyes. The depth of the object is inverted overall. When interpreting the seen image, the mind tries to correct the optical effect. Especially with complicated objects or whole landscapes the inverted effect will not be complete. It works well if simple objects with structured surfaces are placed freely in

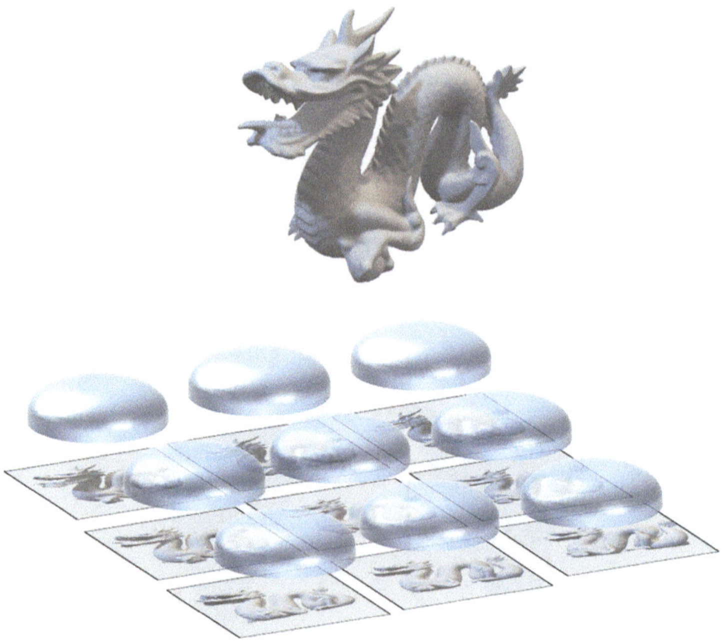

Fig. 4.34 Integral photograph recording, Grasnick (2019)

space and viewed with the pseudoscope. A convex cylinder can then look like a concave gutter.

Notwithstanding the "pseudoscopic defect" inherent in the process, Lippmann's integral photography[86] provided further challenges due to the underlying heliochromy. The need for extreme resolution was accompanied by lower sensitivity of the emulsion and the resulting increase in exposure time. In addition, mercury was used during the exposure, which is known to be toxic and gives off its toxic fumes even at room temperature. The color rendition of a heliochrome is angle dependent and requires good lighting. Since these color photographs could not be duplicated either, Lippmann's color photography did not catch on despite winning the Nobel Prize. Integral photography, on the other hand, has survived. The multi-camera technique that Lippmann used was the basis for various light field cameras.[87] The concept is quite simple. An object is recorded from several positions on some light-sensitive layer or sensor (see Fig. 4.34). Under each lens, the image of the object is formed from a slightly different perspective.

This is roughly equivalent to taking a picture with several cameras from different positions (illustrated in Fig. 4.35). Here it is also clear why a lenticular is more suitable

[86] At least that is what Herbert E. Ives called it (see also [163, 164]).

[87] Well known especially the Lytro [165] developed by Ren Ng (now discontinued).

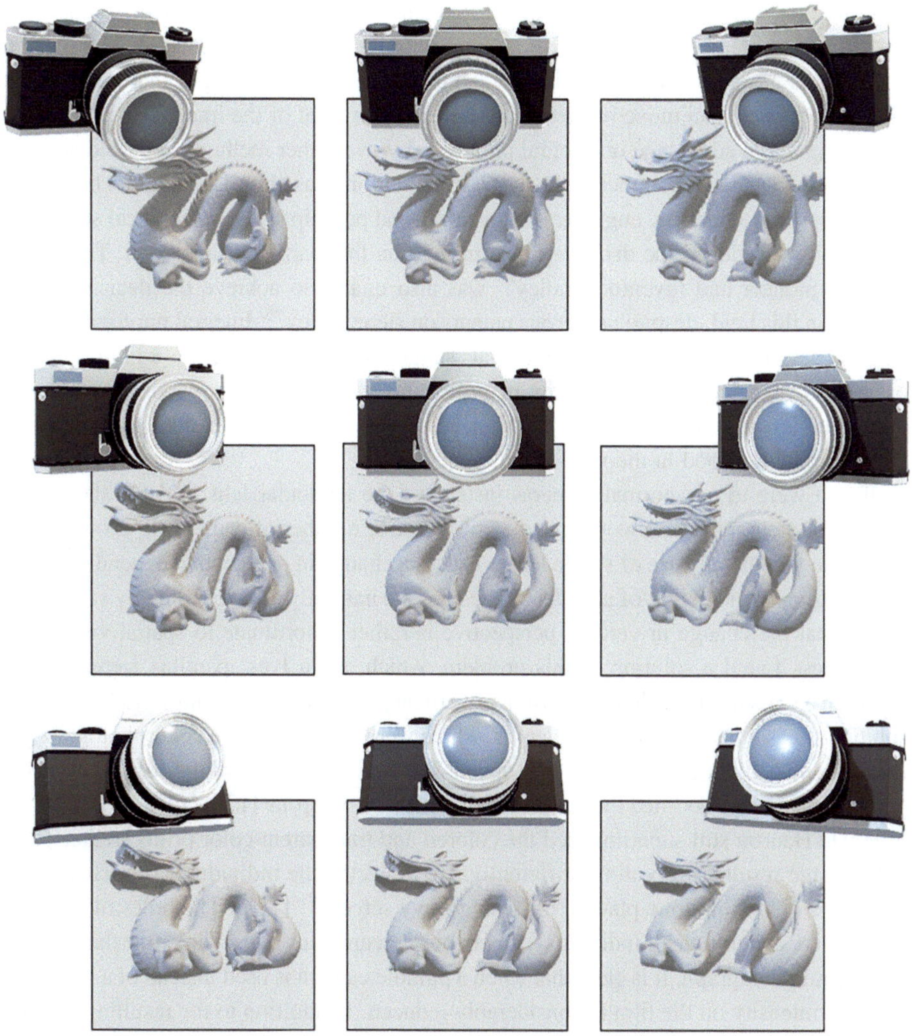

Fig. 4.35 Integral photography with multiple cameras, Grasnick (2019)

(and cheaper) than an array of many cameras. A lenticular is easier to mount in front of each image, and the lenses can be so small that the eye cannot resolve them.

Compared to the autostereoscopic methods of Herbert Ives or Eugène Estanave, integral photography certainly provided a freer 3D impression, since the freedom of movement in front of a barrier is geometrically restricted by the limitation to two perspectives. Because the left eye is only allowed to see the left image and of course the right eye only the right image, the horizontal freedom of movement is limited to the interpupillary distance (only a little more than 6 cm). In front of an integral photograph, the viewer can move comfortably and without restriction as long as he stays within the aperture angle of the single lenses.

Using the well-known structure of a Lippman grid (see e.g. [143], Sect. 4.4.3.1), the freedom of movement in front of an integral photograph can be estimated. The opening angle of the globules is about 45 degrees, so that an observer looking at the image from a distance of 50 cm can move more than 40 cm freely in front of the image. Because of this promising initial advantage of integral photography over other methods, repeated attempts were made to achieve commercially successful solutions based on Lippmann's basic idea. Although some scientists, engineers or inventors had participated in a technical solution,[88] they had not achieved the decisive breakthrough on this particular approach. The British 3D film pioneer and inventor Dudley[89] was also unable to achieve the decisive breakthrough in this field, despite numerous patents on stereoscopy.[90] Integral photography was less successful than its initial potential suggested. Apparently, in addition to the pseudoscopic defect, the difficulties in producing the lenticular grid with a myriad of individual lenses in high quality and the necessary extraordinary resolution of the reproduction material stood in the way of widespread use.

If you were to use normal cameras instead of the lenticular lens grid for shooting in integral photography, you would need a considerable number of cameras to do so. This is due to the fact that the use of single lenses causes a change in perspective depending on the lateral position and height of each lens. The eyes are naturally arranged side by side, which means that the change in vertical perspective is rather subordinate to spatial vision. The Swiss Hess found a solution to this problem, which, as in Ives' parallax stereogram, is based on a striped grid with the omission of vertical perspectives. In contrast to Ives' barrier solution, Hess relied on cylindrical lenses, which were already used in color photography at the time and could therefore also be produced. The basis for this was John Joly's line screen photography. Joly modified the basic idea of Ducos du Haurons [173] color photography a little. Du Hauron still superimposed the colored and transparent color prints so that a full-color image resulted. Joly did it differently. He projected the individual color channels in fine strips onto a halftone plate [174], using Levy screens[91] [176]. The holes of the Levy grid act like a multitude of individual camera obscura and thus in principle like the lenses of an integral photograph. It is clear that when a pinhole camera is used instead of a lenticular, the light intensity on the film is considerably reduced. In addition to the resulting increase in exposure time, the poorer imaging properties of a simple pinhole also cause blurring. This can be improved by the use of many microscopically small individual lenses, but

---

[88]E.g. Kanolt 1930 "Production of Stereoscopic Pictures", Herbert Ives 1935 "Optical Device" [164], Winnek Coffey 1935 "Apparatus for making a composite stereograph" [166], Valius 1949 "Raster Optics" [167], de Montebello 1966 "Integral Photography" (Integram) [168] or Dudnikov 1970 "Autostereoscopy and Integral Photography" [169, 170].

[89]Who designed an autostereoscopic cinema as early as 1935 (see [171], p. 173).

[90]He himself even claimed in 1956 that he held more patents in stereoscopy than any other living person (from [172], p. 56).

[91]Invented by Max Levy, an American photographic pioneer of German descent, after Frederic Ives' cross grid based on Georg Meisenbach's ruler photograph (see p. [175], pp. 913-919).

again increases the difficulties in manufacturing. Berthon addressed precisely these problems and, as a logical consequence, proposed the use of cylindrical lenses (from [177], p. 2):

> One can make do with a smaller number of lenses if cylindrical lenses are applied, for example, by arranging semi-cylindrical filaments of transparent, refractive material parallel to each other on a flat plate.

The production[92] he proposed was implemented by the industrialist Keller-Dorian[93] and later used by Siemens[94] for German color film. Since Hess was familiar with both Ives' parallax stereograms and Lippmann's integral photography, he was able to combine both elements with a cylindrical lens grid [180]. Hess patented his idea worldwide and even founded his own company, but was not commercially successful with the idea itself [181]; however, he later received the Nobel Prize for his research in the field of physiology. A few decades later, with the availability of such lenticular lenses, the scientists Kanolt (as "ribbed lenses" ([182], p. 4) and "lenticular" ([183], pp. 4, 5)) and Ives (as "lenticular" ([184], pp. 1, 2, 4, [185])) also used similar technology for autostereoscopic image rendering, unaware of Hess' earlier development. Both use the term "lenticular", which almost always refers to a lens array with vertical cylindrical lenses. The availability of inexpensive lenticulars[95] for 3D display seemed to spur the development of autostereoscopy. Inventors in the 10s, 20s, and 30s of the nineteenth century were often concerned with lenticular and barrier simultaneously. Kanolt and Ives had certainly noticed that a parallax barrier or cylindrical lenticular, combined with a single stereoscopic image pair, allows only a limited viewing space. If a viewer is to be able to move freely like an integral photograph, more images must be used. Kanolt also recognized this and offered a solution with his "photographic method" [187]. A viewer standing in front of Kanolt's photograph can perceive a three-dimensional image.[96] In addition, the impression of the image also changes with movement.[97] Kanolt moved the camera lens, the barrier or lenses[98] or even the recording film horizontally during the recording for this purpose. If, after development and during viewing, the image taken in this way is again superimposed with the same barrier or lens grid, the viewer sees a slightly different spatial impression from different positions. This motion parallax comes very close to natural spatial vision. In the 1930s, Ives[99] used this

[92] "The cylindrical lenses can be made on film[s] with grooved rollers…" ([177], p. 2).

[93] Under the name Keller-Dorian-Berthon process [178].

[94] As Opticolor or Berthon-Siemens [179].

[95] E.g. as roll stamping of the film material directly in the film casting process for the production of film webs of any length with lens screens (see IG Farben [186]).

[96] "When the picture is viewed through a screen … a stereoscopic image is obtained …" ([187], p. 1).

[97] "… representation of continuous motions …" ([187], p. 1).

[98] "…preferably a line screen, although it may be made up of parallel cylindrical lenses" ([187], p. 2).

[99] This time Herbert Eugene Ives, son of Frederic Eugene Ives [188].

effect for his "parallax panoramagrams" [189]. Ives does not need any movement of the camera components, but maps the scene through a camera lens directly through the grid onto the film.[100] Just like Kanolt, Ives does not reduce himself to the parallax barrier in his ideas, but quite naturally proposes lenticular grids as well [185]. However, he goes even further than Kanolt's idea. In the same patent he mentions the use of his technology for use in moving picture film.[101] The Belgian Noaillon developed at the same time some patents in the field of stereoscopy, of which his barrier cinema system [190] is particularly noteworthy. In this publication, Noaillon departs from the previous parallel barrier strips and develops a radial grid,[102] in order to provide the entire auditorium with an equally good 3D experience.

The Russian Ivanov[103] actually built a whole cinema on this idea, for which he stretched a perspective grille (from [195]) of 150 km of copper wire in a Moscow cinema (see Fig. 4.36). A specially produced film[104] was even shown in this cinema.

The technique has not spread beyond these unique setups, despite the massive size of the representation and even with the improvements that Gabor had contributed [197]. One reason for this is obviously the limited viewing space of the representation, as Ivanov himself admits.[105]

[100]In principle also a light field camera with reduction to horizontal parallax (see e.g. Ng's light field camera of 2005 [165]).

[101]"motion picture film" ([185], p. 3).

[102]This technique was also called "Noaillon theory"in the 1952 Journal of the Society of Motion Picture Engineers [191]. Incidentally, in the same issue Kaplan also published his remarkable essay on the theory of the parallax barrier [192].

[103]Occasionally, two Ivanovs are reported (e. g. in [193], p. 5): besides Semyon Pavlovich Ivanov and was also B. (Boris?) T. Ivanov (and even Valius) participated in the construction of the Moscow grid screen . Since he had turned out to be B. T. Ivanov refers to S. P. Ivanov in the preface of his book on autostereoscopy ([194], p. 4), it must indeed be another Ivanov.

[104]"Semlja Molodosti" (Land of Youth); directed by Alexander Andriyevsky, partly color scenes; the film is sometimes still screened, e.g. in 2013 at the Munich Film Museum [196].

[105]"One drawback of this first stereo-cinema is that each spectator has to find his own viewpoint and stick to it; a movement to right or left, a bend of the head, and the image is lost for the moment." (from [195], p. 213).

Fig. 4.36 Ivanov (in grey suit) in front of the grid screen. (Preslit (1941) from [195], p. 213)

Barrier or Lens?

From the beginning of autostereoscopy, it was possible to achieve the spatial effect equally with barriers or lenses. Berthier's crude idea of the parallax barrier was further developed by F. Ives and Estanave into a functional, free-eyed spatial rendering. Lippmann's integral photograph, again reproduced by Sokolov using a perforated grid, could be simplified by Hess and Berthon to a purely horizontal parallax using cylindrical lenses. Kanolt and H. E. Ives extended the viewing space by reproducing several perspective views simultaneously. In these variants, too, it was possible to achieve the effect with lenses or barriers.

The term "barrier" commonly used today is a shortening of the unwieldy term "parallax barrier", which entered the technical literature at the latest in the 1950s with Kaplan's essay "Theory of Parallax Barriers" [192]. The term "barrier" is quite descriptive: with the help of the barrier, the view of some parts of the image is blocked. The term parallax describes in general the apparent change in position of an object when its own position is changed and thus also defines the perceived difference in position of this object in binocular vision. Father and son Ives used the term "parallax" to describe their technology (e.g. "parallax stereogram" or "parallax panoramagram"). Thus, a parallax barrier is nothing more than a grid of fine opaque lines applied to a transparent material. Like Joly or Sokolov, you can also create a grid of holes on a surface. Since each individual hole acts like a magic lantern when

(continued)

reproduced, one thus obtains a multitude of distributed lanterns over a large surface by simple means. The same principle of operation can be observed in reproduction by Lippmann's lenticular in integral photography. The Hess-Kanolt-Ivesian (vertical) cylindrical lenticular is nothing more than a reduction of circular imaging to one dimension. Since the light is only refracted horizontally by the one-dimensional curved lens surface, one obtains a "focal line" instead of a focal point. If you use a cylindrical lens as a camera lens, the image will show a horizontally compressed image. A second cylindrical lens in front of the image rectifies the image and enlarges it horizontally again. So with a sheet of cylindrical lenses you again have a multitude of camera obscura, but this time only horizontally distributed. A barrier, if the apertures are sufficiently small, will function just like a pinhole camera, but with cylindrical imaging, strictly speaking, therefore, like a "slit camera". If the slits are very wide, there is no optical imaging, but the barrier works purely on position-dependent masking and revealing of image parts. The image parts are then contiguous strips of the different perspective images (see Berthier Fig. 4.31).

As can be seen, barriers and lenses compare quite well in terms of imaging and are quite functional. Barriers can be easily exposed by photographic means or printed in large format. The production of lenticulars is technologically more demanding, as one needs a corresponding embossing tool. On the other hand, lenticulars have the decisive advantage that they allow more light to pass through. A barrier raster is also a barrier to light. If, for example, you want to reproduce a panoramic diagram from 10 individual perspectives, only 1/10 of the light can pass through the barrier during reproduction.

The decision whether to use a lenticular or a barrier raster thus depends on the number of images, the usable illumination of the image (i.e. the brightness of the projection or monitor) and the availability of the manufacturing technology. Barriers, for example, are often used in teaching and research due to their ease of production, while lenticulars are used more frequently in the professional sector.

Once the basic technology for recording and reproducing 3D images that can be viewed without any aids was developed, further development was not long in coming. It was already possible to reproduce short image sequences with the magic lantern. The development of television technology gave new impetus to the subject. In 1931, Winnek Coffey described a "Television Apparatus" (see [198], p. 7), which could project spatial images on the basis of lens grids. With the actual appearance of cathode ray tubes as image generators in television, there were immediately proposals to use picture tubes and lenses simultaneously for spatial image generation. An early testimony to this is Eisler's idea of stereoscopic television, which combined the idea of Hessian vertical lenses and Kanolt's parallax stereogram with picture tubes. Eisler had also proposed simultaneous recording and playback of images from multiple positions in his 1936 patent US 2,209,747. This technique would be called multiview today. Winnek Coffey also had such a system in mind when he designed a "...composite stereograph of the multiple aspect type...",[106] thereby casually addressing a basic problem of autostereoscopy. When two rasters are superimposed, there are always and inevitably superposition phenomena that appear as undesirable patterns and are referred to as interference or moiré patterns.

Moiré Effect

Simple moiré patterns were already used in the "motograph" to create the illusion of movement. Here, however, the effects were desired to create the impression of image movement.

Originally, the term moiré was used for a fabric that had a characteristic pattern. The flowing sheen and the wavy pattern[107] of the fabric gave it its name. Since the seventeenth century, the production of this fabric in France was established by Colbert, a minister of the "Sun King", who with a veritable regulatory zeal[108] strove to increase the prosperity of France in the sense of Louis XIV. Under Colbert's influence the customs barriers between the French provinces were lowered, and the manufacture of silks was protected against foreign countries (see [201], p. 6). The moiré here is the result of a technique of placing two layers of ribbed fabric on top of each other and pressing them against each other. The fabrics, preferably silk, but also taffeta or rep, have a strong surface structure, which is defined by the yarn used and the weaving technique. As an example, a taffeta fabric is shown. Which has a fairly regular, lattice-like surface (see Fig. 4.37).

Now, if you lay two of these fabrics on top of each other, the alignment of the two fabrics to each other will never be perfect. The fabric will be stretched or compressed in one direction or the other, causing a somewhat irregular lattice structure. For illustration purposes, two slightly different layers have been created based on Fig. 4.37 (Figs. 4.38 and 4.39).

If both layers are placed on top of each other, a moiré pattern results (see Fig. 4.40).

This moiré is based on the presence of two layers. However, a moiré fabric does not have to be two-layered at all. Therefore, both layers are pressed together with strong pressure, depending on the fabric with the help of water or heat. At the points

(continued)

[106] from [199], p. 1, but by "aspect" Winnek actually means a certain point of view.

[107] After the French moirer = to water, to marble and the English mohair = mohair wool or fabric (after its gloss).

[108] An idea of this is given by Colbert's letters, instructions, and memoirs [200].

where the two layers lie on top of each other,[109] the threads in both layers are flattened and shine more than the original round threads. This effect is also a desired result when ironing laundry (see [202], p. 235).

The "real" moiré effect is different for every fabric and its actual appearance can hardly be predicted. The "fake" moiré is easier to produce. This is simply embossed onto the textile with a structured roller.

The effect of textile moiré is always somewhat random due to the imprecise spacing of the threads. With a lenticular in front of a display, things are different. The grid of the pixels of a screen is arranged in columns and rows, a lenticular and a barrier grid as well. If you simply superimpose two barrier grids, the moiré effect is self-evident. Figure 4.41 shows two line grids laid one on top of the other. Since the slit widths[110] differ slightly from each other, this alone results in moiré.

Likewise, a moiré results when both grids are rotated relative to each other (Fig. 4.42).

These moiré effects are usually undesirable in print or screen applications. Since these effects can be described mathematically, the undesired effect can be minimized in the design. A good introduction to this is Amidror with his "Theory of the Moiré Phenomenon" [203].

Winnek had proposed an inclination of the lenses for his multi-aspect system to reduce the unwanted moiré stripes [199]. A solution approach that was later adopted by 3D display manufacturers Stereographics [204] and Philips [205]. Just as the slanting of the lenticulars reduces moirés on 3D displays, the positive effect of a slanted structure can also be observed for barriers. Based on this, the company techXpert from Jena used slanted barriers for their autostereoscopic displays [206], as did the Singaporean company Sunny Ocean Studios.[111]

The systems presented so far all had the disadvantage that they were only suitable for use as (auto)stereoscopic displays. However, a barrier or a lenticular always reduces the resolution of the basic device. For example, if one wants to use a 3D monitor for high-resolution 2D output as well, the structure of the lenses or stripes will have a negative effect on the resolution. It is obvious that if the grid were to move quickly across the screen, all pixels would eventually be visible to the viewer after a very short time. The Japanese

[109]These are the places where the warp thread (in the loom the longitudinal threads, quasi the "basic grid" of the fabric) lies above the weft thread (which is woven into the warp threads at right angles to the basic grid).

[110]The spacing of the transparent lines.

[111]Whereby Sunny Ocean used not only parallax barriers but also slanted zone plates [206] and thus could produce screens with up to 64 perspective views [207].

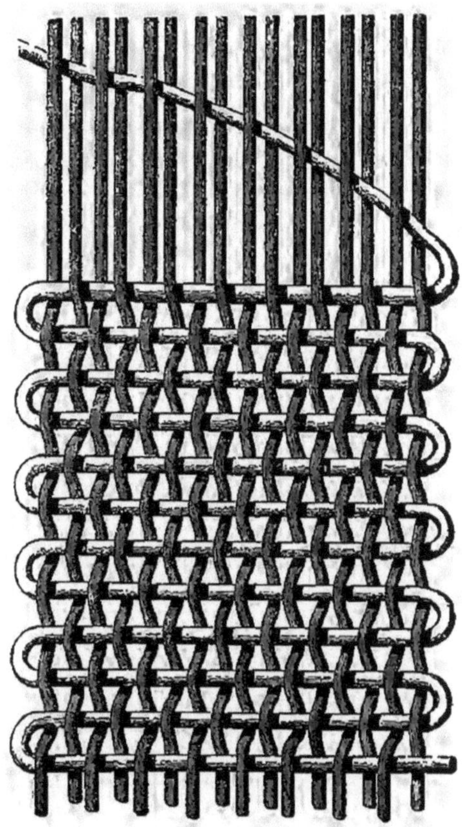

Fig. 4.37 Schéma du taffetas. (Henri Algoud (1912) from [201], p. 41)

company Sanyo[112] developed some remarkable 3D monitors at the end of the 90s of the twentieth century and for this also the bizarre idea of the mechanical barrier transport. However, the captivatingly simple idea of mechanical displacement [208] carries with it the requirement for rapid movement of the screen. By using two LC layers on top of each other, the mechanical challenges can be circumvented. The U.S. military made some effort in the 1980s to develop three-dimensional displays for aircraft and submarines.[113] Mathematician Sombrowsky improved on an idea by Holmes [209] and presented the concept in the early 1990s for a monitor in which "small transparent apertures" appeared sequentially on a

[112]Sanyo was a leading manufacturer of LCDs in the 1990s and was also strong in related development until 2004. Due to the enormous price pressure in the 2000s, Sanyo also got into trouble and tried to save the LC business through a joint venture with Epson and a focus on small-format screens. In the meantime, Sanyo has become part of Panasonic.

[113]Among others, the companies "Electronic Image Systems Inc." (Xenia, Ohio, taken over by BARCO in 1996) and Dimension Technologies (Rochester, New York) profited from this and subsequently came up with several publications and real displays.

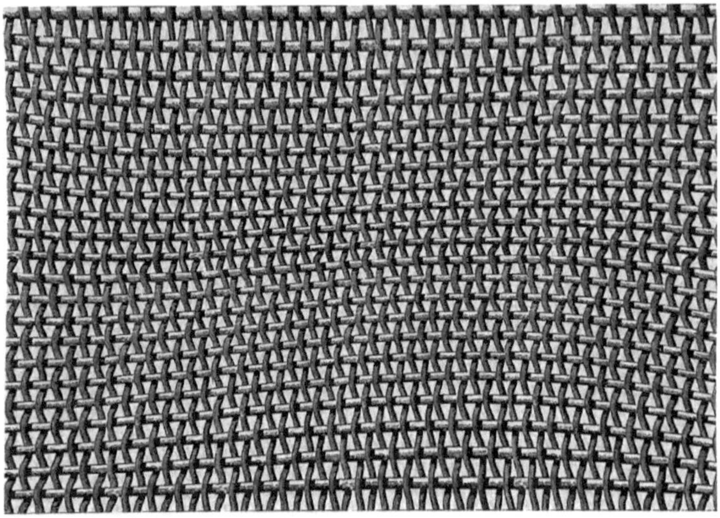

Fig. 4.38 Taffeta, lower moiré layer, Grasnick (2019)

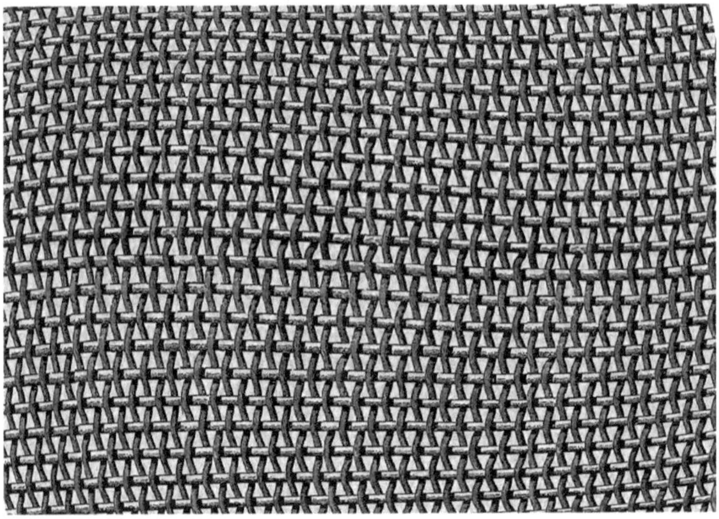

Fig. 4.39 Taffeta, upper moiré layer, Grasnick (2019)

"special glass pane surface" in front of a "special monitor" [210]. The unwieldy neologism "special glass pane surface" of course simply means a switchable barrier structure. The monitor and the barrier must switch synchronized to each other so fast that the image change is no longer perceptible. But this requires a monitor with very fast image repetition. Such a monitor was easier to obtain in the 90s as a Braun tube than such a fast LCD.

Fig. 4.40 Taffeta, two-layer moiré, Grasnick (2019)

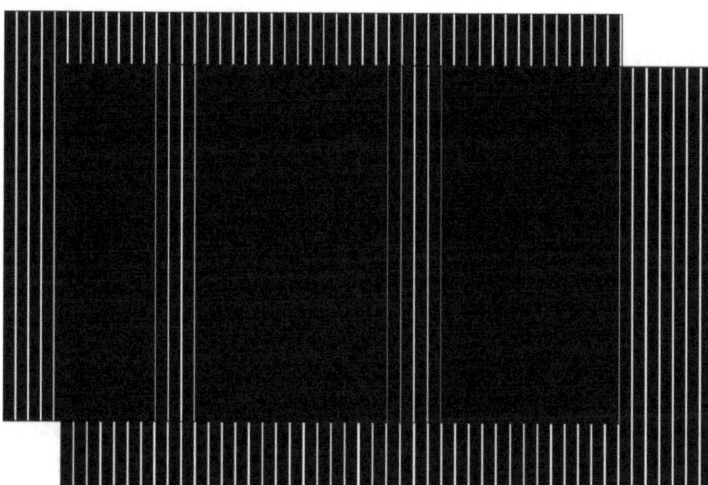

Fig. 4.41 Moiré of two parallel line grids, Grasnick (2019)

Therefore, the idea of switching off the barrier was also investigated. The American Eichenlaub used the barriers as illumination lines of the display, which served to illuminate the "light valves"[114] of the LC panel. This, as well as a barrier in front of the panel, made it possible to view certain pixels depending on the viewing direction. The screens are still

[114] "illuminating lines and light valve" [211].

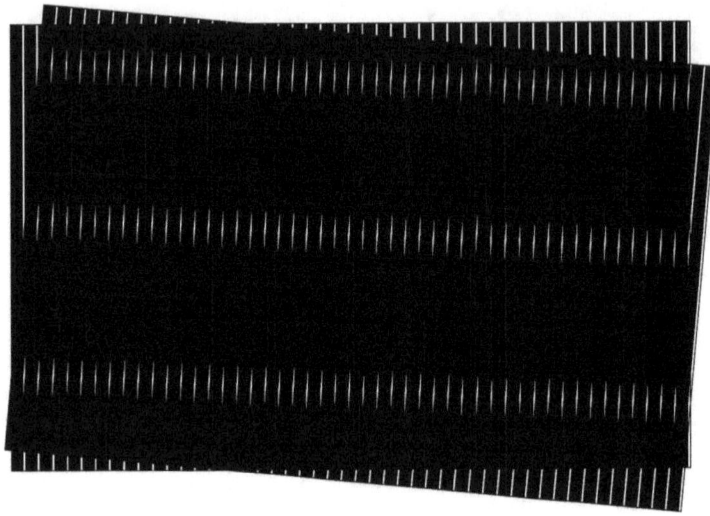

Fig. 4.42 Moiré of two twisted line grids, Grasnick (2019)

produced today.[115] It is also worth mentioning Sharp's 3D displays, which were produced in series and were also switchable between 2D and 3D modes [212]. Sharp preferred the classic 3D setup where the barrier is placed in front of the screen. As a mass-produced LCD manufacturer, Sharp could now design both elements (screen and barrier) as LC. Even with the potential production capability of the Sharp 3D display, it was not a success and the 3D monitor disappeared from the market after the pilot series was sold off. However, Sharp had definitely continued to work on the issue in the background. A Sharp barrier system was used in the 3D display of the Nintendo 3DS [213]. Nintendo itself preferred a simple switchable LC barrier behind an LC display [214]. This makes the display indistinguishable from a conventional 2D display when the barrier is switched off.

Sharp's or Nintendo's 3D displays also used pure black and white barriers (precisely formulated "transparent-opaque"). However, by using the spatial design of an LC pixel, the color of the pixels can also be used. The color filters of the subpixels divide a flat screen into separate areas formed by the primary colors red, green and blue (see Fig. 4.43).

The structure of a pixel from three single elements is the usual structure, but by no means the only one. There are exceptions here as well. Sharp, for example, uses 4 colors[116] in its Quattron technology [215] to improve color perception [216]. The Israeli company

[115] After all, the basic idea dates back to 1980.

[116] RGBY (red, green, blue, yellow).

Fig. 4.43 RGB subpixel, Grasnick (2019)

Genoa Technologies[117] researched 5-color[118] LCDs [217] and Samsung itself even tried 6 colors[119] [218].

If one puts a barrier with different color filters over any of these filter arrangements, interesting constellations result. 4D-Vision had experimented with such color filters [219]. The difference is quickly explained: The transparent surfaces of the barrier are no longer transparent for the entire visible spectrum (white), but only selectively for certain wavelength ranges. These are then quite simply color filters. In its simplest form, a transparent color filter pattern can be printed or even accomplished with a second LC panel. The advantage of such a grid is the subjectively higher resolution. A barrier can be constructed in such a way that it can be permeable over the entire surface, at least for certain colors.

Now that much has been said about the barrier, we will conclude by briefly discussing Kaplan's theory of parallax barriers [192], which has already been mentioned several times. Kaplan had already foreseen the potential use of parallax barriers for picture tubes in the early 1950s. Kaplan could not yet have known the later LCD technology, but nevertheless already pointed out an important fact. It is always necessary for the stereoscopic effect of a display that the pixels are present in a defined arrangement at a certain point in time. Even regardless of the shape of the individual picture elements,[120] all pixels must be arranged in a certain pattern. Kaplan names two indispensable requirements for a 3D display:

[117]Taken over by Samsung.

[118]RGBYC (red, green, blue, yellow, cyan).

[119]RGBYCM (red, green, blue, yellow, cyan, magenta).

[120]Kaplan suggests round, square, hexagonal or rhombic picture elements.

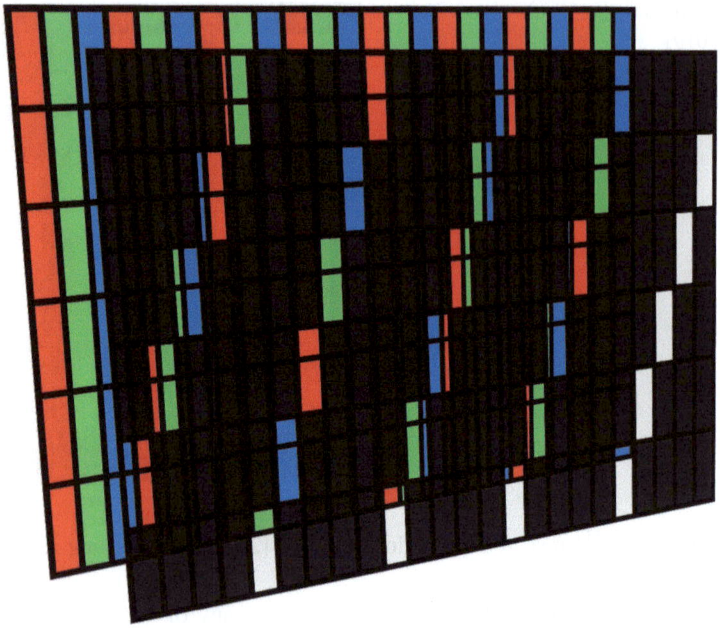

Fig. 4.44 LCD with stair barrier, Grasnick (2019)

1. All elements of a pattern must be of the same shape and size.
2. All elements of a design must be arranged in the same orientation.

Kaplan's conditions are already fulfilled with tube monitors despite the somewhat imprecise arrangement of the pixels.[121] Of course, due to the unchanging barrier restructure, the analogue television signal must be readjusted and aligned with the grid again and again. With a digital flat screen, this is no longer necessary. Each individual pixel is now clearly addressable via its columns and line allocation. As a result, 3D displays are appearing more and more frequently as flat-panel displays become more widespread. Only as an example, the already mentioned Sanyo patent [208] should be cited again, in which a switchable LC barrier was described in addition to the mechanical mask shift. If a second panel is now used as a parallax barrier, in which the color filters have been omitted, a staircase structure[122] can also be used instead of an oblique barrier. In a simplified form without the possibility of switching, 4D-Vision from Jena also had such steps on offer [222]. Kaplan had called this type "open barrier arrangement, staggered" (from [192], p. 16). Usually, the open rectangles of the barrier are matched to the size of the subpixels, so that a colored pattern presents itself to the viewer (see Fig. 4.44). However, if the

[121] Variance of image size, position or distortion.

[122] According to Sanyo terminology, "Step Barrier" [220] or "Step Structure" [221].

structures are small enough so that individual pixels can no longer be resolved, a coherent picture emerges again.

4.8 Light Field

The Light Field in the Illumination Space

To create a spatial impression from a flat image or monitor, the light emitted from the display must be emitted in different spatial directions. The light belonging to a certain perspective must therefore only be seen from that perspective.

The description of light propagation seems to be particularly useful in interior spaces when planning buildings. In 1863, the Berlin painter Magnus was concerned about how to illuminate the works of art on display in the increasing number of museums in Germany. As principles of illumination he named the "unity of light"[123]and the place "where the rays of light fall most intensely" (from [223], p. 204). In his lecture, Magnus describes the intensity of light as a function of the angle of incidence and degree of reflection of the art objects. By appropriate dimensioning of skylight, windows and room dimensions, the art objects can be illuminated evenly and brightly. Magnus's simple geometrical theory was more intensively illuminated by Mehmke on the basis of Lambert's photometry [224] and Wiener's theory of illumination ([225], from p. 390). Mehmke adopted Wiener's concept of the illumination space ([225], from p. 401) and constructed in it the illumination vector.[124] This enabled Mehmke to calculate the illuminance at any point on the wall [226]. Mehmke's diagram of an illumination space with ceiling light can certainly be regarded as a representation of a light field (see Fig. 4.45). The lines of the diagram indicate the locations of equal brightness (isoluminance).

In the 1930s, the Russian physicist Gershun, who had worked intensively on photometry,[125] proposed to describe the propagation of light in the form of a field for which he used the term "light field[126] [228]. The basic idea in Gershun's concept of the light field is a photometric view of the energy distribution of light in space. This allows a propagation of light, measurable by intensity at each individual location, to be described geometrically. The illustration below (Fig. 4.46) shows such a light field, which in this case is caused by the sun (not in the picture). The lines here illustrate what Gershun meant by a light field. The lines in the picture show the radiation density and thus describe the direction-

[123]By which Magnus means lighting from one direction.

[124]The illumination vector describes the predominant light direction.

[125]See, for example, "Selected Works on Photometry and Light Technology" [227] or "Light field", in which he also dealt with Mehmke's work, see [228], p. 159.

[126]"Svetovoye Pole" (light field), but the absolute attribution to Gershun is problematic in that the term light field was used in microscopy and photometry long before.

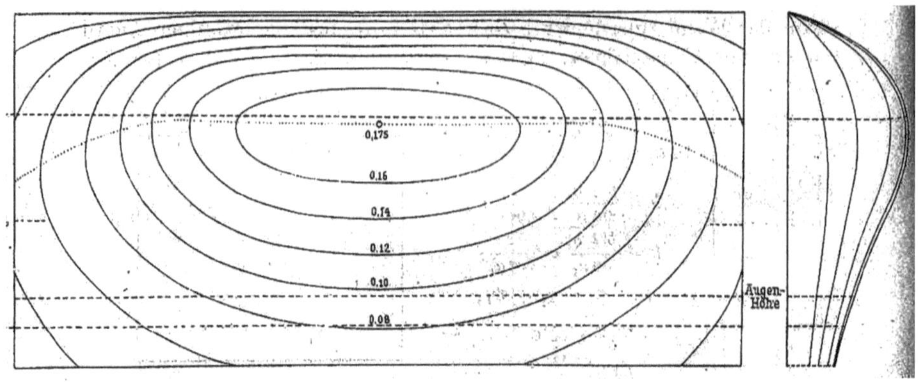

Fig. 4.45 Diagram of an illumination space. (Rudolf Mehmke (1898) from [226], p. 52)

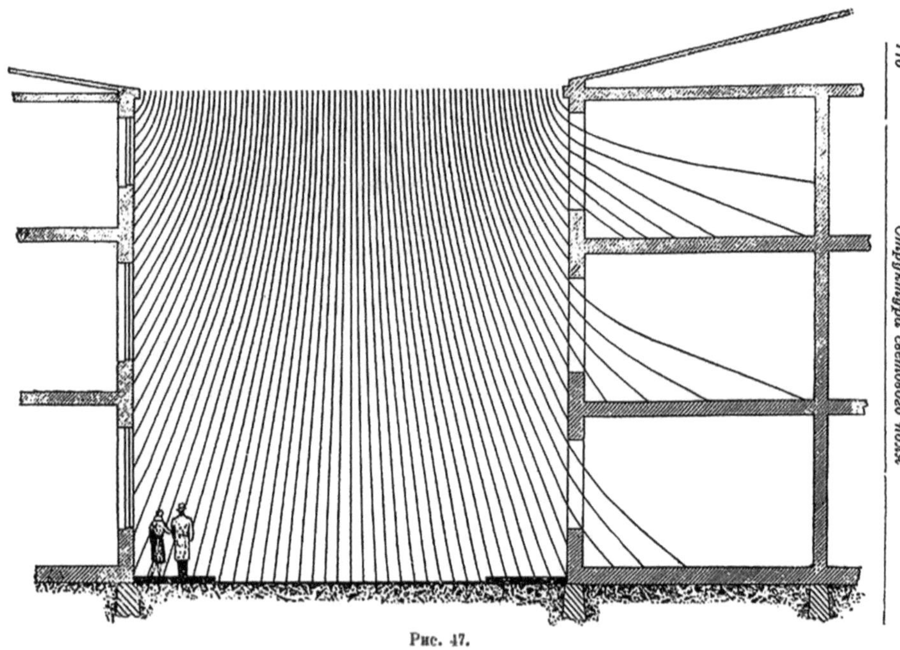

Fig. 4.46 Struktura svetovogo polya, light field structure. (Andrey Aleksandrovich Gershun (1936) from [228], p. 116)

dependent intensity of the light. The greater the distance between the lines, the lower the intensity at that location.

Gershun's light field became known outside the Soviet Union mainly with the English translation of his paper "The Light Field" [229] by Moon and Timoshenko. Moon and Eberle Spencer developed the "photic field" [230] from Mehmke's illumination vectors and Gershun's light field. The term "photic" or "photonic" field did not catch on, but the idea of light field did. Adelson and Bergen later described the light field as a plenoptic function[127] [231]. If the plenoptic function is to describe the entire distribution of light in space, all the parameters of light must be considered for this purpose. First of all, the agreement is that the light radiation is considered at a position and thus dependent on the three spatial dimensions x, y, and z. Geometrically simplified, for each point in space an addition of the vectors of all incoming light sources can be calculated, resulting in a new vector at this very point. The length of the vector indicates the respective intensity of the radiation, two solid angles[128] the direction. Already now one would have a description of the intensity in space, which depends on five parameters[129] and thus becomes five-dimensional. In a dynamic system, the lighting situation can change, adding time as an influencing variable. Since the real world is not monochromatic, the colors or simply the wavelength of the light must be taken into account, which means that a total of seven dimensions are now included in the calculation (and that only if the phase or polarization state of the light is neglected). This doesn't seem like much of a simplification. Levoy and Hanrahan, two American computer scientists, proposed an efficient calculation in the late 1990s [232].[130] Basically, they started from the definition that the radiation from a point of light is directed in a certain direction. This is in accordance with the plenoptic function of Adelson and Bergen. However, the radiant power of light along a straight line changes only if the light is blocked by something. This constraint is certainly sufficient for most situations and now allows the description by four parameters. To do this, simply place two plates in space and draw a straight line from a point on one plate through an empty space to another point on the second plate (see Fig. 4.47).

A light ray passes through a point on the first plate through another point on the second plate. With a single light ray no scene can be described yet, for this one needs many light rays. Levoy and Hanrahan call the space between the two plates, where many light rays travel from point to point, simply "light slab" [232]. This may become a little clearer if one draws several light rays in the light slab (Fig. 4.48). The sketch illustrates a model of the light field as it would appear in a camera shot, whether with real cameras or virtual cameras in computer graphics. The red plane represents the camera plane, from which now two different points (red and blue) are looking at the object (illustrated by the green focus plane). Here, only green points are shown as an example.

[127] Plenoptic—fully grasping what is visually perceptible.

[128] Horizontal angle (azimuth) and vertical angle (elevation).

[129] x, y, z, azimuth, elevation.

[130] A nice summary on the subject was given by Levoy himself in 2006 [233]. Interesting in this article is also Levoy's assumption that in 25 years most photographic cameras will be light-field cameras (2031—we will see).

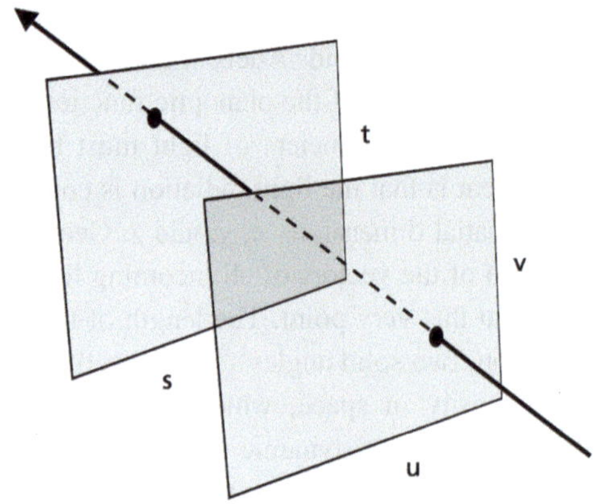

Fig. 4.47 Light beam through 2 plates, Grasnick (2019)

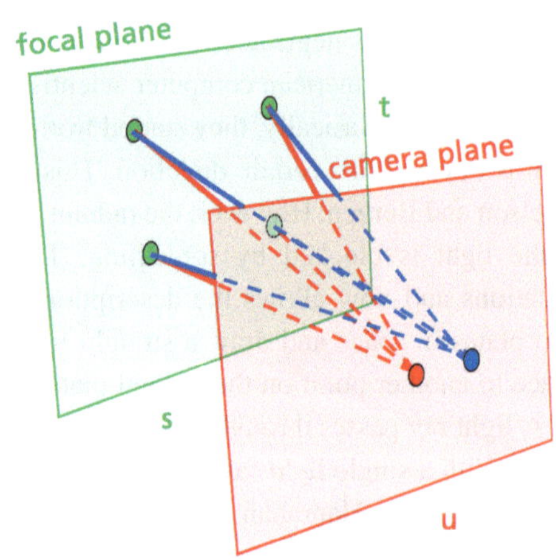

Fig. 4.48 Light slab with multiple light beams, Grasnick (2019)

The practical recording of a real scene as a light field requires a large number of different camera positions, which always involves a certain amount of effort. The recording of moving scenes also requires the synchronous recording of the scene with all cameras simultaneously, which increases the effort even more. At Stanford University, the "Stanford Multiple Camera Array" was set up under Levoy's direction [234]. For this purpose, 100 inexpensive cameras were set up in a two-dimensional array and synchronized, which not only resulted in high data rates, but also in a considerable setup.[131] This setup, which is only transportable to a limited extent, was converted into a highly mobile and practicable camera based on an idea by Ren Ng (who has already been mentioned) [165]. Here, Ng basically uses Lippmann's integral photography [236] in

[131] With a side length of 30 mm for each camera board (see [235], p. 29), the system probably had a minimum size of 40 × 40 cm, and 4 computers were needed to operate 100 cameras.

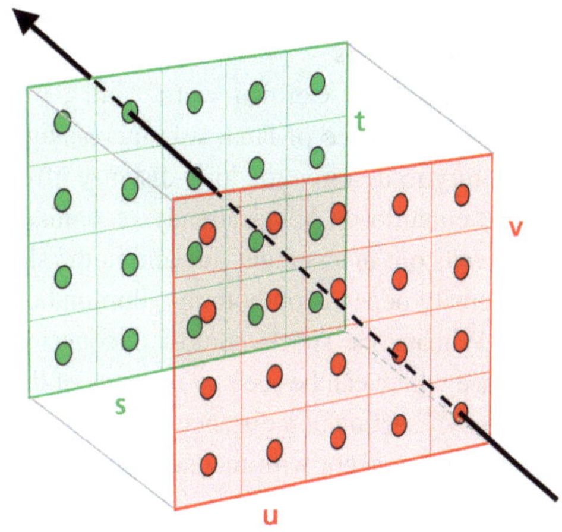

Fig. 4.49 Lumigraph. (By Gortler, Grasnick 2019)

combination with Ives' large parallax lens [189] and accommodates these components in a digital camera.

At the same time as Levoy's "light slab", Gortler presented his "lumigraph". The basic idea is more or less the same as in Levoy's and Hanrahan's work. A nice model in Gortler's work, however, is the cube that surrounds an object to be viewed. If you always look at two opposite faces of it, you get Levoy's "light slab"[132] again. Here Gortler makes an exceedingly practicable suggestion. He divides both surfaces into grids and thus obtains a clear number of light rays to be calculated. Figure 4.49 shows the lumigraph cube. On two opposite sides grid points are drawn and a single ray connecting two defined points can be described.

In principle, the corresponding rays can now be drawn from any point outside the lumigraph through the grid points of two surfaces in each case. It is no coincidence that this is reminiscent of Bettini's construction of the anamorphoses (see Fig. 2.55), for more or less the same process takes place here. From a certain camera or eye point, geometrically always a slightly different perspective is seen. Every position in space results in a slightly different perspective. Once a light field has been recorded, countless perspectives can be calculated from any other position. These are then two-dimensional sections from the four-dimensional light field.

[132]With the restriction that Gortler's surfaces are parallel to each other, while Levoy's surfaces can be positioned anywhere in space.

Physical Fields

The Old High German word "fëld", which originally denoted an open, flat and unwooded piece of land, was increasingly used from the middle of the nineteenth century to describe the long-distance effects of forces.

Leonardo da Vinci already recognised that the light emitted by a light source spreads out in a straight line and in the shape of a "pyramid".[133] The perceived light intensity depends on both the illumination and the position of the observer.

Wheatstone speaks almost unconsciously of a "field of vision" and means the space which can be seen by the eyes of the observer [238]. A direct comparison with forces[134] acting at a distance is however already drawn by Faraday [240], who links the field concept with magnetism [241]. Thus, the idea of the electromagnetic field was born.

Today, a field is understood to be the distribution of a physical quantity in a given space.

A light field can not only be recorded, but also reproduced. The best known device for this is Lippmann's integral photography. As is well known, the images were quite small, so that the spatial impression was limited to the size of a postcard. The conversion to larger displays requires a high resolution, which became possible with the further development of LC technology. Based on this, the possibilities of applying integral photography as a 3D display were investigated (see e.g. [242, 243]) and also some prototypes were produced (see e.g. [244 or 245]). The problem with such technologies lies in the still limited resolution of available display systems. However, by foregoing the use of vertical parallax and interleaving a larger number of perspectives only horizontally, it may be possible to exploit the advantages of integral photography[135] in larger displays. Sunny Ocean Studios had experimented with 64 perspectives for this purpose, Tokyo University with 128 (now called "super multi-view", see [244]), the author with up to several million (called "hyperview" see [245–247]). The concept of hyperview allows to assign to each pixel of an autostereoscopic display (whether lenticular or barrier) a section from a slightly different perspective.

[133]"Light and shadow do their service in the form of pyramids" (from [237], p. 302).

[134]Whereby Faraday here and elsewhere speaks of "magnetic line of forces" [239].

[135]In addition to the large number of individual views and the resulting free viewing point, the absence of the stereoscopic/autostereoscopic accommodation-convergence conflict is particularly positive. One can actually focus on the aerial images of the 3D objects in integral photography (and not only on the screen, as in stereoscopy).

Literature

1. von Helmholtz H. Populäre wissenschaftliche Vorträge. Braunschweig: Friedrich Vieweg & Sohn; 1871.
2. Rosenbach O. Ueber die monokulare Vorherrschaft beim binokularen Sehen. Münchener Med Wochenzeitschrift. 1903;30:1290–2.
3. Buxton CE, Crosland HR. The concept eye-preference. Am J Psychol. 1937;49:458.
4. Bourassa DC, McManus IC, Bryden MP. Handedness and eye-dominance: a meta-analysis of their relationship. Laterality. 1996;1:5–34.
5. Li J, Lam CSY, Yu M, Hess RF, Chan LYL, Maehara G, et al. Quantifying sensory eye dominance in the normal visual system: a new technique and insights into variation across traditional tests. Invest Opthalmol Visual Sci. 2010;51:6875.
6. Carey DP. Vision research: losing sight of eye dominance. Curr Biol. 2001;11:R828–30.
7. Goersch H. Sensorische Abweichungen vom idealen Binokularsehen. Neues Optikerjournal. 1979;2:7–13.
8. Kundert G, Marty I. Welchen Einfluss hat das Führungsauge auf die Abweichung am Valenztest? DOZ. 2005;6:44–8.
9. Stollenwerk G. Möglichkeiten und Grenzen bei der Anwendung des neuen reduzierten Valenztestes. Neues Optikerjournal. 1999;9:26–31.
10. Souman JL, Frissen I, Sreenivasa MN, Ernst MO. Walking straight into circles. Curr Biol. 2009;19:1538–42.
11. Scheiner C. Oculus hoc est fundamentum opticum. Daniel Agricola: Innsbruck; 1619.
12. von Helmholtz H. Handbuch der physiologischen Optik. Leipzig: Leopold Voss; 1867.
13. Bach L. Pupillenlehre. Berlin: S. Karger; 1908.
14. Marg E, Morgan MW Jr. The pupillary near reflex. The relation of pupillary diameter of accommodation and the various components of convergence. Optom Vis Sci. 1949;26:183–98.
15. Zuber BL. The control of eye movements. In: Bach-y-Rita P, Collins CC, Smith-Kettlewell Institute of Visual Sciences, Stockton, Calif, editors. The control of eye movements. New York: Academic; 1971. p. 447–71.
16. Westheimer G, Blair SM. The parasympathetic pathways to internal eye muscles. Investig Ophthalmol. 1973;12:193–7.
17. Schultheis H. Die Pupillengöße als Index zur Online-Erfassung der kognitiven Belastung. 11 GI-Workshop "Adaptivität und Benutzermodellierung in interaktiven Softwaresystemen". 2003.
18. Winn B, Whitaker D, Elliott DB, Phillips NJ. Factors affecting light-adapted pupil size in normal human subjects. Invest Ophthalmol Vis Sci. 1994;35:1132–7.
19. Curcio CA, Sloan KR, Kalina RE, Hendrickson AE. Human photoreceptor topography. J Comp Neurol. 1990;292:497–523.
20. Diels H. Die Fragmente der Vorsokratiker. 3. Aufl. Berlin: Weidmannsche Buchhandlung; 1912.
21. Aristoteles. Problemata Physica. 4., gegenüber der 2., durchges., unveränd. Aufl. Berlin: Akademie-Verlag; 1991.
22. Wilde E. Geschichte der Optik, vom Ursprunge dieser Wissenschaft bis auf die gegenwärtige Zeit. Berlin: Rücker & Püchler; 1838.
23. Siebert H. Die ptolemäische Optik in Spätantike und byzantinischer Zeit: historiographische Dekonstruktion, textliche Neuerschliessung, Rekontextualisierung. Stuttgart: Steiner; 2014.
24. Alhazen W. Opticae thesaurus. Risner F, Editor. Basel: Episcopios; 1572.

25. Kepler J. Dioptrik oder Schilderung der Folgen, die sich aus der unlängst gemachten Erfindung der Fernrohre für das Sehen und die sichtbaren Gegenstände ergeben. Leipzig: Wilhelm Engelmann; 1904.
26. Descartes R. Discours de la methode pour bien conduire sa raison, & chercher la verité dans les sciences: plus la dioptrique et les meteores, qui sont des essais de cette methode. Paris: Theodore Girard; 1657.
27. Smith R. A compleat system of opticks. Cambridge: Cornelius Crownfield; 1738.
28. Smith R. Vollständiger Lehrbegriff der Optik. Richtersche Buchhandlung: Altenburg; 1755.
29. Dornblüth O. Medizin: Horopter [Internet]. Klinisches Wörterbuch. 1927; http://www.textlog.de/14790.html. Accessed on 18. Sept. 2016
30. Aguilonius F. Opticorum Libri Sex. Apud Viduam et Filios Moreti: Antwerpen; 1613.
31. von Baer KE. Vorlesungen über Anthropologie. Gebrüder Bornträger: Königsberg; 1924.
32. Vieth GUA. Versuch einer Encyklopädie der Leibesübungen. Berlin: Carl Ludwig Hartmann; 1795.
33. Vieth GUA. Ueber die Richtung der Augen. Ann Phys. 1818;58:233–53.
34. Müller J. Von dem natürlichen Doppeltsehen. Zur vergleichenden Physiologie des Gesichtssinnes des Menschen und der Thiere, nebst einem Versuch über die Bewegungen der Augen und über den menschlichen Blick. Leipzig: Cnobloch; 1826. pp. 165–204.
35. Hering E. Vom binokularen Tiefensehen. Kritik einer Abhandlung von Helmholtz über den Horopter. Beitr Physiol. 5. Heft/1864;347–358.
36. Hering E. Der empirische Horopter. Beitr Physiol. 1863;3:200–24.
37. Hillebrand F. Die Stabilität der Raumwerte auf der Netzhaut. Z Psychol Physiol Sinnesorgane. 1893;5:1–60.
38. Panum PL. Physiologische Untersuchungen über das Sehen mit zwei Augen. Schwersche Buchhandlung: Kiel; 1858.
39. Richards W. Stereopsis and stereoblindness. Exp Brain Res. 1970;10:380–8.
40. Wheatstone C. Contributions to the physiology of vision. –Part the first. On some remarkable, and hitherto unobserved, phenomena of binocular vision. Philos Trans R Soc Lond. 1838;128: 371–94.
41. Dodgson NA. Variation and extrema of human interpupillary distance. In: Woods AJ, Merritt JO, Benton SA, Bolas MT, Editor. Proc SPIE. 2004. pp. 36–46.
42. Mach E. Untersuchungen über den Zeitsinn des Ohres. In: Moleschott J, editor. Untersuchungen zur Naturlehre der Menschen und der Thiere. Gießen: Emil Roth; 1870. p. 181–200.
43. Edison TA. Caveat 110: Kinetoskop [PT031AAA1; TAEM 113:238] [Internet]. Orange. 1888; http://edison.rutgers.edu/images/ei/ei0235.jpg
44. Wittmann M, Pöppel E. Hirnzeit: Wie das Gehirn die Zeit macht. Kunstforum Int. 2000;151:85.
45. Pöppel E. Lost in time: a historical frame, elementary processing units and the 3-second window. Acta Neurobiol Exp. 2004;64:295–302.
46. Carmack J. Latency mitigation strategies [Internet]. 2014. http://archive.is/udOC9#selection-595.0-595.29. Accessed on 19. Sept. 2016.
47. Howarth PA, Costello PJ. The occurrence of virtual simulation sickness symptoms when an HMD was used as a personal viewing system. Displays. 1997;18:107–16.
48. DiZio P, Lackner JR. Motion sickness side effects and aftereffects of immersive virtual environments created with helmet-mounted visual displays. The Capability of Virtual Reality to Meet Military Requirements. Orlando, Florida verlegt von NATO RTO HFM; 2000;1. pp. 11/1–11/3.
49. Carbon C-C, Hesslinger VM. Da Vinci's Mona Lisa entering the next dimension. Perception. 2013;42:887–93.

50. Grasnick A. Hyperview: Maximalwert binokularer Perspektiven im Bildraster autostereoskopischer Raumbildsysteme. Hagen: FernUniversität Hagen; 2017.
51. Da Vinci L, Della Bella S. Trattato della pittura, Bd. 1792. Florenz: Pagani & Grazioli; 1792.
52. Talbot WHF. Photographic Pictures. Patent, GB 8842; 1841.
53. Dingler JG. Editor. XCVI Talbot's Verfahren die photogenischen Gemälde darzustellen Polytech J. 1839;71:468–77.
54. Talbot WHF. LXXVI. Facts relating to optical science. No. IV. The London, Edinburgh, and Dublin Philos Mag J Sci. 1836;IX:401–7.
55. Daguerre LJM. Geschichte und Beschreibung des Verfahrens der Daguerreotypie und des Dioramas. A. Bielefeld: Karlsruhe; 1839.
56. Niépce JN. Die Heliographie. Das Daguerreotyp und das Diorama, oder genaue und authentische Beschreibung meines Verfahrens und meiner Apparate zu Fixirung der Bilder der Camera obscura und der von mir bei dem Diorama angewendeten Art und Weise der Malerei und Beleuchtung. Stuttgart: Metzlersche Buchhandlung; 1839.
57. Wheatstone C. The Bakerian lecture.— Contributions to the physiology of vision.— Part the second. On some remarkable, and hitherto unobserved Phenomena of Binocular Vision (continued). Philos Trans R Soc London. London: Richard Taylor and William Francis; 1852. pp. 1–17.
58. Moser L. Ueber das Auge. Mechanik Allgemeine Gesetze der Wellenbewegung Literatur des Magnetismus und der Elektricität Ueber das Auge. Berlin: Veit & Comp; 1844.
59. Stoeterogge BL. Recueil von allerhand Collectaneis und Historien, auch Moral-Curieux-Critic- und lustigen Satyrischen Einfällen Zu Entretenirung einer Galanten Conversation. Hannover;1721.
60. Kohlhans JC. Neu-erfundene mathematische und optische Curiositäten. Leipzig: Friederich Lankisch; 1677.
61. Brewster D. An account for a new stereoscope. Report of the nineteenth meeting of the british association for the advancement of science, section mathematics and physics, notices and abstracts. London: John Murray; 1849. pp. 6–7.
62. Brewster D. The Stereoscope: its history, theory, and construction with its application to the fine and useful arts and to education. London: John Murray; 1856.
63. Maxwell JC. The scientific papers of James Clerk Maxwell. Niven WD, Editor. New York: Cambridge University Press & Dover Publications; 1890.
64. Koschnitzke C, Mehnert R, Quick P. Faszinierende Natur, dreidimensional. Stuttgart: DRW-Verlag; 1983.
65. Bernier RV. Three-dimensional motion picture applications. J Soc Motion Picture Eng. 1951;56:599–612.
66. Starkman D, Dennis J. Another chance for over/under viewing. Stereo World. 1992;19:32–3.
67. Rollmann W. Notiz zur Stereoskopie Ann Phys Chem. 1853;165:350–1.
68. Rollmann W. Zwei neue stereoskopische Methoden. Ann Phys Chem. 1853;166:186–7.
69. d'Almeida J-C. Nouvel appareil stéréoscopique. Comptes rendus hebdomadaires des séances de l'Académie des sciences. Paris: Bachelier; 1858.
70. Davanne M. RAPPORT au nom du Comité des Constructions et de Beaux-Arts, sur les IMAGES DITES ANAGLYPHES. Bulletin de la Société d'Encouragement pour l'Industrie Nationale. Paris; 1894. p. 475.
71. Jorke H, Fritz M. Stereo projection using interference filters. In: Woods AJ, Dodgson NA, Merritt JO, Bolas MT, McDowall IE, Editor. 2006. p. 60550G. http://proceedings.spiedigitallibrary.org/proceeding.aspx?doi=10.1117/12.650348. Accessed on 11. Aug. 2016.
72. von Frisch K. Die Polarisation des Himmelslichtes als orientierender Faktor bei den Tänzen der Bienen. Experientia. 1949;5:142–8.

73. Hegedüs R, Åkesson S, Wehner R, Horváth G. Could Vikings have navigated under foggy and cloudy conditions by skylight polarization? On the atmospheric optical prerequisites of polarimetric Viking navigation under foggy and cloudy skies. Proc R Soc A Math Phys Eng Sci. 2007;463:1081–95.
74. Bartholinus E. Erasmi Bartholini Experimenta crystalli islandici disdiaclastici: quibus mira et insolita refractio detegitur. Danielis Paulli: Hafinae (Kopenhagen); 1670.
75. Huygens C. Traité de la lumière. Pierre Vander: Leyden; 1690.
76. Malus M. Sur une propriété de la lumière réfléchie. Mémoires de physique et de chimie. Paris: Bernard; 1809.
77. Malus M. Théorie de la double réfraction de la lumière. Paris: Baudouin; 1810.
78. Anderton J. Method by which pictures projected upon screens by magic lanterns are seen in relief. Patent, US 542,321. 1895.
79. Herapath WB. On the optical properties of a newly-discovered salt of quinine, which crystalline substance possesses the power of polarizing a ray of light, like tourmaline, and at certain angles of rotation of depolarizing it, like selenite. Philos Mag Ser. 1852;4(3):161–73.
80. Herapath WB. On the manufacture of large available crystals of sulphate of iodo-quinine (herapathite) for optical purposes as artificial tourmalines. Philos Mag Ser. 1853;4(6):346–51.
81. Bernauer F. Polarisationsvorrichtung. Patent, DE 547 429. 1932.
82. Wempe J. Eigenschaften und astronomische Anwendungsmöglichkeiten von Polarisationsfiltern. Astron Nachr. 1939;269:331–6.
83. Giesen R. Drittes Reich: Schon zu Nazi-Zeiten wurde an 3D-Filmen gearbeitet – DIE WELT [Internet]. 2011. http://www.welt.de/kultur/history/article12578466/Schon-zu-Nazi-Zeiten-wurde-an-3D-Filmen-gearbeitet.html. Accessed on 14. Aug. 2016.
84. Hagemann PA. Der 3D Film. Stiftung Deutsche Kinemathek, Editor. München, Monika Nüchtern; 1980.
85. Spöhrer M. Die ästhetisch-narrativen Dimensionen des 3D-Films Neue Perspektiven der Stereoskopie. Springer VS: Wiesbaden; 2016.
86. Land EH, Friedman JS. Polarzing refracting bodies. Patent, US 1,918,848. 1933.
87. Kaempffert W. Science: the evolution of the scientific museum; An event in New York points to the trend toward the dynamic exhibit. New York Times. New York. 16. Februar 1936.
88. Norling JA. The stereoscopic art (a reprint). J Soc Motion Pict Eng. 1953;60:268–308.
89. Norling JA. Three-dimensional motion pictures. J Soc Motion Pict Eng. 1939;33:612–34.
90. Wesemann W. Sehlust und Sehfrust – Sehprobleme bei 3D-Kino und 3D-Fernsehen. Teil I SMPTE J. 1994;103:522.
91. Land EH. Polarizing optical system. Patent, US 2,099,694. 1937.
92. Walworth VK. History of polarized image stereoscopic display. In: Proc. SPIE 8648 Woods AJ, Holliman NS, Favalora GE, Editor. 2013. p. 864804.
93. Lipton L, Berman A, Meyer LD. Achromatic liquid crystal shutter for stereoscopic and other applications. Patent, US 4,884,876. 1989.
94. Lipton L, Berman A. Push-pull liquid crystal modulator for electronic stereoscopic display. In: Robbins WE, Editor. 1988. pp. 31–6. http://proceedings.spiedigitallibrary.org/proceeding.aspx?articleid=1251451. Accessed on 20. Aug. 2016.
95. Faris SM. Electro-optical display system for visually displaying polarized spatially multiplexed images of 3-D objects for use in stereoscopically viewing the same with high image quality and resolution. Patent, US 5,537,144. 1996.
96. Victor Company of Japan J. The Xpol Circular Polarizing System used by JVC enables natural 3D pictures [Internet]. JVC's 3D technology latest 3D technologies Xpol® Circular Polarizing System. 2009. http://pro.jvc.com/pro/microsite/3d/technology/xpol.html. Accessed on 21. Aug. 2016.

97. LG Display. Film Patterned Retarder (FPR) 3D Fact Sheet [Internet]. LG Display 2013. http://lgdnewsroom.com/wp-content/uploads/2013/04/FPR_3D_Fact_Sheet.pdf.
98. Lim YJ, Yu JH, Song KH, Lee M-H, Ren H, Mun B-J, et al. Film patterned retarder for stereoscopic three-dimensional display using ink-jet printing method. Opt Express. 2014;22: 22661.
99. Walker G. Display week 2014 review: touch technology. Inform Display. 2014;30:40–5.
100. Woods AJ. Crosstalk in stereoscopic displays: a review. J Electron Imaging. 2012;21:040902.
101. Victor Company of Japan J. DT-3D24G1 DT-3D24G1Bedienungsanleitung. JVC. 2011.
102. Hammond L. Stereoscopic motion picture. Patent, US. 1922;1:435,520.
103. Puchalski S. Remembering The Selwyn [Internet]. Shock Cinema Magazine 2007. http://www.shockcinemamagazine.com/selwyn.html. Accessed on 20. Aug. 2016.
104. New York Tribune. Teleview will bring stereoscopic movie to the Selwyn Theater. New York Tribune. New York. 17. Dec. 1922;8.
105. Lipton L. Large-screen electro-stereoscopic displays. In: Proc. SPIE 1255 Bleha Jr. WP, Editor. 1990. pp. 108–13.
106. ELSA AG. Die neue 3D-Brille ELSA 3D REVELATOR ist der Schlüssel zur dritten Dimension | Pressemitteilung [Internet]. 1999. http://presseservice.pressrelations.de/pressemitteilung/die-neue-3dbrille-elsa-3d-revelator-ist-der-schluessel-zur-dritten-dimension-7853.html. Accessed on 21. Aug. 2016.
107. XPAND. XPAND 3D Cinema System – XPAND [Internet]. 2016. http://www.xpand.me/products/3d-cinema-system/. Accessed on 21. Aug. 2016.
108. Samsung. Samsung TV-Zubehör: Batteriebetriebene 3D-Active-Shutter-Brille SSG-P1002 [Internet]. 2016. http://www.samsung.com/de/consumer/tv-av/accessories/tv-accessories/SSG-P51002/XC. Accessed on 21. Aug. 2016.
109. Wolf M. Katalog von 1053 stärker bewegten Fixsternen. Veroeffentlichungen der Badischen Sternwarte zu Heidelberg. 1919;7
110. Pulfrich C. Die Stereoskopie im Dienste der isochromen und heterochromen Photometrie. Naturwissenschaften. 1922;10:553–64.
111. Beard TD. Low differential 3-D viewer glasses and method with spectral transmission characteristics to control relative intensities. Patent, US 4,836,647. 1989.
112. Free J. 3-D TV. Pop Sci Juni 1988;232.
113. Der Spiegel. Wiederkehr der Pappbrille. Der Spiegel. 12. Oct. 1998.
114. Lipton L. Foundations of the stereoscopic cinema. New York: Van Nostrand Reinhold; 1982.
115. Holliman N. Three-dimensional display systems. Handbook of optoelectronics. New York: Taylor & Francis; 2006.
116. Dove HW. Beschreibung mehrerer Prismenstereoskope und eines einfachen Spiegelstereoskops. In: Poggendorff JC, editor. Annalen der Physik und Chemie. Leipzig: Johann Ambrosius Barth; 1851. p. 183–9.
117. Julesz B. Binocular depth perception of computer-generated pattern. Bell Syst Tech J. 1960;39: 1125–62.
118. Tyler CW, Clarke MB. Autostereogram. In: Proc. SPIE 1256 Merritt JO, Fisher SS, Editor. Santa Clara; 1990. p. 182.
119. Baccei T. Das magische Auge: dreidimensionale Illusionsbilder. Ars Edition: München; 1994.
120. Newton I. A letter of Mr. Isaac Newton, . . . containing his new theory about light and colors. Philos Trans R Soc Lond. 1672;80:3075–87.
121. Brewster D. Notice of a chromatic stereoscope. In: Brewster D, Taylor R, Kane R, Francis W, editors. The London, Edinburgh and Dublin Philosophical Magazine and Journal of Science. London: Taylor & Francis; 1852.
122. Steenblik RA. Stereoscopic process and apparatus. Patent, US 4,597,634. 1986.

123. Steenblik RA. The chromostereoscopic process: a novel single image stereoscopic. In: Proc. SPIE 0761 McAllister DF, Robbins WE, Editor. 1987. pp. 27–34.
124. Chromatek. Chromadepth3D – a brand-new 3D experience [Internet]. 2016. http://chromatek.com/. Accessed on 13. Aug. 2016.
125. Schemali L, Eisemann E. ChromoStereoscopic rendering for trichromatic displays. ACM Press. 2014:57–62.
126. Bailey M, Clark D. Using ChromaDepth to obtain inexpensive single-image stereovision for scientific visualization. J Graph Tools. 1998;3:1–9.
127. Buerger JE. French daguerreotypes. Chicago: University of Chicago Press; 1989.
128. Claudet A. On the stereomonoscope, a new instrument by which an apparently single picture produces the stereoscopic illusion. Proc R Soc Lond. 1857;9:194–6.
129. Schlömilch O, Witzschel B, Cantor M, Editor. Das Stereomonoskop von Claudet Z Math Phys 1859;4:169–170.
130. von Rohr M. Die binokularen Instrumente. Berlin: Julius Springer; 1907.
131. von Rohr M, Englisch E. Der Verant, ein neues Instrument, um Photogramme vom richtigen Standpunkt zu betrachten. Z Wiss Photographie Photophys Photochem. 1904;2:188–9.
132. von Rohr M. Die Brille als optisches Instrument. Leipzig: Wilhelm Engelmann; 1911.
133. Carl Zeiss Jena. Der Doppelverant. Beschreibung und Gebrauchsanweisung. Berhanrd Vopelius Jena; 1902.
134. von Rohr M. Reflecting instruments annulling the perception of depth in binocular vision. Patent, GB 916. 1907.
135. Ames A. The illusion of depth from single pictures. J Opt Soc Am. 1925;10:137.
136. Vishwanath D. Induction of monocular stereopsis by altering focus distance: a test of Ames's hypothesis. i-Perception. 2016;7:204166951664323.
137. Ittelson WH. The Ames demonstrations in perception. Princeton University Press: Princeton; 1952.
138. Euklid, Dee J. The elements of geometrie of the most auncient philosopher Euclide of Megara. Faithfully (now first) translated into the Englishe toung, by H. Billingsley, citizen of London. Whereunto are annexed certaine scholies, annotations, and inuentions, of the best mathematiciens, both of time past, and in this our age. With a very fruitfull præface made by M. I. Dee, specifying the chiefe mathematicall scie[n]ces, what they are, and wherunto commodious: where, also, are disclosed certaine new secrets mathematicall and mechanicall, vntill these our daies, greatly missed Elements. London: John Daye; 1570.
139. Homer. Odyssee. dt. Voss 1781, Nachdruck 1881. Stuttgart: Cotta'sche Buchhandlung; ca. 8. Jhd v. Chr.
140. Koenderink J. Wijntjes M, van Doorn A. Zograscopic viewing i-Perception. 2013;4:192–206.
141. Ritschel T, Templin K, Myszkowski K, Seidel H-P. Virtual passepartouts. International Symposium on Non-Photorealistic Animation and Rendering. 2012;7 pages.
142. Ell JJ, Gresty MA. Uniocular Pulfrich phenomenon: an abnormality of visual perception. Br J Ophthalmol. 1982;66:610–3.
143. Grasnick A. 3D ohne 3D-Brille: Handbuch der Autostereoskopie [Internet]. Springer; 2016. http://www.springer.com/de/book/9783642305092#aboutBook
144. Symons W. Improved movable devices for producing optical illusion changes. Patent, GB 5759. March 14, 1896.
145. Symons W. Improved movable devices for producing optical illusion changes (Transparencies). Patent, GB 21,480. Nov 20, 1897.
146. The motograph moving picture book. London: Bliss, Sands & Co.; 1898.
147. Berthier A. Images stéréoscopiques de grand format. In: Moigno FNM, editor. Le Cosmos: Revue des Sciences et de les leurs applications. Paris: Le Cosmos; 1896. p. 227–33.

148. Ives FE. Kromskop color photography. London: The Photochromoskop Syndicate; 1898.
149. Ives FE. A novel stereogram. J Frankl Inst. 1902;153:51–2.
150. Ives FE. Parallax Stereogram and Process of making same. Patent, US 725,567. 1903.
151. Estanave E-P. Relief à vision directe et images changeantes par plaque autostéréoscopique. Extrait des "Comptes rendus du Congrès des sociétés savantes". Paris: Imprimerie Nationale; 1911.
152. Estanave E. Dispositif de stéréophotographie et de stéréoscopie à l'aide des rèsaux. Patent, FR 392.871. 1906.
153. Estanave E-P. Le relief stéréoscopique en projection par les réseaux lignés. J Phys Théor Appl. 1908;7:878–80.
154. Timby K. 3D and animated lenticular photography: between Utopia and entertainment. Berlin: De Gruyter; 2015.
155. Gaumont L. Un laboratoire des premières fois [Internet]. Les collections de la Société Française de Photographie 2012. http://laboratoiredespremieresfois.tumblr.com/post/30879706814/eug%C3%A8ne-estanave-et-les-images-changeantes-relief. Accessed on 28. Aug. 2016.
156. Lippmann G. Épreuves réversibles donnant la sensation du relief. J Phys Théor Appl. 1908;7:821–5.
157. Scientific American, Editor. Integral photography- a new discovery by Professor Lippmann. Sci Am. 1911;164.
158. Fujifilm. FUJICHROME Velvia 50 | Fujifilm Deutschland [Internet]. Fujifilm. 2016. https://www.fujifilm.eu/de/produkte/analoge-fotografie/p/fujichrome-velvia-50. Accessed on 10. Sept. 2016.
159. THORLABS. Holography plates and supplies [Internet]. https://www.thorlabs.com/newgrouppage9.cfm?objectgroup_id=6716. Accessed on 9. Juni 2019.
160. Kellerbauer A. Farbbilder aus gefrorenem Licht. Interferentielle Fotografie Phys Unserer Zeit. 2010;41:16–22.
161. Lavédrine B. Photographs of the past: process and preservation. Getty Conservation Institute: Los Angeles; 2009.
162. Sokolov PP. Avtostereoskopiya i integral'naya fotografiya po pr. Lipmanu/Avramova E, Übersetzer. (Autostereoscopy and integral photography by professor Lippman's Method) Reetz D, Editor [Internet]. Z Ges Naturforscher. 1911. http://futurepicture.org/PDF/P_P_Sokolov_1911_Autostereoscopy_and_Integral_Photography_by_Professor_Lippmans_Method.pdf. Accessed on 10. September 2016.
163. Ives HE. Optical properties of a lippmann lenticulated sheet. J Opt Soc Am. 1931;21:171.
164. Ives HE. Optical device. Patent, US 2,174,003. 1935.
165. Ng R, Levoy M, Brédif M, Duval G, Horowitz M, Hanrahan P. Light field photography with a hand-held plenoptic camera. Computer Science Technical Report CSTR. 2005;2:1–11.
166. Winnek Coffey DF. Apparatus for making a composite stereograph. Patent, US 2,063,985. 1935.
167. Valius NA. Растровая оптика (Rasteroptik). Moskau: Allunionsinstitut für Fotokinotechnik (NIKFI); 1949.
168. De Montebello RL. Integral photography. Patent, US 3,503,315. 1966.
169. Dudnikov YA. Autostereoscopy and integral photography. Soviet J Opt Technol. 1970;37:422–3.
170. Dudnikov YA, Rozhkov BK. Растрове систем для получения объемнх изображений (Rastersysteme für volumetrische Bilder) [Internet]. Maschinenbau, Niederlassung Leningrad. 1986. https://books.google.de/books?id=YO2jxgEACAAJ.
171. Dudley LP. Stereoscopy in the telekinema and in the future. British Kinematogr. 1951;18:172.
172. Technical notes, 3-D television. Radio, TV and recording technician-engineer. 1956;5:14–5.

173. du Hauron LAD. Les Couleurs en Photographie – Solution du Probleme. Paris: A. Marion; 1869.
174. Joly J. On a method of photography in natural colours. Scientific Transactions of the Royal Dublin Society. Dublin: Royal Dublin Society; 1898. 127–139°S.
175. Eder JM. Geschichter der Photographie. 4. Aufl. Halle: Wilhelm Knapp; 1932.
176. Ahriman. Von der zukünftigen Photographie. Photographisches Archiv. Düsseldorf: Liesegang; 1896. pp. 249–251.
177. Berthon R. Verfahren zur Herstellung von mehrfarbig wiederzugebenden photographischen Bildern. Patent, DE 223,236. 1908.
178. Ede F. Un épisode de l'histoire de la couleur au cinéma: le procédé Keller-Dorian et les films lenticulaires. 1895. 2013;187–202.
179. Gretener E. A brief survey of the physics and technology of the berthon-siemens color process. J Soc Motion Pict Eng. 1937;28:447–63.
180. Hess W. Unmittelbar wirkendes Stereobild. Patent, CH 61475. May 30, 1912.
181. Lorenz D. 3-D-Linsenrasterbilder, eine Erfindung aus der Schweiz. Arbeitskreis Bild Druck Papier. Münster: Waxmann; 2004. S 20–7.
182. Kanolt CW. Production of stereoscopic pictures. Patent, US 1,882,648. Juni 18, 1930.
183. Kanolt CW. Projection of pictures for viewing in stereoscopic relief. Patent, US 2,075,853. 1937.
184. Ives HE. Projection of stereoscopic pictures. Patent, US 1,883,290. 1932.
185. Ives HE. Making stereoscopic parallax panoramagrams from pseudoscopic parallax panoramagrams. Patent, US 1,905,716. 1933.
186. Eggert J, Heymer G. Verfahren zur Herstellung von Filmen mit linearen lichtbrechenden Elementen. Patent, DE 558,365. 1932.
187. Kanolt CW. Photographic method and apparatus. Patent, US 1,260,682. 1915.
188. Buckley OE, Darrow KK. Herbert Eugene Ives 1882–1953 [Internet]. National Academy of Sciences 1956. http://www.nasonline.org/publications/biographical-memoirs/memoir-pdfs/ives-herbert.pdf.
189. Ives HE. Parallax panoramagrams made with a large diameter lens. J Opt Soc Am. 1930;20:332.
190. Noaillon EHV. Art of making cinematographic projection. Patent, US 1,772,782. 1928.
191. Jennings WW, Vanet P. New direct-vision stereo-projection screen. J Soc Motion Pict Eng. 1952;59:22–7.
192. Kaplan SH. Theory of parallax barriers. J Soc Motion Pict Telev Eng. 1952;59:11–21.
193. Blundell BG. On aspects of glasses-free 3D cinema ~70 years ago. Auckland University of Technology 2013.
194. Ivanov BT. Stereoskopicheskoye Kino (Stereoskopisches Kino). Goskinoisdat: Moskau; 1951.
195. Ivanov SP. Russia's third dimensional movies. American Cinematographer. 1941;22(5):212–3.
196. artechock film. FILMINFO: Konzert: Das Land der Jugend (Konzert: Semlja Molodosti) [Internet]. 2013. http://www.artechock.de/film/text/filminfo/k/ko/kodala.htm. Accessed on 3. Sept. 2016.
197. Gabor D. Three-dimensional cinema. The New Scientist. 1960:141–5.
198. Winnek Coffey DF. Apparatus for projecting pictures in relief. Patent, US 2,100,634. 1931.
199. Winnek Coffey DF. Composite stereography. Patent, US 3,409,351. 1968.
200. Colbert J-B. Lettres, Instructiones et Mémoires de Colbert. Clément P, Magne P, Editor. Paris: Imprimerie Impériale; 1861.
201. Algoud H. Grammaire des Arts de la Soie. Paris: Jean Schemit; 1912.
202. Bergmann J. Handbuch der Appretur. Marschik C, Editor. Berlin: Julius Springer; 1928.
203. Amidror I. The theory of the moiré phenomenon. Bd. 1: Periodic layers. 2. Aufl. London: Springer; 2009.

204. Lipton L, Feldman MH. New autostereoscopic display technology: the SynthaGram. In: Proc. SPIE 4660 Woods AJ, Merritt JO, Benton SA, Bolas MT, Editor. 2002. pp. 229–235.
205. van Berkel C. Image preparation for 3D LCD. In: Proc. SPIE 3639 Merritt JO, Bolas MT, Fisher SS, Editor. 1999. pp. 84–91.
206. Grasnick A. Autostereoskopische Anzeigevorrichtung. Patentanmeldung, EP 1 895 782. 2007.
207. Zippel T. Jenaer zeigt Weltpremiere auf Cebit – Räumliches Bilderlebnis ohne Brille möglich. Thüringer Allgemeine [Internet]. Onlineversion. Erfurt; 24. Februar 2010. https://www.thueringer-allgemeine.de/wirtschaft/jenaer-zeigt-weltpremiere-auf-cebit-raeumliches-bilderlebnis-ohne-brille-moeglich-id225967659.html. Accessed on 27. Oct. 2019.
208. Nakayama E, Hamagishi G, Yamashita A, Mashitani K, Sakata M, Furuta Y, et al. Two-dimensional/three-dimensional compatible type image display. Patent, US 5,831,765. Nov, 1998.
209. Holmes R. E. 3-D TV without glasses. Proceedings of the IEEE 1991 National Aerospace and Electronics Conference NAECON 1991. Dayton, OH, USA: IEEE; 1991. pp. 914–20.
210. Sombrowsky R. Verfahren zur Erzeugung stereoskopischer Darstellungen. Patent, EP 0 584 752 B1. 1993.
211. Eichenlaub JB. Autostereoscopic display with illuminating lines and light valve. Patent, US 4,717,949. 1988.
212. Jacobs A, Mather J, Winlow R, Montgomery D, Jones G, Willis M, et al. 2D/3D Switchable displays. Sharp Technical J 2003;4.
213. Rassweiler A. Nintendo 3DS Carries $100.71 Bill of Materials, IHS iSuppli Physical Teardown Reveals – IHS Technology [Internet]. 2011. https://technology.ihs.com/389045/nintendo-3ds-carries-10071-bill-of-materials-ihs-isuppli-physical-teardown-reveals. Accessed on 4. Sept. 2016.
214. Oyagi Y. Storage medium storing display, controlling program, display controlling apparatus, display controlling method and display controlling system. Patent, US 8,599,246. 2013.
215. Sharp. 4-Color Innovation, Quattron Technology | TVs AQUOS Global Website [Internet]. 2010. http://www.sharp-world.com/aquos/en/product/4_color_innovation.html. Accessed on 9. Sept. 2016.
216. Asano Y, Fairchild MD, Blondé L. Observer variability experiment using a four-primary display and its relationship with physiological factors. 21st Color and Imaging Conference Final Program and Proceedings. Albuquerque, New Mexico, USA: Society for Imaging Science and Technology; 2013.
217. Cheng H-C, Ben-David I, Wu S-T. Five-primary-color LCDs. J Disp Technol. 2010;6:3–7.
218. Yang Y-C, Song K, Rho S, Rho N-S, Hong S, Deul KB, et al. Development of six primary-color LCD. SID Symposium Digest of Technical Papers. 2005;36:1210.
219. Grasnick A, Relke I. Verfahren und Anordnung zur räumlichen Darstellung. Apr, 2002.
220. Mashitani K, Hamagishi G, Higashino M, Ando T, Takemoto S. Step barrier system multiview glassless 3D display. In: Proc. SPIE 5291 Woods AJ, Merritt JO, Benton SA, Bolas MT, Editor. 2004. pp. 265–272.
221. Mashitani K, Takahashi H, Aida T. Multi-view glass-less 3-d display by parallax barrier of step structure. Memoirs- Faculty of Engineering, Osaka City University. 2007;48:1–8.
222. Schmidt A, Grasnick A. Multiviewpoint autostereoscopic displays from 4D-Vision GmbH. In: Proc. SPIE 4660 Woods AJ, Merritt JO, Benton SA, Bolas MT, Editor. 2002. pp. 212–221.
223. Magnus E. Ueber Einrichtung und Beleuchtung von Räumen zur Aufstellung von Gemälden und Sculpturen. Erbkam G, Editor. Z Bauwesen 1864;14:202–220.
224. Lambert JH. Lambert's Photometrie (Photometria sive de mensura et gradibus luminis, colorum et umbrae). (1760). Leipzig: Wilhelm Engelmann; 1892.
225. Wiener C. Lehrbuch der darstellenden Geometrie. Leipzig: B. G. Teubner; 1884.

226. Mehmke R. Über die mathematische Bestimmung der Helligkeit in Räumen mit Tagesbeleuchtung, insbesondere Gemäldesälen mit Deckenlicht. Z Math Phys. 1898;43:41–57.
227. Gershun AA. Izbrannyye trudy po fotometrii i svetotekhnike (Ausgewählte Arbeiten zur Photometrie und Lichttechnik). Verlag für Physik und Mathematik: Moskau; 1958.
228. Gershun AA. Svetovoye Pole (Lichtfeld). Leningrad: ONTI. Edition allgemeine Technik; 1936.
229. Gershun A. The light field. J Math Phys. 1939;18:51–151.
230. Moon P, Eberle SD. Theory of the photic field. J Frankl Inst. 1953;255:33–50.
231. Adelson EH, Bergen JR. The plenoptic function and the elements of early vision. In: Landy MS, Movshon JA, editors. Computational models of visual processing. Cambridge: MIT Press; 1991. p. 3–20.
232. Levoy M, Hanrahan P. Light field rendering. ACM Press; 1996. pp. 31–42. http://portal.acm.org/citation.cfm?doid=237170.237199. Accessed on 10. Juli 2016.
233. Levoy M. Light fields and computational imaging. Computer. 2006;39:46–55.
234. Wilburn B, Joshi N, Vaish V, Talvala E-V, Antunez E, Barth A, et al. High performance imaging using large camera arrays. ACM SIGGRAPH 2005 Papers on – SIGGRAPH '05. Los Angeles, California: ACM Press; 2005. p. 765.
235. Wilburn B. High performance imaging using arrays of inexpensive cameras. Stanford: Stanford University; 2005.
236. Lippmann G. Épreuves réversibles. Photographies intégrales. Comptes Rendus de l'Académie des Sciences. Paris: Gauthier-Villars; 1908. pp. 446–451.
237. da Vinci L. Traktat von der Malerei. Herzfeld M, Editor (Ludwig, H. Übersetzer). Jena: Eugen Diederichs; 1909.
238. Wheatstone C. Contributions to the Physiology of Vision. No. 1. J R Inst G B 1830;1:101–117.
239. Faraday M. On the magnetization of light, and the illumination of magnetic lines of force. Philos Mag Ser. 1845;3(28):294–317.
240. Faraday M. Thoughts on ray-vibrations. Philos Mag Ser. 1846;3(28):345–50.
241. Faraday M. On new magnetic actions, and on the magnetic condition of all matter. Philos Mag Ser. 1845;3(28):396–405.
242. Kim Y, Hong K, Lee B. Recent researches based on integral imaging display method. 3D Res. 2010;1:17–27.
243. Hong J, Kim Y, Choi H-J, Hahn J, Park J-H, Kim H, et al. Three-dimensional display technologies of recent interest: principles, status, and issues [Invited]. Appl Opt. 2011;50:H87.
244. Takaki Y. Development of super multi-view displays. ITE Trans Media Technol Appl. 2014;2: 8–14.
245. Grasnick A. Concept of an autostereoscopic system containing 29 million of stereoscopic image pairs. In: Tian A, Asundi A, Liu W, Zhang C, editors. Xi'an. China: SPIE; 2015.
246. Grasnick A. Creating of autostereoscopic displays with random lenticulars by hyperview approach. DGaO Proceedings 2016. Erlangen: Deutsche Gesellschaft für angewandte Optik e. V.; 2016.
247. Grasnick A. Super resolution from hyperview image stack by spatial multiplexing. 3D Res. 2016;7:1–6.

5 Virtual Spaces

The Creation of Illusory Worlds

Summary

For a truly virtual impression, it is not enough to look into a small peep-box or stare at a screen. The captivating impression of a virtual narrative in Weinbaum's sense cannot be achieved in this way, and one is arbitrarily far removed from Lem's phantomatics. It therefore seems necessary to stimulate even more senses. The audience expands its scope of action and turns from a mere voyeur into an actor. The chair on which a viewer sits becomes a motorcycle, a car or even an airplane. Wind blows around the pilot of the virtual airplane, he smells the odor of exhaust fumes and holds a control stick in the hand instead of a computer mouse. Interaction with the virtual world requires knowledge of the direction in which the interactor is looking or moving. For this purpose, suitable sensors are attached to different parts of the body, which can now be used to determine the position of the hands, fingers, legs or head. The desire to realize the virtual world has made people dream of the materialization of thought since the time of Ovid. With the enlivenment of Pygmalion's beloved statue Galatea, there is also an alleged testimony to this. Today, ideas can be turned into real things by using a 3D printer. It is not yet possible to actually make virtual objects real, as on the holodeck of the Starship Enterprise. But who knows...?

5.1 Panoramas

Right in the Middle Instead of Just There

As it seems, you always need a monitor or a screen for a 3D presentation. If you want to do without, you will have to use a stereoscope. The impression of being in an alien world will then certainly evaporate quickly.

A. Grasnick, *Basics of Virtual Reality*,
https://doi.org/10.1007/978-3-662-64201-6_5

The desire for greater involvement of the viewer in the artificial illusion has always existed. An obvious idea is the large-scale presentation of scenes in the form of a panorama.

La Nature a Coup d'Œil: Nature at a Glance

The idea of completely depicting the surrounding world is not new. As early as the twelfth century, the Chinese painter Zhang Zeduan depicted life in the city of Kaifeng on a silk scroll over five meters long.[1] Perhaps the unrolling of a long painting is entertainment enough in itself. But it is certainly more impressive when such a painting is also set up. A panorama with a length of several meters can then have a much more intense and realistic effect on the viewer than a small picture.

Large-format mounted panoramas were known and popular since the late eighteenth century. The creator of the word "panorama"is considered to be the Irish painter Robert Barker, who applied for a patent for it in 1787 (from [2] p. 165):

> ....for his invention of an entire new device, called by him La Nature a Coup d'Natureil, for the purpose of displaying Views of Nature at large, by Oil-Painting, Fresco, Water-colours, Crayons...

Barker had stipulated that these representations of nature be fixed in a circular space or frame, and that the viewer must always be at a certain distance from the image in order to achieve the desired effect.

The term "panorama"[2] does not yet appear in Barker's application of 1787, but it does in the publication of 1796, where the explanation of the French term "La Nature a Coup d'Œil" is given in English as "This invention has been since called the Panorama" (from [2], p. 165). Barker had mounted his panoramas in rotundas (buildings with a circular floor plan) and presented them to the public in exchange for money. The "Waterloo-Panorama" brought in 10,000 pounds with only one shilling admission (1/20 pound) [3].

The panorama painter Daguerre created with his colleague Bouton in Paris a diorama,[3] which increased the reality of the scene. It was the Daguerre who would later establish a precursor to modern photography with the daguerreotype, achieving even greater realism.

Initially, however, Daguerre painted a canvas, as large as possible, on both sides and, depending on the desired effect, illuminated the canvas from the front, from behind, or only in certain areas ([4] from p. 61). This also allowed for variations in the scenery despite having a single large image. Daguerre describes it like this (from [4] from p. 60):

---

[1] The Qingming scroll is now in the Palace Museum in Beijing [1].

[2] From Greek pan = all- + horama = sight.

[3] From Greek dia = through + horama = sight.

> In all these paintings an alternation of day and night illumination is carried out, with which the disappearance of figures is still connected, whereby, for example, in the mass at midnight human figures appear in the place where before church chairs were to be seen...

These changeable paintings were set up in their own buildings from 1822 in Paris [5], from 1823 [6] in London and from 1827 in Berlin [7][4] and were exceedingly large in size.[5] The dioramas were constructed as theatres and the performances were carefully staged in each case. The performances were not limited to a single screen. A look into the auditorium reveals the "theatrical" intention (Fig. 5.1 left). The auditorium was 35 feet[6] in diameter and seated 250 people inside (according to [8] CCCXXXIX). After about a quarter of an hour, the entire hall rotated, giving the audience a view of one of the other large screens.[7] Figure 5.1 on the right shows the floor plan of the Paris diorama. The three screens are each shown there as the stretches ***f-g***, the rotating auditorium is the circle in the middle (marked ***a***).

Dioramas were usually large, stationary, and completely unsuitable for home applications or for showing around. Therefore, this technique was also packed in a small box and sold as a toy or souvenir after visiting a diorama.

The 1820s saw the construction of a number of dioramas in Britain,[8] so it is not surprising that as the technique became more widely known, corresponding souvenirs and peepshows began to appear. The London publisher William Spooner had a type of transformation pictures on offer which he called "Protean Views",[9] contemporary with the "Dioramic Views"[10] of the London publisher William Morgan ([9], p. 76). Like the large dioramas, these pictures were painted or printed on both sides. Under normal lighting, one saw an ordinary scene, such as a view of a landscape. If one then held the picture against a bright light, the back picture became visible and one saw here, for example, a volcano suddenly erupting.

The Frenchman Lefort built a small box for the pictures, which he called "Polyorama panoptique" and for which he received a patent in 1849 ([9], p. 78). One looked through an eyepiece at the diorama, which could be illuminated from above or behind via two flaps. Both flaps were connected by a mechanism. If one flap was closed, the other one opened. This meant that only one effect of the two that could be displayed was visible.

---

[4] Built and operated by the brothers Carl, Ferdinand and Georg Gropius ([7], p. 147).

[5] In the Berlin diorama, the paintings occupied an area of 2500 square feet (ibid).

[6] According to the Weimar measure (1 foot = 28.2 cm) about 10 m.

[7] 65 × 42 feet, or about 18 × 12 m.

[8] e.g. in London, Liverpool, Manchester, Dublin and Edinburgh [6].

[9] After the sea god Proteus from Greek mythology, Proteus could take on various guises and as a legendary quick-change artist is also eponymous for "Proteus' perspectives", also known as "Spooner's perspective views".

[10] Also marketed as "Improved Protean" or "Transformations".

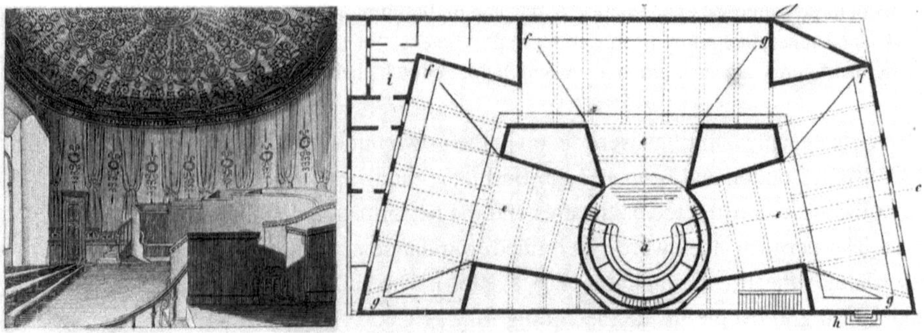

Fig. 5.1 Parisian diorama with rotating auditorium, Friedrich Justin Bertuch (1830) from [8] Melanges. CCCXXXIX

This was not enough for the stained-glass artist Seguin. Seguin saw the polyorama as a "magical" instrument and it demanded movement from him (from [10], p. 237)[11]:

> The "polyorama animé" is a visual instrument of the genre "fantasmagoria". The effect produced by this instrument consists in imitating, in picturesque views, the movements of nature: such as the movements of water, flames, clouds, steam and animated people. . .

By moving a metal screen, several images are projected one after the other at the same position on a screen. In principle, several images could already be displayed in a sequence, so that an impression of natural movement could be created.[12]

Seguin also worked for the lantern magician Robin,[13] who gave him the idea of the polyoscope[14] ([13], p. 95). A polyoscope is basically a scaled-down version of the Pepper ghost illusion (see description from [14], p. 22):

> In the Séguin polyoscope there is a 45-degree mirror. It serves to reflect images into the eyes of the 'voyeur' and to cause a change between tableaux. These, which are in the box, transform into each other and seem to merge.

Although polyoscopes and polyorama animé offered interesting possibilities for creating the illusion of image changes or even movement, Seguin's ideas did not prevail. The desire for a realism enhanced by movement remained, however.

[11] Own translation.

[12] The Paris drugstore Menier & Cie also lists a Polyorama animé with 12 pictorial plates in its catalogue of 1854 ([11], p. 216).

[13] The magician Henri Robin presented his "living spirits" at the same time as Pepper in London (1863) (see [10], p. 236).

[14] For which Séguin also applied for a patent (see [12], p. 112).

The architect Langhans took up a suggestion by the court theater painter Gropius, who, inspired by his own panoramas and dioramas, wanted to realistically imitate a water journey (see [15], p. 548). Langhans invented the pleorama[15] for this purpose and thus created an artificial reality (from [16], Preface):

> The onlooker will be able to get into a barque, and, if the execution is at all successful, to sail a real surging sea, to dream of sailing round real islands and promontories, also to enjoy quite different views every time he chooses a different seat on renewed visits, as in nature itself.

The screens were moved past the viewer so that the impression of movement was created.

With the triumph of the cinema, the heyday of large-format still images came to an end and the effect of the screens of pano- and pleoramas was replaced by the moving images on the screen of the cinemas.

But here, too, they wanted more realism. For this, the American Waller had a "big" idea in the true sense of the word. With the invention of his "Vitarama" Waller wanted one thing above all (from [17]): "... to produce the effect or illusion that the spectator is actually in and surrounded by the environment depicted."

In Waller's Vitarama cinema, several projectors[16] were used to project onto a curved screen. The screen could completely surround the viewer (360° projection) and also be curved vertically ("from horizon to zenith"). For a seated viewer, 180° was considered sufficient due to the human field of vision,[17] and thus a quarter hollow sphere (quasi a semi-hemisphere) was considered a suitable projection screen. Public performances of projections with ten projectors were given by Waller at the 1939 World's Fair inside the "Perisphere", a huge sphere more than 50 m in diameter.[18]

With the outbreak of the war, the potential of the Vitarama cinema for the training of bomber gunners was recognized. These sat for the training in the middle of a Vitarama behind their weapon in a gun turret and fired at approaching aircraft. Everything was real—except the planes. Since this "Gunnery Training" [17] was additionally equipped with 3D sound, a very realistic air combat could be trained without danger for the shooter. In an article about Waller the interviewer describes the reaction of the air gunner quite vividly: "... he's spinning and firing and it's no longer a game—the sweat is real sweat, the fear is real fear." (from [20]). As a reward for sweat and fear, the combatant had the decisive advantage of being out of any danger.

---

[15] From Greek pleo = ship- + horama = sight.

[16] In Waller and Walker's 1937 patent [17], four projectors were illustrated, although it is also described that there may well be more or fewer.

[17] Although the horizontal field of vision of the human being actually comprises almost 220° and can be completely extended to 360° by turning the eyes and the head, as Hueck already communicated in 1840 ([18], p. 84).

[18] Waller was responsible there for the "Living Murals", a film of only 83 seconds by Eastman Kodak, which was projected onto the sky of the Perisphere [19].

A Vitarama cinema was also set up on the "Chicago Railroad Fairground" in 1949. It showed a panoramic video of Niagara Falls [21]. To fill the screen, no less than ten synchronized projectors were needed. This was a bit costly to be practicable for a large number of cinemas. So Waller reduced the number of projectors to three, set up a movie theater with them at the Broadway Theater in New York in 1952, and invented the sonorous name "CINERAMA" [22] for the whole system. Although the CINERAMA, a combination of "Cinema" and "Panorama", was an impressive technology, the installation and operation of multiple synchronized projectors remained not easy. In addition, it was difficult to superimpose the three parts of the image without distortion and with an exact blending. As a result, the edges of the individual images could be visible and thus destroy the impressive overall impression. Waller himself had the idea for a scanning projector as early as 1944, in which certain film sections were projected onto the screen one after the other [23]. It then took almost two decades until the first CINERAMA film[19] was shown with a single projector in 1963.

The idea of equipping a large projection with a smaller number of projectors was the basis for the foundation of the IMAX Corporation. Immediately after the 1967 World's Fair in Montreal, four Canadians got together with the idea of creating a panoramic projection using only a single projector. The CINERAMA cinemas have largely disappeared, but the IMAX cinemas are represented in large numbers and are particularly successful with the showing of 3D films (IMAX 3D) including surround sound (IMAX Sound). IMAX is now considered the de facto "gold standard" of cinema.

The Teaching Swing of the Aviators

It is certainly appropriate to call Waller's shooting cinema a simulator. Yet flight simulators in the 1930s were by no means something fundamentally new.

The "Link Trainer" by Ed Link was well known at that time and used by air forces and airlines in different countries (see [24], Chap. 2).

Pilot training was the original purpose of flight simulators. Already with the beginning of military aviation in World War 1, the need to train pilots for the new combat system increased. In order to do this effectively, ground-based and non-flying aircraft replicas were also constructed among the belligerents. These constructions, as simple as they may have been, can certainly be seen as the forerunners of today's flight simulators. The British "Rocking Nacelle"[20] by Major Hawker, the German "Lehrschaukel" by the engineer Drexel [26] (set up at the Döberitz airfield at the time, (see [27], p. 52) and the French "Piloteur" by Lieutenant Maurice Lender and Paul Heidelberg [28] have passed down, for example. From the French "Piloteur" even a film still exists, on which the two soldiers

[19]The hit comedy "It's A Mad, Mad, Mad, Mad World".

[20]Rocking gondola, also rocking fuselage [25], p. 151.

Fig. 5.2 Le Piloteur, La Section de Cinématographie Technique de la Direction des Inventions (1919) from [29], courtesy of CNRS Images

show the simulator in full function and with amazingly fast movement. A still from this film is shown in Fig. 5.2.

Of course, the Americans also had such a device on offer with the "Dry Shooter", with which the machine gunner could also carry out his target practice on the ground (see Fig. 5.3).

5.2 Duplicated Reality

Copied Worlds for All Senses

The flight simulators and shooting trainers used the stimulation of the sense of motion and the sense of sight to create realistic impressions. Stimulation of other senses can enhance the spatial impression and create the sensation of actually being in a different, distant place.

At the beginning of the 1950s, the American film artist and cinematographer Heilig was inspired by the colour 3D films of the time, spatial sound and the new CINERAMA cinema to combine these different "sensory materials" into one work [31]. He believed that the movement of a person is inferior to his other senses in perception and formulated four "laws" of cinematic art from this:

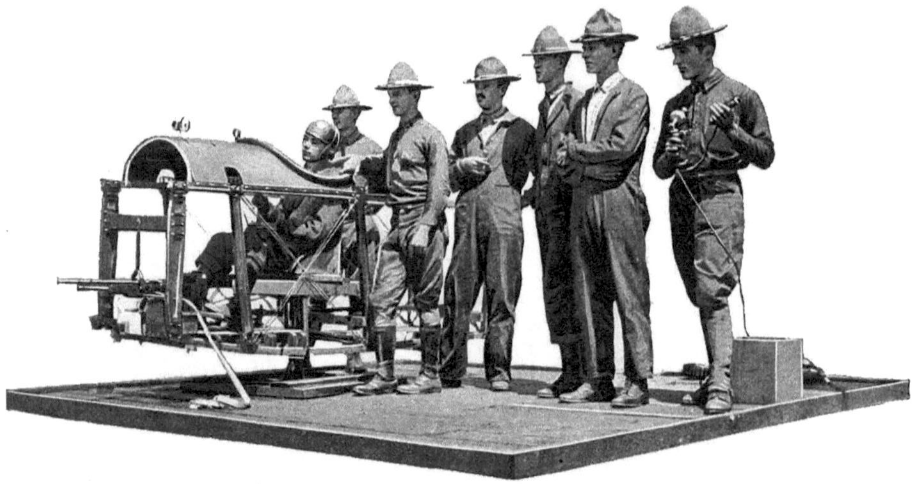

Fig. 5.3 "Dry Shooting" for Airplane Gunners, Captain E. C. Crossman (1919) from [30], p. 13

1. "The nature of man's art is fundamentally rooted in his peculiar psychic apparatus and is limited by the material means at his disposal" (from [32], p. 244).
2. "Consciousness is a composite of all the sense impressions conveyed to the brain by the sensory part of the nervous system which can be divided into the great receiving organs - the eyes, ears, nose, mouth, and skin." (ibid. p. 245)
3. "The brain of man shifts rapidly from element to element within each sense and from sense to sense in the approximate proportion of sight, 70%; hearing, 20%; smell, 5%; touch, 4%; and taste, 1%, selecting one impression at a time according to the needs of individual and racial development." (ibid. p. 248)
4. "In his creative process, man is imposed on by outer impressions. He learns the secrets of their basic principles through imitation and then subjects these to the needs of his own expression. He goes from reception to imitation to creation, i.e., from portraying the outer to portraying the inner world." (ibid. p. 251)

It is perhaps a bit daring to speak of laws here, rather we are talking about statements. The first two statements are immediately clear and actually require no explanation. Expressiveness is undoubtedly dependent on the senses, perception and imagination. The third statement is somewhat more difficult. The classical 5 senses in Aristotle's sense make varying contributions to perception. The division into qualitative proportions is correct insofar that vision is indisputably the largest contributor to perception. In all epochs, sight was widely regarded as the most important sense, mostly followed by hearing and taste (see also "History of the Senses" [33]).

Aristotle had divided the sensory faculty into five categories: The visible, the audible, the olfactory, the gustatory, and finally the tactile. Each sense is thereby quite peculiarly receptive only to certain sensations. The vision only for the colours and the light, the hearing for the sound of the tones and the voice, the smell for the odorous particles, the taste for the evaluation of the food and the feeling for the condition of the things (warm and cold, dry and damp, soft and hard) and also for the excessive [34]. The number five, like the number four,[21] has been of symbolic significance since antiquity and thus has been rule-giving. Until modern times, the five senses were sufficient to describe perception.

The Sixth Sense

In the middle of the eighteenth century, the persecution of witches came to an end and with it the official belief in magic and ghosts. In fact, however, this end was also the beginning of a quasi-scientific study of supersensory phenomena, i.e. phenomena that go beyond the 5 senses. One of the first modern stories about telepathic perceptions is that of the Swedish mystic Swedenborg, who was allegedly haunted by the vision of a fire in Stockholm at the same time in Gothenburg, hundreds of kilometers away.

The philosopher Kant examined the story in the treatise "Träume eines Geistersehers" ("Dreams of a spirit-seer") [35]. He begins his criticism with the fact that Swedenborg's (whom Kant calls Schwedenberg) entire occupation is reduced to the fact "...that he ... has been in the most exact contact with spirits and separated souls for more than 20 years..." (ibid. p. 84).

Kant is extremely clear in his assessment of Swedenborg when he writes (pp. 98-99):

> "The great works of this writer contain eight quarto volumes of nonsense, ...".
>
> "His stories ... seem to have sprung from fanatical views, and give rise to little suspicion that speculative fantasies of a confused mind should have induced him to invent them and use them to deceive."

The ghostly appearances now supposedly meet Swedenborg's inner, hidden sense (which Kant rather considers madness), which today one would call the sixth sense. This means that even things that cannot be grasped with the usual 5 senses are still somehow noticed. Today, parapsychology deals with phenomena such as clairvoyance, fortune telling, telepathy or ghostly apparitions.

However, with serious scientific investigation, such phenomena regularly prove to be nonsense or explainable. Often the sixth sense is the involvement of the subconscious in perception. You don't notice things consciously, but you still get something. A good example of this is when someone is looking at your back. You turn around and notice that you are actually being stared at. The sixth sense is hardly responsible for this. The most likely explanation is that one has subconsciously picked up certain cues with the other senses (especially hearing and smell in this case). And let's not forget that the sense of sight with head movement covers 360°—including looking backwards.

---

[21] See also Tetractys.

If one assumes that Aristotle meant more by feeling than the absolute sense of touch, as well as the sense of temperature and humidity, then the perception of balance and of one's own body can also be grasped with the five senses. One can assume this interpretation of Aristotle, since he wonders whether there is only one organ of feeling or several, whether the feeling is something internal, and thereby speaks of several feeling-sensations (11th major, of the tangible and the feeling, from [34]). Today, sight, hearing, taste, smell, bodily movement, balance, as well as the sensations of the organs and the skin are counted among the senses (see, e.g., [36], p. 76). One would thus come to 8 senses.

But back to Heilig. The fourth law shows a simplified theory of knowledge, which is essentially determined by external impressions and a primitive learning process by imitation. His consideration of the representation of the world according to one's own desires already sounds a little like Schopenhauer's "World as will" ("Welt als Wille") [37, 38], but also reveals the artist.

It is significant that the practitioner Heilig did not leave it at his mere theory. He built the Sensorama, a device in which he stimulated four of the five senses through simulation. The device was designed to provide full 3D vision, including peripheral vision, and to deliver stereo sound while simultaneously generating smell, evoking tactile sensations, and producing wind. This "environmental simulator" allowed Heilig to fulfill his desire to shape the external world to his will [39]. The actual device was not very elegant, but a large box in front of which one had to sit and press one's head under a hood against two eyepieces. In order to provide several visitors with the illusion at the same time, Heilig had designed a complete sensor cinema a few years later, in which the chairs were also moved: The "Experience Theather"[22] [41].

Admittedly, Sensorama and Experience Theather were immobile, as the devices were fixed to a specific location. If you wanted to enjoy the sensory experience, you had to visit the respective panorama.

Heilig had, however, developed another device that, in contrast to the Sensorama, was highly mobile. The device, for which Heilig first used the name "stereoscopic television set for individual use", was later called "Telesphere Mask" [42]. This amazing device was clearly reminiscent of today's VR goggles[23] in design (see Fig. 5.4), but was much more extensively equipped compared to those. The 3D impression was complete, the stereo image and stereo sound were supplemented by stereo sound.

[22]The "Experience Theater" is roughly equivalent to what is now called 4D or 5D cinema, which can be experienced, for example, in the Ostsee-Welten 5D in Warnemünde [40] (www.ostsee-welten.de) or in Ripley's 5D Moving Theater in Gatlinburg, Tennessee (https://www.ripleys.com/gatlinburg/moving-theater)

[23]Some of the current representatives have already been mentioned in the introduction.

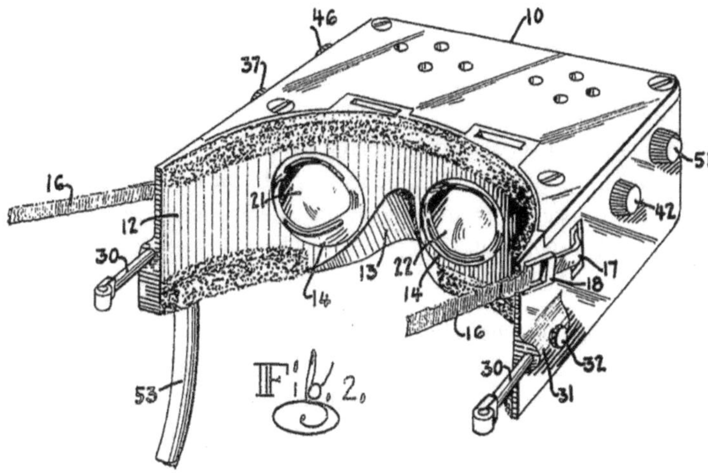

Fig. 5.4 Telesphere Mask, Morton L. Heilig (1957) from [43]

Olfactory Cinema

It is known that cinema performances have been accompanied by scents since the beginning of the twentieth century. One of the first appearances of an intended smell in the cinema is the performance of the "Pasadena Rose Bowl Game"[24] by the showman "Roxy" Rothapfel. For this, Roxy simply placed cotton wool soaked in rose water in front of a powerful electric fan ([45], p. 172). With this approach, the accompaniment of a longer movie was hardly possible.

In 1932, for his novel "Brave New World", the writer Huxley also devised a scent organ with which entire olfactory pieces could be composed (from [46], p. 170)

> The scent organ was playing a delightfully refreshing Herbal Capriccio–rippling arpeggios of thyme and lavender, of rosemary, basil, myrtle, tarragon; a series of daring modulations through the spice keys into ambergris; and a slow return through sandalwood, camphor, cedar and newmown hay (with occasional subtle touches of discord–a whiff of kidney pudding, the faintest suspicion of pig's dung. . .)

A few years after Huxley's vision, the Swiss inventor Laube actually succeeded in combining film and scent. His scented film "My Dream" premiered at the 1940 World's Fair in the Swiss pavilion [47]. The complex system, in which the various scents were synchronized with the film, initially had the unwieldy name O.T.P. ("Odorated-Talking-Pictures" [48]), later Scentovision. When in 1960 the

(continued)

[24] A legendary American college football final ("The Granddaddy of Them All" [44]).

film "Scent of Mystery" was brought to the screen in full Cinerama and with programmed smells, a new name also came into play. The film was presented, "...in Glorious Smell-O-Vision!"[25]

The film was not a success, nor was the smell film "Behind Great Wall" with "AromaRama" ([50], pp. 19–20), which was performed around the same time. In addition to the technical difficulties with the quality, intensity and timing of the smells, the cost of installing the equipment was of course also an obstacle.

The Sensorama, like the other olfactory experiments, was commercially unsuccessful. In principle, however, Heilig had achieved one thing with it. He had brought Huxley's vision of "An all-super-singing, synthetic-talking, coloured, stereoscopic feely. With synchronized scent-organ accompaniment" (from [46], p. 171) into reality.

5.3 Interaction

Perspectives of the Artificial Space

The prefabricated films of the panorama, 3D or experience cinemas always have a built-in limitation. The events are fixed in sequence and angle of view. An active intervention in the events is not possible. The viewer is therefore bound to a fixed perspective. Even if you put on 3D glasses or attach the "Telesphere" mask to your head, the image impression does not change in the slightest when you move your head. This contradicts what one would perceive when watching a scene in real life.

An interactive experience like in Weinbaum's "Pygmalion's Spectacles" [51] or a counter-eye mirroring reality like in Lem's "Summa technologiae" [52] cannot be achieved without interaction. The general wishful thinking corresponds to a complete telepresence. This is an attempt to record reality in all its facets and to reproduce the copy of the true scenery as comprehensively as possible.

Computer scientist Sutherland described a kinesthetic display in 1965, at the time when computer displays were just able to draw letters, dots and lines. Kinesthetics is the study of the body motion and the perception of movement. By controlling all movements, the necessary forces for the input systems could be controlled (as was already done on the joystick of a link trainer) and the position of all muscles, including eye movements, could be detected. But that's not all. In Sutherland's vision, the ultimate display is a room where

[25] From a 1960 Jagres movie poster, "FIRST (1893) They Moved THAN (1927) They Talked NOW (1960) They Smell; MICHAEL TODD, Jr. presents SCENT OF MYSTERY IN GLORIOUS SMELL-O-VISION! The Process to end all Processes" [49].

the computer is also able to control matter. One could sit in a displayed chair, virtual handcuffs would shackle, and shots fired from a simulated gun would be lethal [53].

This seems a little strange and is almost reminiscent of the holodeck on the Starship Enterprise,[26] in which the virtual things are not only displayed, but actually materialize. The idea of the holodeck comes from the hologram and 3D enthusiast Dolgoff. Dolgoff inspired StarTrek producer Roddenberry to come up with the idea of the holodeck, in which matter is created from energy under computer control [54]. In fact, since 2014, Imperial College in London has been trying the idea of turning light into matter [55] based on the Breit-Wheeler effect described 80 years ago [56]. However, it does not work like in the movie, even less in the 1960s, at the time of Sutherland's paper.

In reality, however, Sutherland was quite practical. He developed a head-mounted stereoscope in which the user is shown a 3D image adapted to the position by changing the position of the head.[27] This is now simply called a "head-mounted three dimensional display" [59]. Due to its weight, the device had to be attached to a suspension in addition to the mount on the head and thus hung above the user, which explains the later nickname "sword of Damocles" (see e.g [60]., p. 360). In the fifth book of the "Tusculan Disputations" ("Tusculanae disputationes") Cicero tells the story of the "Sword of Damocles", in which the tyrant Dionysius teaches his courtier Damocles a lesson by attaching a flashing sword only to a horsehair above Damocles during a splendid banquet. In this way Dionysius wanted to show that there can be no happiness for one over whom a horror constantly hangs (after [61], p. 315).

Although the displayed images were only quite simple geometric figures, the system already pointed in the direction of interaction with the illusory world. According to the computer artist Lanier,[28] the term "virtual worlds" goes back to the philosopher Langer and was later adopted by Sutherland [64]. However, Langer uses the term "virtual" quite universally in connection with world, space, scene, memory, presence, and many others in her theory of art "Feeling and Form" [65]. In principle, art always creates something virtual.

Lanier's virtual reality, however, is more comprehensive (from [66], p. 15):

> To me, what virtual reality originally meant was moving beyond the headset experience to include some other elements, which would include your own body being present, so to have an

[26]From 1974, first in the StarTrek Animated Series, episode "The Practical Joker", where Uhura, Sulu and McCoy have to deal with unbridled matter (in the form of pitfalls and blizzards) on the holodeck.

[27]Whereby, besides the stereoscopic effect, the kinetic depth effect is also mentioned as providing space (see e.g. [57, 58]).

[28]The computer artist Lanier is credited with the invention of the term "virtual reality", which was, however, already used by Broderick in 1982 in his book "The Judas Mandala" [62]. The term "virtual reality" can also be found as early as 1938 in the book "The Theater and Its Double" by the extraordinary director and theater theorist Artaud ([63], p. 49).

avatar where you could pick up things, and also where there could be multiple people, where it could be social.

In order to interact with the artificial world, one needs a form of input in addition to the image. One of the first ideas of this kind came from Sutherland himself, who in 1964 designed the "light pen", a graphic input system for his "Sketchpad" drawing computer [67]. The "light pen" was the size of a fountain pen and consisted essentially of a photocell with a switch. The photocell measured the light of a pattern displayed on the screen and could determine the movement of the pen based on the measurements.

The RAND tablet by Davis and Ellis from the same year [68] was the low-cost variant of the Sketchpad.[29] In addition to the price advantage, the RAND tablet was an almost practical device, unlike the room-sized Sketchpad system. However, Sketchpad and RAND tablet were only capable of capturing two-dimensional information.

In 1966 Roberts, who like Sutherland was employed at MIT's Lincoln Laboratory, invented the "Lincoln Wand", a kind of magic wand to add a third dimension to Sutherland's "light pen" and the RAND tablet. To do this, Roberts used four ultrasonic transmitters and one receiver. Depending on the distance between the wand receiver and the transmitters, the transit time of the signals emitted periodically by all the transmitters in succession changed. According to Roberts, it was possible to determine the position of the wand in a space of $1.2 \times 1.2 \times 1.6$ m with an accuracy of about five millimeters [70].

Lanier had the vision of a social interaction with virtual reality, the creation of a meeting space of one's own telepresence with other avatars. Now this is exactly the opposite of the often proclaimed isolating effect of a head-set and the associated danger of social exclusion ("digital autism", see also [71] from p. 69). For the computer scientist Minsky, interaction was also a tool for creating telepresence. In Minsky's imagination, during telepresence the user wears a comfortable jacket with sensors and muscle-like motors. Every movement of the arms, hands and fingers is recorded and reconstructed elsewhere on a mechanical counterpart. This allows any heavy or dangerous work to be done, even in impossible environments. Of course, a remote eye is also needed, for which Minsky proposed Philco[30]'s solution in 1980 [74].

The American student Fisher already stated in his master thesis that a pure duplication of a scene cannot lead to a real interaction and consequently proposed a position-dependent image display [75]. As a NASA engineer, Fisher, together with the Fake Space founder Bolas, wanted to actually realize Minsky's idea of telepresence. They envisioned something bigger than a "personalized television safari" as described by Clarke in his essay on the second century of the telephone [76]. To this end, they first enlisted the help of another

[29] Still $18,000 (see [69]), but comparatively cheap against the Lincoln TX-2 computers used by Sutherland, an MIT development that were among the largest and most powerful computers in the world at their time.

[30] American electrical and electronics manufacturer, also made televisions and computers from the 1950s ([72], pp. 7–8), developed remote glasses called "Headsight"in 1961 [73].

science fiction writer, the American Heinlein. In the 1942 short story "Waldo", Heinlein has his physically weak protagonist perform complicated and difficult tasks over long distances. In the process, Waldo has a disembodied self-image [77]:

> He saw himself not as a crippled human being, but as something higher than human, the next step up, a being so superior as not to need the coarse, brutal strength of the smooth apes.

In Heinlein's fiction, Waldo can overcome his physical disability by, for example, putting his hands into a pair of mechanical gloves that transmits his movements to a distant second pair of gloves. In doing so, Waldo can control his distant movements on the "colour-stereo television receptor".

A key component in the remote control of telepresence on other planets is vision and thus interactive remote camera control. Such a solution was developed by Bolas and Fisher based on a remotely controlled stereo camera. This camera can be controlled by head movement of the head-mounted display [78] and provides 3D images that correspond to the viewer's line of sight. The project was named VIEW lab (Virtual Interface Environment Workstation) by NASA.

The computer artist Krueger had a somewhat different view of the environment. He wanted the environment to interact with the user, the virtual space to react in an unpredictable way to the user's movements. This was technically achieved in the first installation, called GLOWFLOW, by sensors scattered on the floor so that the computer could switch the light and sound in a dark room depending on the user's position. The METAPLAY installation was a bit more sophisticated and allowed graphical interaction on a 2.5 $\times$ 3 m screen. Most famous, however, is certainly Krueger's VIDEOPLACE, in which several remote locations are coupled together. On the screen inside the darkened room, the visitor sees himself or the remote persons in a composite life-size image. The fact that all the rooms involved are of the same kind can create the impression that all the people are moving in the same room [79].

A single screen cannot simulate a realistic environment. Of course, the screen can be enlarged[31] or composed of individual smaller screens.[32] But the problem always remains that the space formed by a display does not completely surround the viewer—as a real room or cave would. In fact, when naming the CAVE,[33] the developers refer to Plato's ancient allegory of the cave [84]) and build an artificial cave, so to speak. However, the cave is

[31] e.g., University of Minnesota's PowerWall (approximately 2.4 m wide [80]) or the University of Chicago's Infinity Wall (approximately 3.6 m wide [81]).

[32] Or the NASA hyperwall (about 3.2 m [82]), but constructed from 49 individual screens, with the 7 $\times$ 7 limitation based on Miller's "Magical Number Seven" (only about 7 things can be stored in short-term memory at a time, according to Miller 1956 [83]).

[33] "CAVE Automatic Virtual Environment", the recursive acronym (an abbreviation that when written out again contains the abbreviation as a word) describes a cave with automated and virtual environment.

actually a cube in which, moreover, all six surfaces can be projection surfaces. In combination with systems for detecting body, head and hand movements, such a system can be used to create a real virtual reality.

The ideas for the input devices almost arise by themselves. The primary glove of Heinlein's Waldo is almost a concept. In the development of novel input devices for the "humanization of digital computers", DeFanti and Sandin had proposed a special sensor based on an idea by Sayre, which is stimulated differently by the hand movements (from [85], p. 6):

> The basic idea is to put a light source in one end of a flexible tube and a photocell in the other end. When the tube is bent or compressed, the amount of light hitting the photocell decreases evenly.

What was meant by humanizing digital computers was that the input devices should adapt more to the human physique and psyche—not the other way around. Sandin was already somewhat familiar with virtuality and was already involved with Krueger in GLOWFLOW. DeFanti and Sandin were also later involved together in the development of CAVE and Infinity Wall, making them true VR veterans. Richard "Rich" Sayre, however, had the decisive idea ([85], p. 6, see also [86]) and so it comes that the glove as "Sayre glove" is named only after him.

In the Sayre glove, the tension of the photocell could be used to infer the curvature of the tube and thus the fingers. The bending of the fingers alone is not very informative about the position of the hand. Only a few years after the Sayre glove, Grimes of Bell Labs presented a much improved solution. Grimes also initially proposed optical curvimeters as well. In his patent specification [87] such a sensor is shown, which may also be considered as an explanation for the function of the Sayre glove. Figure 5.5 shows the finger in the glove (100). In the upper part of the drawing the finger is straight, in the lower part it is curved. Mounted on the back of the finger is the knuckle bend sensor (102). The tube in Grimes is made of two parts (in the Sayre glove it was one continuous tube). The smaller tube (405) is inserted into a larger diameter tube, which allows the tube to extend in the simplest way when the finger bends. The wall of the tube is flexible and opaque. On the left side there is an infrared diode (403) and on the opposite side a suitable detector (406).[34] This sensor is irradiated with more or less light depending on the bending of the finger. When the amount of light reaches the desired threshold, the Schmitt trigger switches and the KBS signal (Knuckle Bend Signal) is sent when the finger is sufficiently bent.

[34]In Grimes' case, an "optoschmitt," a photodetector that includes a photodiode, an amplifier, and the namesake Schmitt trigger (after Otto Schmitt's "thermionic trigger" [88]).

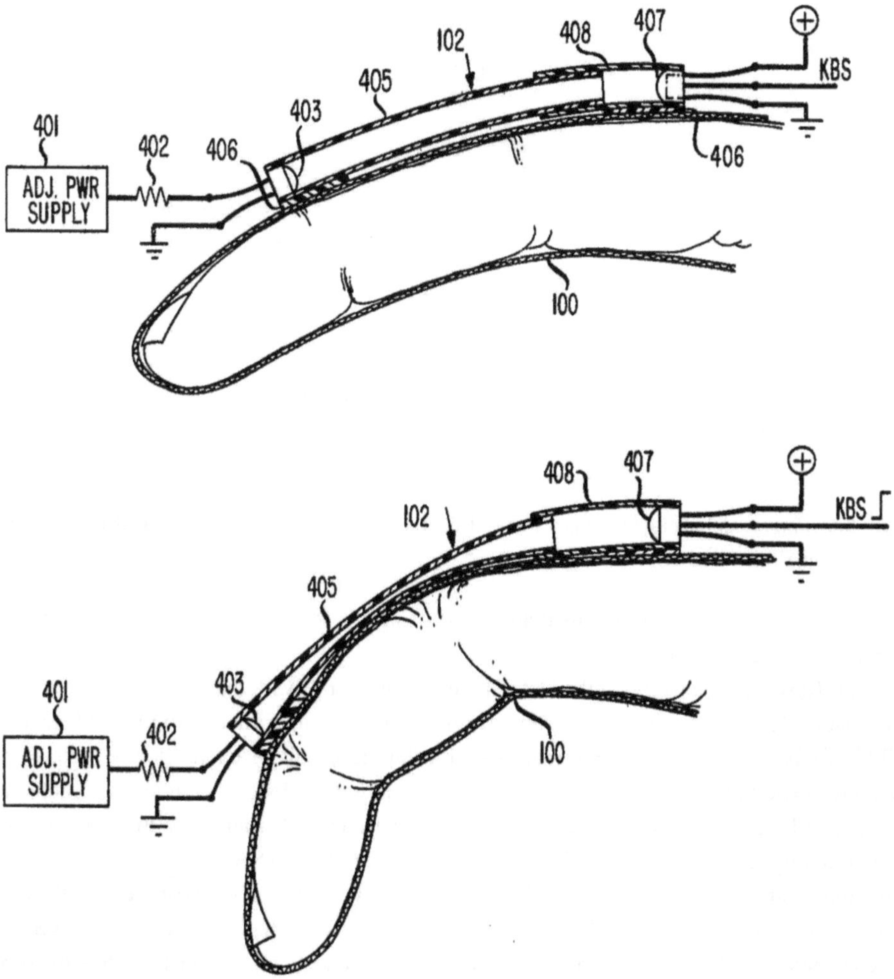

Fig. 5.5 Knuckle-bend sensor, Gary J. Grimes, 1981 from [87]

One intended application of the gloves was to be the input of data using the one-hand finger alphabet.[35] Since certain parts of the hand also touch in the finger alphabet (e.g. thumb and index finger for "O", see Fig. 5.6), the glove had to be equipped with touch sensors[36] that were suitably positioned on the glove. The orientation of the hand in

[35]The basis for the finger alphabet is the publication of the Spaniard Bonet from the year 1620 "Reducción de las letras y arte para enseñar a hablar a los mudos" (Summary of the letters and the art to teach the mute to speak) [89].

[36]Grimes used conductive rubber pads filled with silver.

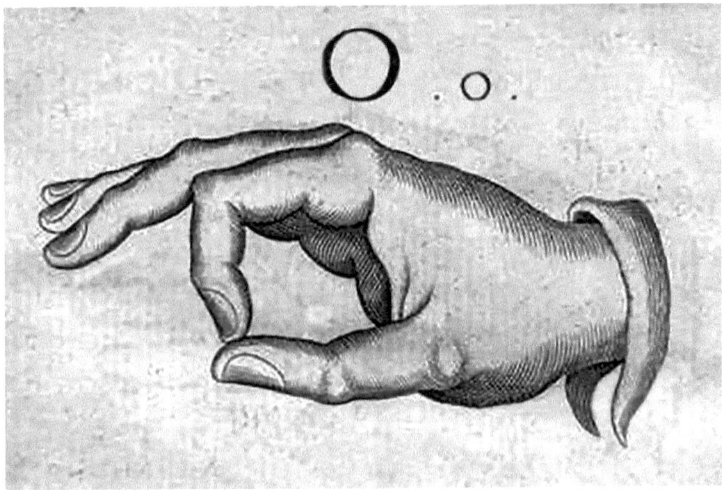

Fig. 5.6 Finger alphabet "O", Juan Pablo Bonet (1620) ABECEDARIO (Spanish for ABC) Demonstrativo, p. OPQ

space with respect to gravity was determined using tilt sensors,[37] and the relative movement was evaluated using inertial sensors.[38]

VPL Research[39] developed the DataGlove, a glove for NASA that would be suitable for astronauts. The idea for it actually came from engineer Zimmerman, who met with Lanier, VPL's founder, in 1984. Zimmerman was developing a pair of cable gloves because he wanted to use them to play a virtual air guitar like Jimmy Hendrix. Lanier, who actually wanted to be a musician, was intrigued by Zimmerman's idea and saw the potential for virtual reality (from [90], p. 90). Eventually, VPL won the NASA contract and developed functional data gloves that could work in conjunction with Fisher's head-mounted displays to create Fisher's telepresence VIEW system. The basis for VPL's gloves were the "optical flex sensors" developed by Zimmerman [91]. The flex sensors also consist of a flexible tube like Grimes', but here now the inner wall is reflective. Here, too, the light source is attached to one side, and on the other side a light sensor that changes its resistance as a function of the incident light (e.g. a phototransistor or a photocell). This makes it possible not only to detect whether a finger has been bent at all, but also to determine the strength of the bend by the resistance of the gloves. In addition, VPL's DataGlove used Polhemus' magnetic "3SPACE" sensors to determine the position and orientation of hands in space, which worked on the basis of Kuiper's tracking and attitude determination [92].

[37] Here mercury-filled tubes (mercury switches).

[38] Inertia switch; the contact closes when a moving mass (here a ball) overcomes a holding force (here a magnet) by acceleration.

[39] Founded by Lanier, VPL is short for "Virtual Programming Language".

Nintendo's Power Glove

The glove developed by Zimmermann and Lanier was very expensive on its own,[40] but in a set with its own head-mounted display "EyePhone", the data suit "DataSuit" and the necessary computer and applications (the "RB2 Virtual Environment") it was completely unaffordable for private users.[41] A New York toy developer named Abrams/Gentile (AGE) acquired the rights to use VPL DataGlove idea. VPL's second generation gloves were very elaborate, with attempts to use fiber optics and even more sensors to reduce the error rate of the gloves. This was done, for example, by roughening the surface of the fiberglass in some places and placing it over these sensors. This way, when the fingers were bent, the light could escape at these spots (see, e.g., [93]). AGE tried to get within the cost range for a toy and focused on simplifying the technology. To make the data gloves significantly cheaper, they developed a simple sensor. AGE's solution was based on a flexible, conductive material[42] whose resistance changed when bent. The toy manufacturer Mattel, in collaboration with AGE, used it to build the "Power Glove", which was to serve as an input device for Nintendo's game console and was offered for a price of US$ 89 [96]. Despite the acceptable price, this glove was also not a success (Fig. 5.7).

At the beginning of the 1970s, the engineer and mathematician Kuipers developed a technique on the basis of which the relative position of a distant object could be determined using a magnetic field. For this purpose, a coordinate frame, so to speak, was set up with the three coils.[43] The receiver is also formed by three coils that detect the magnetic field of the transmitters [97]. In simple terms, the direction and position of the object can be determined from the fields of the transmitter via the coils of the receiver. Kuipers later described the process again in detail in his book "Quaternions[44] and rotation sequences" [99] (from p. 303). Kuiper's invention was patented by the military pilot Polhemus and marketed as a product under the name ROPAMS (Remote Object Position Attitude Measurement System). This small device with a side length of only about two centimeters[45] was used by Bolt in a 3D space installation.[46] The room was comfortably furnished like a private office. The

[40] US$9000 (from [90], p. 91).

[41] Approximately US$400,000 for 2 users (ibid).

[42] Which in the simplest case could be an electrically conductive paint. AGE's patent [94] is the basis for today's flex sensors e.g. from Spectra Symbol or Flexpoint Sensor Systems (see [95]).

[43] Coils that emit a magnetic field.

[44] Quaternions are a mathematical "simplification" by extending the real numbers to four dimensions, which find application in computer graphics and robotics. Quaternions were invented by the Irish mathematician Hamilton [98].

[45] This is the size of the receiver, the transmitter was slightly larger.

[46] The Media Room of the Architecture Group at the Massachusetts Institute of Technology.

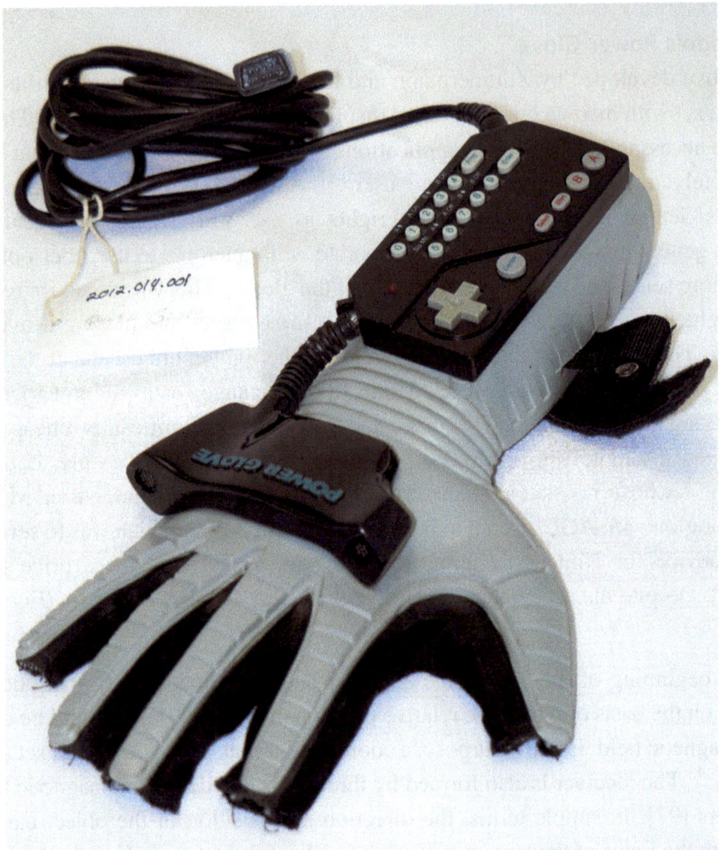

Fig. 5.7 "Power Glove", Grant Goddard and Samuel Cooper (1989) for Abrams Gentile Entertainment (AGE), Digital Game Museum Collection, retrieved September 8, 2019, from www.digitalgamemuseum.org/collection/items/show/693

user sat in an Eames lounge chair with vinyl upholstery,[47] on whose armrests joysticks and touchpads were mounted [100]. The user picked up the Polhemus space tracking sensor and the computer knew exactly what the user was pointing at on the large display in the media room. This allowed for familiar operation of the screen [101]. You point to a spot, said what you wanted, and the computer did it. So if the user said "Create a yellow circle there", the program would paint a yellow circle exactly where the user had just pointed.

A certain amount of interaction was possible in the media room. The reduced input by a finger pointing does not yet correspond to what an engaged user can imagine under his own interaction in and with the virtual space. Determining the position of hands and fingers alone is not sufficient for this. It would also be necessary to determine the position of other

[47] A club chair designed by the American designers Charles and Ray Eames, which is still sold today.

limbs. A first step in this direction was made by Harrison in 1960, who had the idea to move the electron beam of a CRT screen "like the brush of an artist" (from [102]). His approach of transmitting the movement of a real person directly to a figure on the screen was interesting. First, Harrison developed the "Bone Generator", which created simple lines that could be moved in a defined way via electronics. This formed a simple stick figure that could move as Harrison wished. The trick, however, was to transfer the movement from a real person[48] to the stick figure. Using simple potentiometers attached to the original's joints with materials from toy construction kits[49] and rubber bands, the stick figure counterpart performed their dance steps in real time (called "ANIMAC").

The two students Ginsberg [103] and Maxwell [104] from the Massachusetts Institute of Technology worked together with Negroponte, who was significantly involved in the "Media Room" at MIT with Bolt, on a project for a "graphical marionette". The idea was to record human movement as completely as possible in order to reproduce it, as in the case of a marionette, on the basis of the input on a graphic system. For this purpose, a simple body model was created that divides the body into areas from whose position in space the arrangement of the entire body could be inferred (Fig. 5.8).

LEDs were attached to specific points on the body, preferably at joints or other prominent locations, which blinked in a defined manner. The wearer of the LED suit was recorded by two cameras, which now sent the images to the Op-Eye system.[50] The whole system operated in the infrared range, so that it remained somewhat unaffected by ambient light.

On the display of the connected Apple II, one could finally see the dancing strokes moving to the rhythm of the original. Although capturing the spatial position of LEDs with cameras was not a groundbreaking development on its own back in the 1980s,[51] the two ladies had found an application that has since been widely used as "motion capture". It is possible to use passive reflectors as markers[52] or optically active elements,[53] to use a larger number of cameras or to work in other spectral ranges[54] in order to obtain a higher detection accuracy. The principle is always the same. One captures a point with at least

[48]In addition, a dancer in a blue suit with all kinds of wires and various fastening aids for the potentiometers is regularly depicted.

[49]Like "Tinker Toys" or "Lincoln Logs".

[50]From United Detector Technologies (now OSI Optoelectronics).

[51]The principle is based on stereometry and triangulation and has been demonstrated previously with the Op-Eye mentioned earlier, as well as by Woltrings system called SELSPOT (SElective Light SPOT recognition) [105, 106] or Burton and Sutherland's "Twinkle Box"[107].

[52]E.g. retroreflective materials that reflect a certain spectrum.

[53]Emitters emitting in any part of the electromagnetic spectrum.

[54]Both within the electromagnetic spectrum (e.g. visible light, infrared, ultraviolet) and outside it (e.g. mechanical sound waves, especially ultrasound).

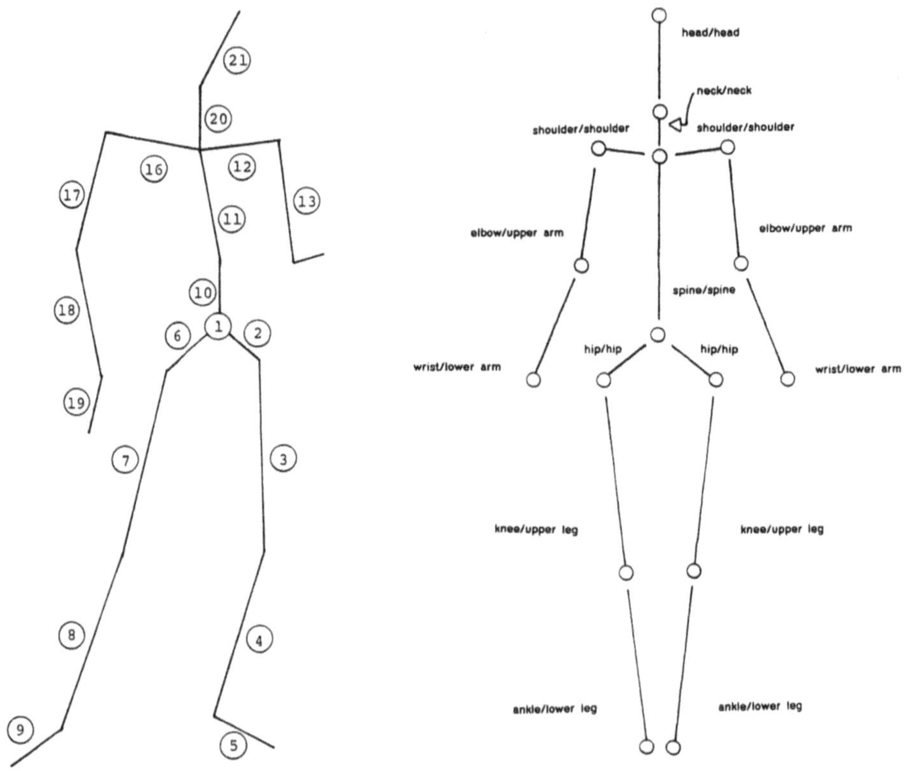

Fig. 5.8 Node numbering, Delle Rae Maxwell (1983) from [104], p. 46 (l) and p. 76 (r)

two cameras and can triangulate it at any point in time.[55] With these and the aforementioned sensors for acceleration, position and joint position, complete sensor suits can also be sewn. Minsky's idea of the comfortable sensor jacket becomes reality. Such Mocap suits (also "Mo-Cap", Motion Capture Suit) were already sold by VPL as DataSuit in the 1980s (see [90] p. 90).

Today's sensors are usually highly integrated and use several technologies in interaction. It has proven useful to determine the position of an object by identifying the location,[56] acceleration[57] and rotation[58] of various measuring points. In order to detect all possible movements, each sensor must be present for each axis. In three-dimensional space, the position is determined by the x-, y- and z-axis, so that the three different sensors

[55]To determine the position of the targeted point from the two angles of the cameras to the point and the base length (the distance between the two cameras).

[56]When referring to the earth's magnetic field, a magnetic field sensor is used.

[57]Acceleration, translation or inertia sensor with reference to the earth's gravitational field.

[58]Gyroscopic sensor (from Greek gyros = gyroscope).

must each be present three times, i.e. exist nine times for each measuring point. For further improvement, GPS[59] or compass are sometimes added. The collected data are processed intelligently[60] and thus allows a very accurate prediction of the position. The nine individual sensors are usually housed in a common housing and thus appear to the user as a single sensor. Now, to connect the sensor to the user's body, the sensors can be attached to the limbs and head with rubber bands or Velcro. Such "ANIMAC" solutions are preferred, for example, by the Dutch company "XSens Technologies", the American "Yost Labs" or the Chinese company "Noitom Perception Neuron".

Today, however, the VR suits can also look quite professional. The use of such suits now no longer makes the wearer seem like a weird nerd,[61] but can also give them a superhero feeling. The "Teslasuit" suit from the English company VR Electronics not only has sensors for motion detection, but also provides tactile feedback via electrical stimulation.[62] In addition, the suit can be selectively heated, so that physical stimulation can also be achieved here. For example, the effect of heat can create a real-life sensation when virtually walking through a fire. Similar suits are also available from the Danish company Rokoko Electronics (as a Smartsuit without feedback) or from the Indian startup Kaaya Tech (as a Holosuit with feedback).

In addition to the trousers, jacket and gloves of the real suit, you also need a head-mounted display to immerse yourself in the virtual world. Of course, no one wants to have a sword of Damocles hanging over them, but there is quite naturally the desire for a lightweight and inexpensive system.

The young American Luckey took pleasure in buying the early, formerly unaffordable head-mounted displays at auction, but at the same time he cherished the desire described above for something better. By taking them apart, he gained an understanding of how such systems were built. In the small VR community of the early 2010s, game developer Carmack[63] became aware of Luckey's idea of lightweight glasses and eventually developed a version of the game Doom for Luckey's glasses [111]. Through Carmack's support, the idea became popular and the Oculus company became successful. The first available glasses were the Oculus Rift in the developer version (Fig. 5.9).

[59] Global Positioning System, a navigation satellite system originally developed by the U.S. military to determine position, officially "Navigational Satellite Timing and Ranging - Global Positioning System" (NAVSTAR GPS).

[60] e.g., the Yost Labs 3-space sensor with an "on-board quaternion-based orientation filtering" algorithm [108].

[61] In the sense of being a highly intelligent and computer fanatic nerd, a wonderful example the Canadian computer science professor Steve Mann, who for decades has appeared in public wearing the quaintest homemade VR goggles [109].

[62] Neuromuscular electrical stimulation (NMES, see e.g. [110]).

[63] Known for the games Wolfenstein and Doom, both of which made it to the index in Germany, but are also extremely well known.

Fig. 5.9 Oculus Rift—developer version, Sebastian Stabinger "Petheon" (2013), Wikimedia Commons

The Oculus Rift and certainly the takeover of Oculus by Facebook[64] triggered a real VR glasses hype.[65] A prominent example of this is the "Magic Leap" idea. The engineer Abovitz already had success with his company economic MAKO.[66] The innovative company had filed numerous patents in the field of robot-assisted surgery and developed robotic arms that could be used in surgeries. One of the company's specialties was the generation of haptic feedback for the surgeon during surgery (see, e.g., [114, 115]). Abovitz himself was involved in some patents and through these developments was familiar with interaction in space and virtual reality. By selling his company, Abovitz was financially independent and could follow his vision of virtual reality. The eccentric Abovitz was invited to speak at the TEDx conference in Sarasota in 2012 as a successful medical entrepreneur and company founder. He did this in an unusual way, by stepping on stage in a spacesuit, accompanied by a green and a red hairy monster, and saying only one sentence after the music of Space Odyssey[67] played:

> A few awkward steps for me, a magical leap for mankind.[68]

[64]Sold to Facebook in 2014 for a staggering US$2 billion [112].

[65]E.g. Sony PlayStation VR, Valve Index or HTC Vive.

[66]Sold to Stryker in 2013 for 1.65 billion [113].

[67]Written by Alex North for "2001: A Space Odyssey" a 1968 film based on the screenplay by Stanley Kubrick and Arthur C. Clarke.

[68]See, e.g., [116, 117].

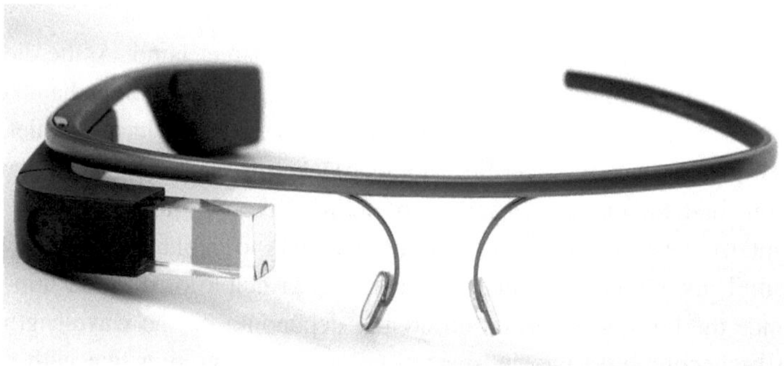

Fig. 5.10 Google Glass - front view, Tim Reckmann (2014) Wikimedia Commons

Such an appearance is certainly not something that is taught in the business start-up seminar of the Chamber of Industry and Commerce. But if Abovitz's aim was to attract attention, he undoubtedly succeeded. Although he had communicated little of substance beyond this sentence, various investors[69] have already invested more than two billion dollars in the company. The highlight of "Magic Leaps" data glasses, which were only introduced in 2018, is the display, which, unlike the Oculus glasses, is transparent and so does not close the view of the environment. Thus, reality is not completely virtual, but augmented by the virtual part. However, these "augmented reality" glasses were previously known by the available data glasses "Google Glass" and "Microsoft Hololens" in a similar way. The "Google Glass" is strictly speaking not a pair of glasses, but a frame that holds a prism in front of the eye (Fig. 5.10). Via this prism, a small projector reflects the images of a small display directly into the eye of the wearer.

The "Hololens", like the "Magic Leap One", is a real pair of glasses. The basic function is roughly the same for both systems. The display no longer sits directly in front of the eye, as it does with Oculus and the like, but rather a light guide. However, this light guide is not a simple plate, but a waveguide with a somewhat more complicated structure. Basically, these waveguides have the decisive advantage of being able to be placed in front of the eye like a pair of glasses. The image is coupled into the waveguide at one side, corner or edge and coupled out again directly in front of the eye. The significant technologies in recent years have been "Surface Relief Gratings" (SRG) and "Volume Bragg Gratings" (VBG). Both technologies are based on the principle of diffraction gratings.

Optical Gratings

Diffraction gratings are not a new invention, but already known for centuries. The American judge Hopkinson was a man of many interests. One evening, for whatever

(continued)

[69] Google, Qualcomm, Alibaba and Axel Springer, among others.

reason, he had looked through his handkerchief at a street lamp. As he did so, he noticed black stripes which he thought were the enlarged threads of the handkerchief. However, these stripes did not move with the movement of the handkerchief, which puzzled Hopkinson beyond measure. He approached Rittenhouse in 1785 with his observation, and he made an analysis of it. Rittenhouse built a grid of hair to experimentally verify the phenomenon and was able to explain it with Newtonian diffraction. Thus, Rittenhouse had described the principle of a diffraction grating [118]. Since the light is diffracted differently depending on the wavelength, white light can be decomposed into its spectral colors with a grating like with a prism. Around 1820, Fraunhofer also observed this color separation on the thread grating he used. Just as he had previously done with the prism [119], Fraunhofer was now also able to detect the black absorption lines in the spectrum of sunlight with the thread grating [120].

In principle, an ordered structure does not have to be two-dimensional. The French mineralogist Haüy had noticed in 1784, when accidentally breaking a calcite crystal, that the fragments had a smooth, regular shape. Haüy packaged this observation into a geometric theory of crystal structure [121]. From this basic geometry Bravais then developed in 1848 into a whole system of points distributed in space. From a simple series of points, he developed a two-dimensional mesh. Laying several of these meshes with identical spacing on top of each other, one finally obtains a uniform spatial grid [122]. Laue used these Bravais grids to prove that X-rays[70] are also electromagnetic waves.[71] In 1913, Bragg[72] had studied and mathematically described the reflection of X-rays from various crystals [123].[73] The Bragg reflection now behaves differently from the known reflection at a mirror. In a mirror the laws of geometrical optics apply, the angle of incidence is equal to the angle of reflection. In Bragg's diffraction-reflection, the light is reflected at all only if it is at a certain angle.[74] This reflection[75] can be used when passing through the volume grating to deflect a beam of light. When the light hits the surface, it is reflected or absorbed depending on the wavelength.

[70]The crystal lattice would then have to act like an optical lattice and force the X-rays to interfere; this is what happened, and Laue was awarded the Nobel Prize for this in 1914.

[71]Discovered in 1895 by Röntgen, who received the very first Nobel Prize in Physics for it in 1901.

[72]Physicists William Henry Bragg and William Lawrence Bragg, father and son, were jointly awarded the Nobel Prize in 1915 for their discovery.

[73]Based on Laue's evidence.

[74]The glancing angle (from [123], p. 436).

[75]Whereby this is not really a reflection, but constructive or destructive interference (such as resonance and absorption).

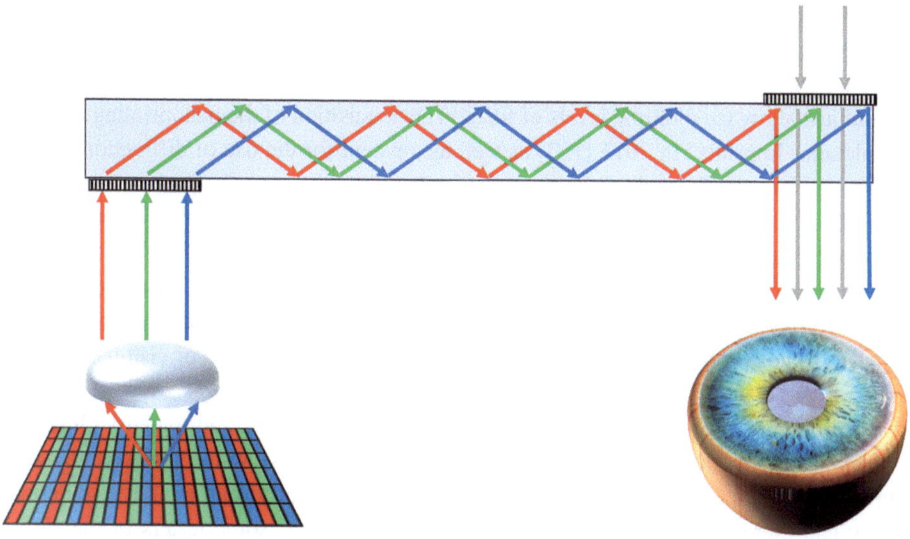

Fig. 5.11 Waveguide with diffraction grating, Grasnick (2019)

Microsoft's Hololens uses Nokia's idea of a structured waveguide ("Surface Relief Grating" SRG, see e.g [124].). For this purpose, tiny structures are engraved on the surface of a glass plate, which bends the image projected onto the glass plate at this structure. This causes the light to be deflected to an angle greater than the critical angle of total internal reflection (see Fig. 5.11). The glass plate thus becomes a waveguide. At the position of the observer (illustrated by the eye in Fig. 5.11), another diffraction grating is installed, which couples the coupled light rays out again by diffraction. The light coming from outside (grey arrows) can pass through the grating, so that the viewer can see the image and the surroundings at the same time. The virtual image overlays the real world.

The attentive observer will not have failed to notice that in Fig. 5.11 the different diffraction angles for different wavelengths are not shown for the sake of simplicity. Long-wave red light is diffracted more strongly than short-wave blue light, which is why several gratings[76] or several glass plates[77] are sometimes used.

Magic Leap calls its technology "Photonic Lightfield Chip" and has filed several patents for its technology.[78] It is not known exactly which of the many variants described therein

[76]Nokia, for example, uses another grating to change the direction of diffraction again so that the direction of the coupled-in and coupled-out beams remains identical [125].

[77]As different layers for wavelength-dependent coupling and decoupling of the light, not shown in Fig. 5.11.

[78]Among other things, a very extensive patent application with various configurations for the use of virtual and augmented reality with an incredible 490 pages [126]. Apparently, blocking future competitors is an important part of the corporate strategy.

will be used. However, it can be deduced from the various patents that the "chip" will actually be a passive optical element that should function similarly to the light guide of the Hololens. The difference is likely to be the use of multiple layers of the light guides, which would just allow for multiple levels of focus. The acquisition of the nanoprinting company "Molekular Imprints" in 2015 [127] as well as a patent for the use of diffraction gratings and zone plates [128] could point to the technology.

Both glasses fulfill the basic expectation of a system with superimposed reality. The overlay of the environment seen through the glasses with the virtual 3D objects succeeds. However, it becomes clear in the test that especially the small field of view and the weak contrast are problematic. The "Netzpiloten" magazine comes to the conclusion that "market maturity is not yet in sight" [129] and "c't" criticizes the narrow field of vision despite the wow effect [130].

Even the "Magic Leap One", which was announced for years and touted with revolutionary features, does not make the magical leap. The field of vision of these glasses is almost as limited as that of the "Hololens", the calculation of virtual reality is sometimes faulty and there is hardly any content or applications. Thus, some testers of the system judge disillusioned "For US$2 billion we would have expected more" (golem.de [131]), "Magic Leap's glasses, which are supposed to enable the magical leap of digital content into reality, still fail at times because of the technology." PC Games Hardware [132] judges the "unfortunately, the hoped-for revolution in hardware failed to materialize and the technology is not really ready for consumers yet" (VR Nerds [133]).

"Hololens" and "Magic Leap One" are also quite expensive due to the complicated structure.[79]

In order to still be able to offer the virtual reality set reasonably inexpensively, some manufacturers[80] therefore rely on the use of smartphones. This not only saves the display, but also provides another advantage for free. The necessary position and motion sensors are already built into the current devices.

With the headset, the virtual reality set would actually be complete. But you have to keep in mind that every hardware set also includes the interaction with the associated application. For a simple interaction with the 3D object,[81] a 3D mouse is sufficient, which has one of the described 3D sensors and the necessary buttons and keys for operation. This can be a stylus,[82] a trackball[83] held in the hand or a somewhat wacky device such as the

[79]Prices in 2019 were US$3500 for the Microsoft Hololens and around US$2300 for the Magic Leap Personal Bundle.

[80]E.g. Samsung Gear, Zeiss One, LG 360 VR or—as a cheap cardboard variant—Google Cardboard.

[81]e.g. rotate, move, zoom.

[82]By analogy with the Lightpen or Lincoln Wand.

[83]The input device developed at the German Aerospace Center was named "SpaceMouse".

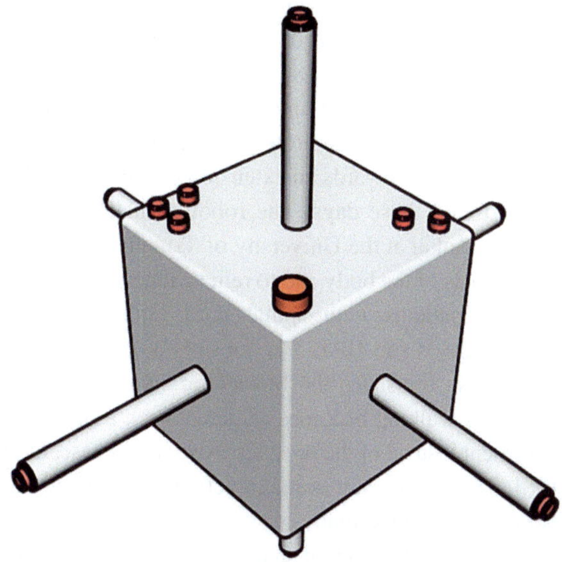

Fig. 5.12 Cubic Mouse, Hannes Grasnick (2019), illustration acc. to [134]

Cubic Mouse from GMD.[84] The computer scientist Fröhlich had developed a device there that could perform an extremely defined movement in virtual space by rotating the cube and dragging the axes [134] (see Fig. 5.12).

Such a device points to the need to develop special input devices for special requirements. It is obvious that for a virtual shooting training a data shotgun or another weapon-like device for data input will be excellently suited. A cyclist will prefer a bicycle simulator, while a runner will prefer a treadmill. Such treadmills[85] interact with the virtual world, generate real effort and, with a well-designed virtual environment, certainly provide a very realistic impression of the virtual action.

In addition to the aforementioned interaction possibilities, an additional external structure can also act on the user and thereby transmit forces. An obvious use case is the amplification of one's own force by an external metal skeleton,[86] which supports the movement of the wearer with servo motors. An early example of this is the "Handyman" and the "Hardiman", two exoskeletons developed by General Electric engineer Mosher.[87]

[84]The GMD Research Center for Information Technology was a large German research center in Sankt Augustin with a focus on mathematics and computer science, today divided into 8 institutes of the Fraunhofer-Gesellschaft.

[85]The word "treadmill" is used for a historical treadmill as well as for a modern treadmill; basically for anything where you move and don't get out of your seat.

[86]Exoskeleton, from Greek exo = outside + skeleton.

[87]Mosher had previously developed the "walking truck", a four-footed device weighing over a ton that could march at walking speed over rough terrain [135].

The Handyman was a two-armed master-slave device[88] that allowed a remote user to perform work in a dangerous location.[89] This mirrors the function of Waldo's glove and is a useful form of telepresence. Called the "Hardiman", the exoskeleton manipulator was a complete suit containing an operator who could lift loads weighing about 700 kg[90] in the suit [135]. Although the prototype was not fully operable (see Final Report [136]), it was still a great idea towards an extension of human capabilities.[91] A practical implementation can be found these days. The robotic suit HAL[92] was first developed by the Japanese scientist Sankai at the University of Tsukuba to understand the signal flow of the bioelectrical signals of the body and to relieve the wearer during heavy work. Today, it is also used to assist patients with spinal cord injuries or after a stroke during rehabilitation [137]. Today's suit HAL-5 is not only a data suit that records the positions of the limbs, a HAL also translates the bioelectric signals of the body into movement patterns. The thought of walking becomes an actual step, as the suit's servo motors perform the body's movement instead of the muscles. Also in this case, the HAL can be used to lift heavy loads [138]. The robotic suit is manufactured by Sankai's company "Cyberdyne" [139] and is also available in Germany.

One can well imagine that the servo motors of the HAL are not controlled by the thoughts of the wearer, but by an external program. The program, in turn, could be controlled by the wearer's movements in a virtual space and thus react according to the situation. When running uphill virtually, the suit could block the movement of the leg motors a bit, so it would be sweaty not just virtually but in real life. A virtual arm wrestle would be just as strenuous as a real one.

However, it doesn't always have to be a whole suit. Sutherland's idea of the ultimate display also includes a virtual chair on which one can sit in real life. In fact, a mechatronic device has been developed at the Institute of Robotics and Intelligent Systems at ETH Zurich that comes quite close to Sutherland's vision [140]. The "chairless chair",[93] which is manufactured according to this idea by a Swiss start-up called "Noonee", is in principle an exoskeleton. However, the skeleton is designed in such a way that when you sit down, you have the real support of a chair. If you now add an electronic triggering of the chair functionality, you would have fulfilled Sutherland's dream—at least in this point.

[88]The term "master-slave" refers to the form of control: a single "master" controls one or more "slaves" who carry out the master's instructions; conversely, the slave cannot take control.

[89]The plan was to use it in the processing of radioactive materials (see [135], p. 3).

[90]1500 lb.

[91]Referred to in the title of the report as "Machine Augmentation of Human Strength and Endurance".

[92]"Hybrid Assistive Limb".

[93]see [141].

5.4 Real Virtuality

From the Virtual into Reality

The Self-Replicating Interstellar Spaceship "Oumuamua"

When the asteroid "1I/2017 U1[94]" flew past the Earth from the depths of space in 2017,[95] initially only its interstellar origin aroused the interest of the discoverers. After all, this asteroid was the first ever observed object that definitely came from outside our solar system. An aspect ratio of 1/10 was determined for the cosmic chunk shortly after its discovery [142], an unusual elongated shape not exhibited by any other known object in the solar system. This flying cylinder,[96] about 200 m in diameter, thus has a shape that has been described repeatedly by science fiction writers[97] for decades. In the publication in the science magazine "Nature" the summary refers to a fantasy novel by Arthur C. Clarke.[98] This led to the assumption in the press that the object was an artificial body, perhaps even a kind of alien spaceship (e.g. [145–147]). In some cases, the story was expanded to include the idea of a self-replicating spaceship, a so-called Von Neumann probe ([148, 149]). A Von Neumann probe is a self-replicating system that works analogously to Neumann's theory of self-producing automata [150].

This probe would therefore be able to produce copies of itself, which in turn would be capable of further self-copies. These daughter probes could now set off in a different direction than the mother probe and replicate themselves again at some point during the journey. In principle, a new path can be taken with each replication and thus a larger part of the universe can be explored.

In fact, the asteroid, now under the Hawaiian name "Oumuamua[99]", was eavesdropped on during the flyby by the "Breakthrough Listen Initiative" using the massive Green Bank Radio Telescope[100] [152]. However, no evidence of extraterrestrial intelligence was found.[101]

---

[94] Discovered by the Pan-STARRS telescope in Hawaii (Pan-STARRS = Panoramic Survey Telescope and Rapid Response System).

[95] Approximately 24 million kilometers away.

[96] In the press, also often referred to as a cigar.

[97] Also in several issues of the Perry Rhodan series (e.g. [143, 144]).

[98] "Science fiction readers will remember Rendezvous with Rama by Arthur C. Clarke, where a cylindrical spacecraft from outside the Solar System passes through it at high speed" [142].

[99] The name is composed of 'ou (scout or scout) and mua (the very first, messenger), the second mua merely emphasizing and reinforcing the meaning of the first mua [151].

[100] Approximately 100 m in diameter.

[101] "No evidence of artificial signals..." [153].

The opening statement in the subtitle is intended as an illustration of what is already possible today on a small scale. A real object can emerge from an idea, a fiction or a construction plan. It is not mandatory that virtual objects always only seem to exist. New technologies enable the transition into the real world.

Since antiquity, artists' ideas have been transferred from the world of thought into real works as images or sculptures. The virtual entity thus became a real thing, something that can certainly be called "real virtuality". The name is meant to describe that in this case the real object is a copy of the virtual world.

If one disregards the purely manual implementation of ideas by artists, a similar process can also be found in the manufacture of everyday objects. The first hand axe was initially created in the imagination of prehistoric man, who then made it out of stone as a real object. A pottery did not exist until the idea was transformed into form and colour by the potter. Throughout the history of mankind there have always been new things that were not based on copying what had gone before, but were created according to the ideas of individuals. This creative process was no longer necessary for the manufacture of the utilitarian object. If the object existed or its manufacture could be described, others than the creative person were also able to make a copy.

For thousands of years, this remained the only way to produce real things. With the increasing mechanization of the working world and the resulting knowledge of mechanical processes, complicated devices could already be constructed in the Middle Ages.

The story of Pygmalion had repeatedly motivated artists and scholars to do the same as this sculptor and create human-like figures. The Dominican Albertus Magnus is said to have at least partially achieved this goal. He is said to have made a talking machine of metal,[102] which his pupil Thomas Aquino smashed out of fear (after [155] p. 500). Whether or how the device might have actually worked has not been handed down.

In 1615, the Heidelberg master builder de Caus had described all kinds of apparatuses that functioned on the basis of mechanical, hydraulic or pneumatic power transmissions [156]. By means of water power he was able to drive clocks and create fountains, but also to create moving figures that imitated bird song[103] or appeared to drink.[104] De Caus designed entire scenes in which figures moved as if by magic and melodies sounded. One example is a scene in which a nymph in a boat pulled by dolphins passes a music-making cyclops (Fig. 5.13).

Now this is indeed something that never existed in reality, but was made real for the viewer by the artist. However, de Caus' engineering art was not limited to water features in the Heidelberg Palatine Garden. De Caus also invented quite useful machines such as water

[102]Or, an ornamental column ([154], p. 164).

[103]"How to do all kinds of bird songs with water and wind" (from the "Gewaltsamen bewegungen" ("Violent movements") [156], First Book).

[104]"How to do to this a swan or other bird, which drinks out all the water so held out to it" (ibid.).

Fig. 5.13 Machine for Galatea with dolphins and cyclops playing shawms, Solomon des Caus (1615) from [156]

pumps, barrel organs, sawing machines and even lathes—an outstanding achievement for the time.

Such artful apparatuses were built again and again in the course of the time, whereby one fixed oneself thereby essentially on the art of the imitation of the reality.[105]

[105]Exemplary are the automata of the Swiss watchmaker family Jaquet-Droz, which amazed the audience especially with the three "androids", the writer, the draughtsman and the musician. The scribe wrote real texts of up to forty characters, the draughtsman could draw four different pictures and the musician played a repertoire of few melodies on her small pipe organ [157].

The philosopher Descartes had a mechanistic view of life and thus consequently saw animals only as complicated machines. In his reflection on the nature of the human soul,[106] he also ponders the possibility that humans are mere automata ([158], p. 36).

> ... alone I had often seen people pass by in the street from my window, of whom I ... used to say that I saw them myself. But what had I seen but hats and clothes, under which might be automatons, but which I took for people?

Descartes negated the "unchristian" idea of an equality of humans and animals, by denying the animals the soul and reason, which seemed to manifested itself especially in the missing language.[107]

During Descartes' lifetime, de Cau's machines were the most complicated automatons that could be made. It is undisputed that he surpassed his contemporaries in technical skill, yet hardly anyone would have mistaken any of his apparatuses for real living beings.

This changed, however, when the Frenchman Vaucanson demonstrated his amazing automata [159]. His mechanical flute player was quite capable of impressing the members of the Royal Academy of Sciences. The player could play twenty different pieces on a Provence pipe (a flute) and was at least of such good quality that even the renowned flutist Quantz[108] took up the game. The music was not played live, but "programmed" on a roller via various plates inserted into the roller. Thus, in principle, purely virtual pieces could be played that no composer had ever written down before.

As impressive as Vaucanson's life-size flute player was, his smaller work was even more so. Vaucanson had created a mechanical duck "which does all that a living duck can do" [160]. The duck could flap its wings and quack, eat and drink, and even digest and excrete food. This extremely realistic imitation of a real duck was quite capable of conveying the vision of an automaton world to Vaucanson's contemporaries[109] (Fig. 5.14).

---

[106] "Meditationes de prima philosophia, in qua Dei existentia et animae immortalitas demonstratur" (Zweite Abtheilung, Untersuchungen über die Grundlagen der Philosophie) [158].

[107] "While reason is a general instrument that can express itself in all kinds of excitations, these organs also require a special device for every special action, and therefore it is morally impossible that there should be so many of them in one machine to act in all the incidents of life as we can by reason. By these means we may also discern the difference between man and beast. For it is very strange that even the most dull-witted and stupid man, yes, even the insane, can combine individual words and make a speech out of them, by which they communicate their thoughts, while even the most perfect and best-begotten animal is not able to do this." (from "The Animals Have No Soul" ("Die Thiere haben keine Seele") in [158]).

[108] Johann Quantz, the flute teacher of Frederick the Great, however, also criticized the mechanical way of playing, since the high notes become "rough and unpleasant" due to the restricted movement of the player's lips (from his textbook "Versuch einer Anweisung die Flöte traversiere zu spielen" ("Attempting an instruction to play the traverse flute") of 1752 p. 46).

[109] And as Helmholtz remarks, to implant in them the desire for a perpetual motion machine ([161], p. 7).

Fig. 5.14 Vaucanson's automata, Jacques de Vaucanson (1738) from [160], frontispiece

In fact, the world had become increasingly mechanized in the 18th and 19th centuries, and automata for entertainment had become useful machines. Vaucanson, who was also an inspector of French silk manufactories, had invented an automatic loom. Like the fluting android, his loom was based on programming by a rotating roller. However, here no sheets were used, but different holes were punched in a cardboard and this was wound around the roller. Scanning these holes with needles resulted in the programming of the pattern [162]. Although Vaucanson's automatic loom was not successful, programmable looms were a real revolution and a cause of rising resentment among weavers. This was mainly due to the fact that since the introduction of programmable looms, simply less manpower was needed to operate them, and cheap factory-made goods (based on "automatic looms") displaced domestic looms, leading to the impoverishment of weavers and subsequently to the weavers' revolts. The idea of punched card control, made famous especially with the Jacquard loom, later spread outside the textile industry. With the use of punched cards in Babbage's calculating machine "Difference Engine" (see [163]), which he had started to develop in the 1920s [164], punched cards became a popular input system for the first computers (e.g. the Zuse Z3, the world's first programmable computer [165]).

A weaving machine creates a flat pattern, a computer only outputs the result of the calculation. The real translation of the virtual idea into a three-dimensional counterpart is not yet. With the development of photography, however, photographic sculptures were created.

The French writer Gautier was known to be a devotee of hashish consumption[110] and in this respect was certainly a friend of non-real perceptions. In fact, he was interested in the modern technology of photography and, in an essay for the "Moniteur universel", he described photosculpture, a hitherto unknown form of reproduction (from [168], p. 4).[111]

> That this age-old dream of fixing the fleeting image on the glass should have been realized was quite wonderful. The sun as a draftsman ! done; but the sun as a sculptor ! The bewildered imagination refuses to believe such a miracle.

He then describes, hardly less floweriły, the photographing of a person in a rotunda with 24 cameras simultaneously. The images are developed in Monsieur Willème's darkroom and assembled into a 3D sculpture. The principle is based on tracing the silhouette of the photographs with a pantograph. The pantograph is a simple mechanical device for transferring drawings of different scale, invented and described by Scheiner in 1603 [169]. With the help of the pantograph, an enlarged silhouette was created for each camera position. Cutting out these individual profiles and then assembling them into a cylinder yielded the 3D sculpture of what was being photographed [170]. In this way Willème created a real avatar of the person photographed, which looked amazingly similar to the real person (see Fig. 5.15).

In Hinduism, the avatar is the physical manifestation of a god; in contemporary understanding, it is the virtual equivalent of a real person, in Gautier's novel [171] more comparable to the soul-swapping Jake Sully from the movie Avatar, who becomes his own avatar at the end of the film.

Willème's form of lifelike representation quickly established itself in other countries. In Berlin, it was mainly Pötschke [172] and later Selke [173] who improved Willème's idea around the turn of the century and achieved some fame with their technique. Pötschke optimized edge sharpness of the silhouettes to be cut out by a vertical screen in front of the

[110] With the doctor Moreau founder of the club of hashish eaters, to which belonged besides Gautier and also Dumas, Balzac or Flaubert, Gautier has written about it a history [166], Moreau a scientific treatise [167].

[111] Own translation.

Fig. 5.15 François Willème, self-portrait, photosculpture, (1865) George Eastman House International Museum of Photography and Film, Rochester, NY, Collection G. Cromer/276

light source. The dividing line between light and shadow on the model then gives the outline of the silhouette [174]. Selke had even invented his own photosculpture apparatus [175] and he had also devised a method by which the shape of an object could be determined from stereo images by comparing corresponding points [173]. Selke had also "photosculpted" the chronophotographer Anschütz [176], so that Anschütz has survived in three dimensions.

These sculptures are not by chance reminiscent of the production of today's lifelike copies. Even in a current body scanner, many cameras are sometimes used to create a copy of reality.[112]

But this is not always necessary. In 1890, the former Austrian officer Blanther had the idea of producing three-dimensional maps from layered images. A map with contour lines was produced several times, cut out at the respective contour line and glued on top of each

[112]E.g. from the Austrian company 3Dcopysystems as "Big Alice" with 64 cameras [177] or from the German osensus GmbH as "3D Scamera One+" with up to 125 cameras.

other [178]. It is clear that not only realistic objects can be created on the basis of real areas of land, but that virtual landscapes and objects can also be created with this method.

Overlaying individual layers is a basic principle of 3D printing, but also in itself a process that became popular again in the 1990s.[113] One advantage is the ability to work with different materials such as paper, wood, ceramics and even metal. If the material is very thin, it can be glued on top of each other layer by layer from the roll, cutting it with a knife or laser [180].

This present-day automatic production possibility was not yet available in the nineteenth century. In addition, the generation of the individual layers of a layered image from a real object was hardly possible with the technology of the time. From the 1930s onwards, the conventional tomograph "Introskop"[114] existed, a device with which it was possible to create images of the inside of bodies. In the 1960s, the South African Cormack laid the theoretical foundations of computed tomography[115] (see [182, 183]), which Hounsfield translated into reality [184].[116] The result of a computed tomography is a multitude of shadow images belonging to certain spatial directions. From this, analogous to Willème's technology, a 3D object could already be created from the outer contours. For the reconstruction of the objects one uses today mathematical procedures,[117] in order to reconstruct from the shadow pictures the well-known section pictures. These could then be combined again into a Blanther 3D object.

In principle, in 3D printing the objects are always built up from layer images, which are gradually transferred, layer by layer, from the virtual 3D object to the real workpiece. The transfer of Willème's angular sections or Blanther's vertical sections, however, had the disadvantage of a strong manual component. Even if, as in Willème's photo sculptures, the originals were photographed, it was still necessary to trace the images and transfer them to the angular section. At the beginning of the twentieth century, Baese, an Italian engineer based in Berlin, proposed a much simpler option that did not require manual transfer. Baese illuminated the original from both sides with graduated light [186]. First, a photograph was taken with the rear parts of the object (as seen from the camera) illuminated most strongly, and the front parts of the object only dimly. A photograph taken from the front then shows an image in the negative as depth information. In Fig. 5.16, the illumination is shown in Figs. 1 and 2 (above). The object is illuminated with light from the side (shown with the

---

[113]Introduced as "Laminated Object Manufacturing" (LOM) by the American company "Helisys"in 1990 and later used especially in industrial mold making [179].

[114]The German name "Verwischungstomograph" (blurring tomograph) is based on the function of the device. During the exposure, the X-ray film (or sensor) and the X-ray tube are rotated around the object. A sharp image is thus produced only at the centre of rotation, the rest seems to be "blurred" (see [181], p. 9). The Siemens name "Introskop" is a compound of Latin intro–/intra- = into, inwards + – skop from Greek skopein = to look at.

[115]Tomography from Greek tome = cut + graphein = to write.

[116]However, in ignorance of Cormack's preliminary work.

[117]Especially the transformation described by Radon in 1917 [185].

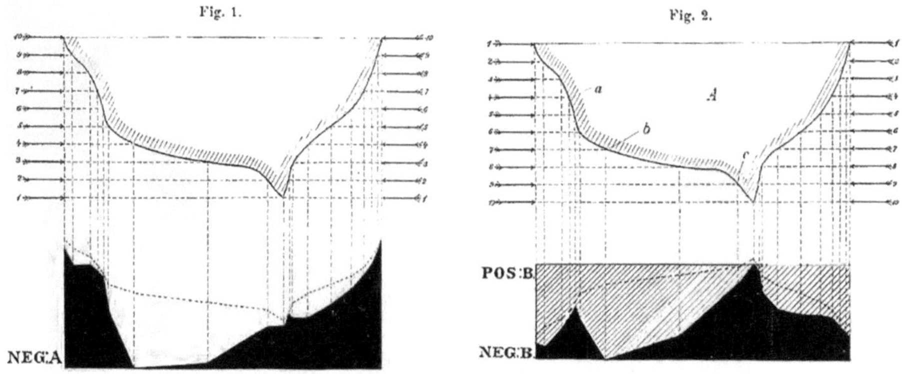

Fig. 5.16 Real photosculpture, lighting and negatives, Carlo Baese (1904) from [187], pp. 393 and 394

arrows). The numbers of the arrows represent the light intensity: level 10 is the strongest light, level 1 is the weakest. As can be seen, in Fig. 2 the lighting situation is reversed. Among the images of the object, Fig. 5.16 also shows the corresponding blackening on the photographic negative. As expected, greater blackening results with stronger illumination.

As can be clearly seen, neither of the two negatives gives the true shape of the object. A positive B is now produced from the negative B. In Fig. 5.16 this is illustrated in Fig. 2 as a hatching (marked POS:B) over the negative (NEG:B). It becomes clear what Baese intended by superimposing Negative A and Positive B. This works, of course, only if both images are superimposed. Of course, this only works if both images are mounted on glass plates, which was a common way in Baese's time. Surprisingly, the resulting blackening then corresponds to the contour of the object. By reversing the illumination with simultaneous inverse blackening in the second image, the resulting blackening is only a function of the outer contour[118] (Fig. 5.17).

With this new negative, a thick chromium gelatine layer is now exposed. Since the swelling capacity of the gelatine depends on the duration of the exposure,[119] the gelatine layer in the strongly exposed areas will swell less in water than in the less or non-exposed areas. From this, a relief could finally be moulded, e.g. with plaster. Since the gelatine layer can naturally only have a small height, a real sculpture must again be composed of several

[118]"All factors, therefore, which affected both photographs in the same way, to which also the effect of the color of the model and even possible reflections of the lens, errors in the condenser, etc. are to be counted, are annihilated by the compensation, in order to allow only the gradation to be valid." (from [187], p. 397).

[119]The light sensitivity of chromium gelatine and its suitability for relief photography was already known at the end of the nineteenth century (see e.g. [188]) and was generally used for collotype printing [189], later the influence of temperature was noticed and used for finer reliefs [190]. In general, substances made light-sensitive (here by chromates), which are present in the gel state (egg white, glue, shellac or celluloid), are suitable for this purpose.

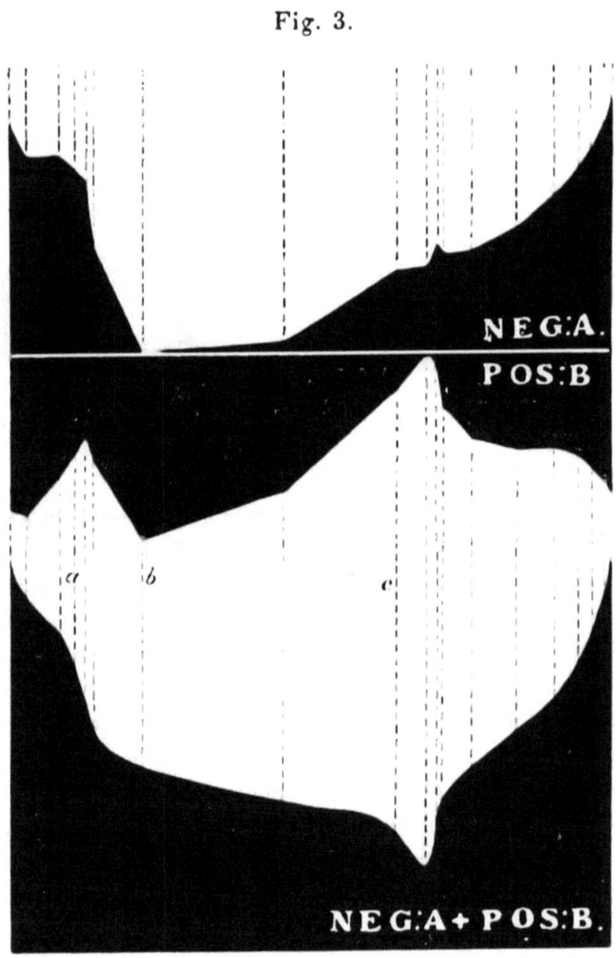

Fig. 5.17 Image superposition as depth image, Carlo Baese (1904) from [187], p. 395

layers. Baese recommends four layers for half of a portrait bust in half natural size ([187], p. 398), which would require a layer thickness of several centimeters after all. The technique of making photographic reliefs seems quite promising, but despite a patent by Baese [191], it did not catch on as a method for making sculptures. Photographic reliefs, however, were used long before and after as collotype[120] mainly for the realization of small editions in art printing.

It is difficult to discern from Baese's sketches whether or not his process could really deliver sufficient depth. Therefore, the test is repeated here once again with modern means.

[120] Also collotype (because of the colloids), heliogravure (engraving with sunlight).

Fig. 5.18 Depth image acc. to Baese, Hannes Grasnick (2019), 3D model of Three D scans [192]

A photosculpture by Pötschke,[121] representing the poet Goethe and available as a 3D scan [192], was illuminated in a 3D program according to Baese's instructions and the resulting images were superimposed (Fig. 5.18). In the figure, the illumination is front light/rear dark on the left, front dark/rear light in the middle, and the superposition of the two images[122] is on the far right. The result indeed shows clear characteristics of a depth image. The front areas are bright, those further back are dark.

From Depth-Maps, Time-of-Flight and Structured Light

The innovative way of obtaining a depth map of real objects had no practical applications in Baese's time except for the photo-sculpture he described. Today, the creation of depth maps is an important area in the acquisition of space. The usual way to do this is to use stereo cameras. From the disparity of the stereo image pair, the horizontal displacement (disparity) can be determined for most points of the stereo image and the depth map can be calculated from this.

Time-of-flight cameras are often used to register a player in a virtual space. The best-known example of time-of-flight technology is the Microsoft Kinect sensor [193]. This uses a pulse of light to illuminate the scene and determines the travel time of the reflected light at each pixel. The technique was developed towards the end of the 1970s [194] and, when equipped with a laser scanner, is capable of capturing scenes with extents greater than one kilometer to millimeter accuracy [195]. The result is a point cloud corresponding to a depth map from the measurement position.

(continued)

[121] Its manufacture is described as early as 1893 (in [172], p. 497).

[122] One of which was inverted as requested, the exposure was adjusted a little (just as Baese would have been able to do in his photo lab).

With several measurement positions around the real object, the complete 3D model can be assembled from several point clouds.

As Baese had shown, an object can be captured in its spatial extent by suitable illumination. The Japanese Morioka had an even better idea in 1933. He projected a number of lines onto the object, photographed them, and was able to produce a large number of sectional images on the basis of the course of the lines, which he assembled into a copy [196]. Morioka did not yet have any need for mathematical reconstruction of the objects, but confined himself to the purely mechanical realization in portrait sculptures.[123] Today, complete 3D objects are captured by means of fringe projection with accuracies of less than one millimeter (see e. g [197].).

One of the main reasons for the lack of commercial success of the light relief was probably the shallow depth of the 3D objects, which were less sculpture than bas-relief with low stiacciato.[124]

In his patent of 1922, the Australian Monteath objected to the low thickness that could be achieved with gelatine plates in general [198]. Monteath first proposed as a solution an illumination of the object exclusively from the front side and positioned the object to be photographed in a chamber lined with black velvet. The solution certainly does not lead to a true image of the external form, but reproduces the edges and colours of the object through the illumination. It is clear from the writing that Monteath is only interested in bas-reliefs and that his process is essentially an optimization of the production of the gelatin layer. However, his casting process in a mould also only achieves a layer thickness of about three millimetres.

That is too small for a real 3D object. It was not until 1951 that the American lawyer Munz succeeded in producing larger formats. His process, called "Photo-Glyph Recording[125]" [199], resembles the processes of Baese and Monteath in that reliefs of layered images are also created here. In contrast to Baese's process, the individual layers no longer have to be produced individually and joined together by hand in order to produce larger sculptures. In Munz's case, this is done by the Photo-Glyph-Recorder. His drawing from the German patent specification (Fig. 5.19) shows the apparatus. It is basically a large container with the lifting table adjustable in height. A thin liquid photosensitive layer is poured onto this lifting table from a reservoir and illuminated by light source. This light source can also be a cathode ray tube, which is focused on the layer by means of a lens, so that the thin layer can be exposed entirely without a negative. Then the necessary photochemical processes of developing, fixing and hardening take place. The lifting table is then

[123] Morioka's company "Rittaishashinzo" made more than 30,000 sculptures on this basis until 1990, including Presidents Johnson, Ford, Carter and Reagan.

[124] A bas-relief of shallow depth, such as Donatello in particular made.

[125] A glyph, Greek for something engraved, refers to the graphic representation of a sign.

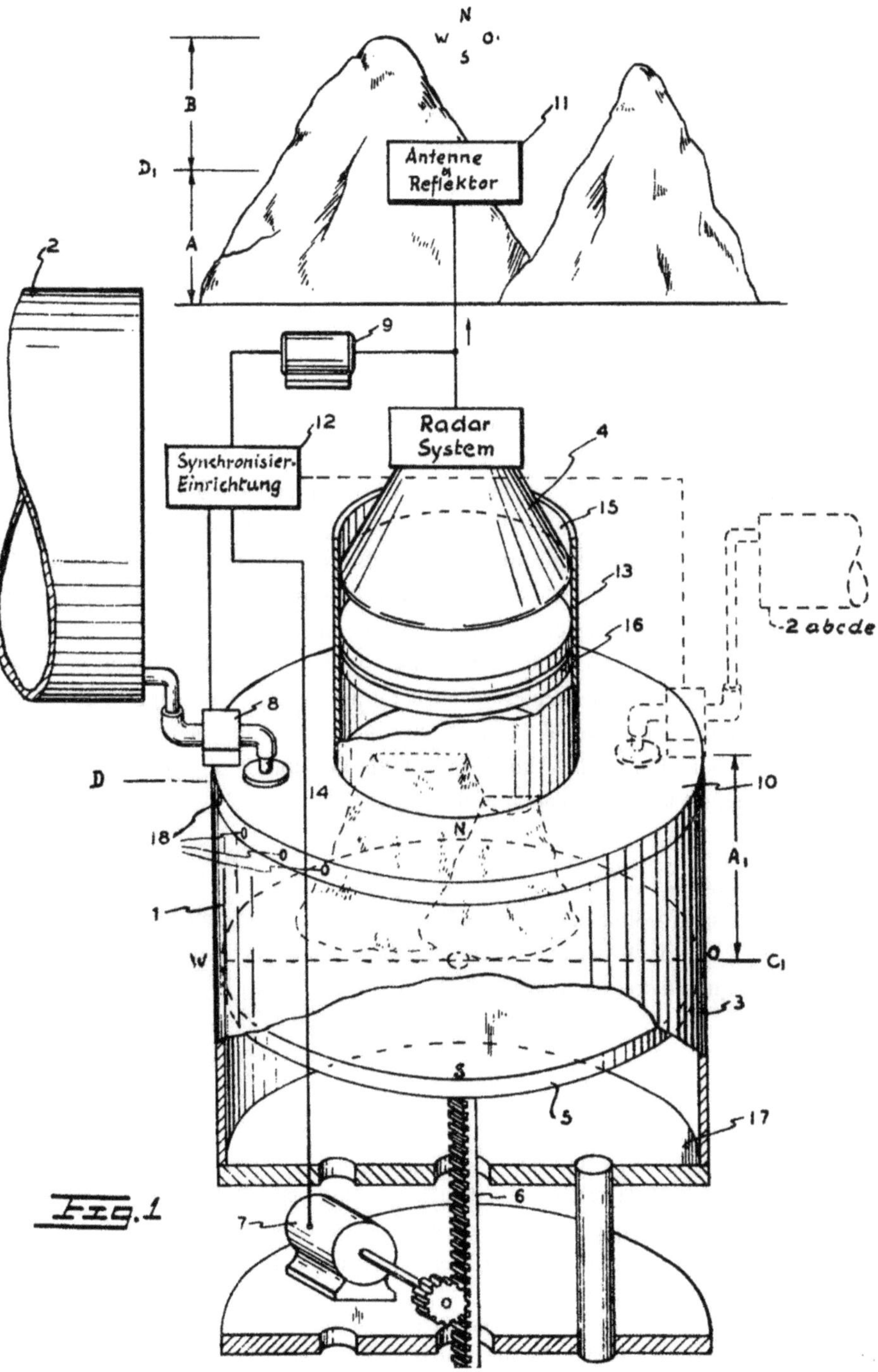

Fig. 5.19 Apparatus for making photoreliefs, Otto John Munz (1956) from [200]

lowered by one layer thickness and the next layer is applied. The process is repeated; exposure, development, fixing, hardening...

At some point the desired total thickness is reached and the printing process is finished. The last step is to wash out the unexposed parts. After that, the 3D object is ready.

The process Munz describes here is very similar to that of modern stereolithography.[126] There, too, a photosensitive material is exposed layer by layer.

It therefore seems surprising that Munz's invention met with little response. The reason here is probably the lack of 3D objects that were suitable for 3D printing in the first place. Until now, the objects to be printed were always created as a copy of an original or had to be created layer by layer. It was only with the advent of computer technology that the real possibility arose of having the virtual 3D models created in the computer also come into being in reality.

A much more advanced approach was pursued by the Dane Swainson in the late 1960s [201]. Like Munz, he also used an electromagnetic radiation source to generate higher intensities at defined points in space, but can do without transporting the material by means of a lifting platform. With Swainson, the entire block of material, which can be liquid, solid or gaseous, is now the projection surface, which also requires a three-dimensional projector. The basic idea is to intersect two rays at a specific point in the image, resulting in a higher intensity at the intersection point and the intended eye-visible[127] or material-changing effect. The elegance of the solution lies in the image source. Swainson proposes holograms as the image source, reconstructed at the appropriate wavelength[128] to introduce real intensities into the material block.[129] Swainson combines at this early stage his knowledge of holograms with knowledge of the possibility of generating computer-generated programs and the radiation of the laser.

Holograms

When Swainson filed his patent in 1967, holograms were still something very new. The Nobel Prize for the discovery of holography had not yet been awarded, the laser was still a solution seeking a problem.[130]

(continued)

[126] From Greek stereo = physical + lithos = stone + graphein = writing, also stereolithography apparatus (SLA), photo-solidification, resin printing or simply optical fabrication.

[127] Swainson is already thinking about three-dimensional television here.

[128] Visible light, but also infrared, ultraviolet, microwaves, ultrasound and even X-rays (which argues a little against its use as a television).

[129] For the sake of completeness, it is worth mentioning the possibility of directly crossing two beams to generate energy points, which is also found in Swainson's patent.

[130] "A laser is a solution looking for a problem"; a frequently quoted statement by laser inventor Maiman.

Gabor, who was to receive the Nobel Prize only in 1971, made his discovery in 1947 while attempting to improve the optics of the electron microscope, whose up to then best possible objective he compared with the imaging quality of a raindrop [202]. His basic idea was to record the image with the quality that was practically achievable, but with "all" information that could be recorded. Thus, it was immediately clear to the electron optician Gabor that, in addition to the intensity of the image, its phase would also have to be recorded. To do this, light must first be considered as a wave, which has been a well-known model since Huygens. The light spreads like a water wave evenly in all directions and takes the known form (see Fig. 5.20 above).

Below the picture, the waveform of the wave is illustrated in a cross-section (indicated by the red arrow in the wave photo). The red wave has a certain wavelength[131] and amplitude[132] The phase of a wave describes its position at a certain point in time compared to a reference. If the red wave is the reference, the blue wave is shifted to the red wave by a certain phase difference. The two waves superimpose in certain ways, as Young recognized as early as 1802: "When two Undulations, from different Origins, coincide either perfectly or very nearly in Direction, their joint effect is a Combination of the Motions belonging to each.".[133] This superposition is the essence of interference.[134] An explanatory sketch illustrates the effects (Fig. 5.21). In the upper diagram, the blue wave is offset from the red by about half a wavelength. In the addition of the values, the resulting amplitude (green) falls almost to zero. The result would be a weakening of the intensity[135] in the image (shown as a faint gray-black change above the diagram). In the lower diagram, the two wave trains (red, blue) are almost on top of each other and thus produce a higher amplitude[136] (green) in the addition.

Fresnel had already observed and described these interference patterns in 1815 [204].

Knowing the research of his predecessors, Gabor was aware that he needed a reference wave to determine the phase, thereby determining the phase shift. For this to succeed, the wave field must have the property of coherence,[137] by which is meant

(continued)

[131] The distance between two crests - or troughs.

[132] Not intensity, but quadratically proportional to it, so one can be derived from the other.

[133] From [203], p. 34.

[134] From lat. Inter = between + ferire = hit.

[135] Destructive interference.

[136] Constructive interference.

[137] Lat.: cohaerere = interrelate.

that the light is emitted in a specific wavelength (monochromatic) and that there must be a fixed phase shift. This coherent reference light now interferes on the recording film with the light modified by the object. The interference patterns created on the film by the reference and object rays produce the hologram. Gabor's technique first imaged a tiny plate less than 1.5 mm in diameter with the names Huygens, Young and Fresnel on it. In the hologram, the letters disappeared and were replaced by the typical interference pattern. However, if the plate was illuminated again with the reference beam, the names became (somewhat) readable again [205].

At first, however, Gabor himself had little joy in his discovery, since the quality of the image was quite low. The reason for this was the low quality (in terms of coherence) of his light source, a simple mercury vapour lamp. Several more years passed before Maiman developed a working visible light laser in 1960 [206]. From the beginning, the laser was capable of emitting coherent light at high intensity.[138] Again, the technology was earlier than the application and the laser was by no means a blockbuster.

A really useful application for the laser was presented by Leith and Upatnieks in 1964 [208]. By means of Gabor's holography, they had produced a three-dimensional photograph of a real object,[139] which also allowed motion parallax and supported the possibility of focusing on objects at different distances in the image. Since the early 1960s, the two physicists had been working on Gabor's holography (see [209]) and had introduced two important improvements to it. Gabor had held the object to be recorded directly into the reference beam, creating a hologram known as an in-line hologram. When viewed, the reference beam produces both a real and a virtual image simultaneously, which is of course distracting. Leith and Upatnieks separated the beam into an illumination beam and a reference beam. This made the double image disappear and the hologram looked even more convincing. For this purpose they had a powerful coherent light source at their disposal, the laser [210].

Holography became quite popular in the 1970s and many researchers participated in the invention.[140] It seemed for a while that holography could solve all the problems of 3D imaging. As with any hype, it turned out that the technical realization

(continued)

[138] See e.g. Maiman's patent of 1961 [207], Maiman uses Gould's term laser "light amplification by stimulated emission of radiation", but also "optical maser".

[139] The object itself was not very spectacular: a toy train and a toy bird.

[140] Without judgement or preference, at least some personalities are mentioned here: two authoritative researchers (Yuri Denisyuk [211] and Steve Benton [212]) and two well-known holography artists (Lloyd Cross [213] and Fred Unterseher [214], with the joint company "Multiplex" producer of multiplex holograms like "The Kiss" or "The Brain of Alice Cooper" for Salvador Dali).

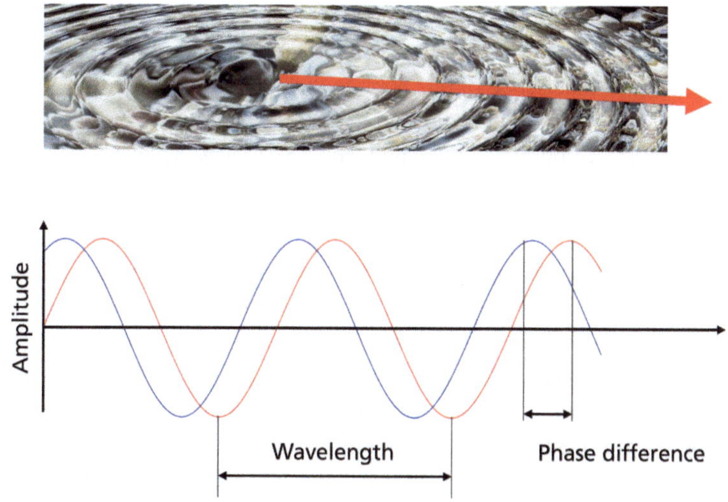

Fig. 5.20 Water wave with undulation, Grasnick (2019), photo by Gerd Altmann (Pixabay)

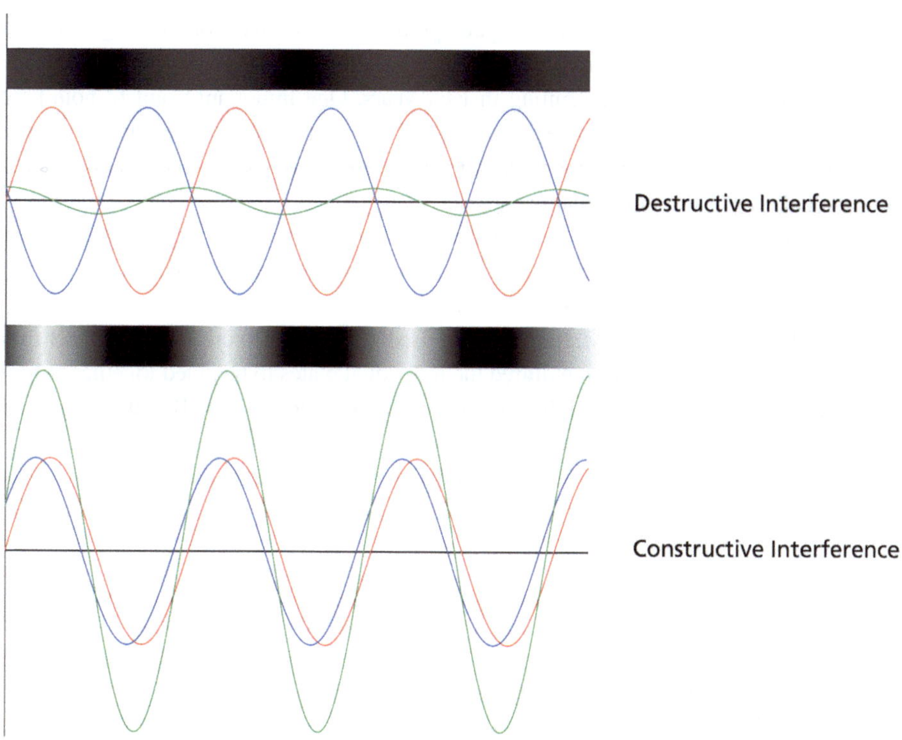

Fig. 5.21 Interference for amplification and attenuation, Grasnick (2019)

could not keep up with the expectations. A holographic television is still an illusion. It would go beyond the scope of this book to include a complete history of holography or to discuss every single development. For those with an honest interest, Johnston's book "Holographic Visions: a History of New Science" [215] is highly recommended.

It is clear that for a conversion of the latent (volatile) spatial image, a material must also exist that can react with the radiation of the hologram. In the case of illumination with light, this can be photosensitive materials that are fixed by heat. Thus, one beam can carry the light radiation and a second can carry the heat radiation. At the point of intersection, heat and light meet and the point is fixed. Swainson also outlines some possible materials,[141] which can be used to create the desired figure. Perhaps the most interesting part is that which deals with the use of plastics and their transition from the soluble to the insoluble state, e.g. by polymerisation or melting. Successful experiments were made with polyester resin and ultraviolet illumination, which gave a "fully polymerized and rigid figure" (from [201], p. 31). A phosphorus liquid[142] gave a greenish spot under ultraviolet light which disappeared by stirring, various other liquids produced light effects or solid figures under different radiation. More generally, Swainson's patent had already described much that would be implemented in 3D printing in later years. One important point to note here: Swainson did not distinguish between durable and ephemeral objects, the way 3D printing and 3D display are distinguished today. In any case, it is the production of a three-dimensional object; just either real or virtual.

Swainson contributed further developments,[143] but was probably a little ahead of his time and not very successful with his company Formigraphic. In general, however, he was one of the first to use computers to create 3D objects.

The American engineer DiMatteo had also used a computer to create a 3D object from individual plates, but this in turn required the individual plates to be glued together to form a complete object [218]. In principle, the process differs little from the Blanther or Helisys processes already described.

Like the Dane Swainson, the Japanese Kodama also described a manufacturing process based on curable polymers [219]. To Kodama's credit, he developed a thoroughly

[141] Actually, more their properties: photosensitive, fluorescent, phosphorescent, photochromic or thermochromic, so that the device is not only a printer, but also a volume display.

[142] An unhealthy mixture of samarium, cerium and strontium sulphide phosphorus powder (see for example the chapters on fluorescence and phosphorescence in "The Theory of Radiant Energy" ("Die Lehre von der strahlenden Energie") [216].

[143] Of these, only the paper "Method for generating changes in the properties of a storage medium in three dimensions" shall be mentioned here [217], in which—in short—a 3D volume memory based on multiphoton absorption as well as a variable lens via 3D change of the refractive index is described—fascinating!

practicable device. From his description it is clear that Kodama intended a cost-effective conversion of a three-dimensional model from the computer into reality. For this purpose, he used a commercially available light-curing polymer.[144] A light guide was used to direct an ultra-violet lamp onto a commercially available xy plotter, which was fixed above the liquid polymer. A shutter allowed the computer to control the coupling of the ultraviolet light into the light guide so that the polymer was selectively cured. As with Munz's apparatus, Kodama lowered the lift by the layer thickness, applied new polymer, and exposed the next layer. This 3D printer was already fine and fast enough[145] to print smaller models. So the Japanese physicist printed a few models and presented his idea at conferences, to machine manufacturers and in trade journals. After finding no interest in his idea and missing the deadline for his patent application, Kodama ended his career as a researcher and became a patent attorney (from an interview with Kodama [220]).

The American physicist Hull, who was working on the same topic at the same time, was more successful. He had spent months in his laboratory experimenting with his idea to create objects from photopolymer and UV light. Here, too, enthusiasm was initially limited. Hull himself recounted in an interview that when he called her into his lab at bedtime to show her his first result, his wife exclaimed, "This had better be good!" [221].

Hull applied for a patent in 1984, which is extremely similar to Kodama's design [222]. This patent already contains the designation "stereolithography", for which Hull can thus be considered the originator. Hull combined the technology for producing a three-dimensional copy with the processes for generating 3D files in "Computer Aided Design[146]" (CAD) and "Computer Aided Manufacturing[147]" (CAM), which were already known at his time, to form stereolithography. The 3D body description of the print files "STL" developed by Hull's company 3D Systems was as an abbreviation for "stereolithography" initially the file name extension (*.stl, see e.g [225], p. 31), but also from the beginning the description of the body surface by individual triangles.[148]

[144] "Tevista" from Teijin (a Japanese chemical company).

[145] With a dot size of 0.1 × 2 mm and a layer thickness of 2 mm, the printer required a printing time of 4.5 h for an object of 70 × 50 × 54 mm.

[146] Computer-aided design, known at the latest since the development of the programming language APT (Automatically Programmed Tool), from 1956—at the Massachusetts Institute of Technology (see [223], p. 35), Douglas Ross [224] is regarded as the "inventor" of the term "computer-aided design" (and "-manufacturing").

[147] Computer-aided manufacturing, Douglas Ross is also mentioned for this, who actually milled an ashtray out of aluminium in 1959 with the aid of a computer (via APT) [224].

[148] The so-called "tessellation" of surfaces by smaller geometric figures has been known in principle since Plato's description of the "Platonic" solids; Plato's octahedron consists of 8 triangles, but Hull can process up to 14,000 triangles with his printing apparatus SLA-1 (see [226], p. 50).

Although Hull may not have been the only inventor with his 3D printer,[149] he established a new market with the commercialization of three-dimensional printing technology. Of course, other materials were also expressed. The introduction of selective laser sintering (SLS[150]) by the American Deckard starting in 1986 [230] enabled the fabrication of functional components from the powder of various metals and plastics. The finished objects can be very hard due to the materials. This means that they can be used immediately and can also be manufactured with undercuts[151] without support structures. The objects do not necessarily have to be described from the common materials plastic or metal. Swainson had already pointed out at the end of the 1960s that ultimately only two things are needed: Directed energy radiation and a medium in which a physical change is induced (in [201] from p. 37). If in Swainson's case the aggregate is still between two crossed beams of energy, at the intersection of which the physical effect occurs, a single beam is sufficient for Hull.[152] Hull's "solidification" (here the transition from the liquid to the solid state of aggregation) is replaced by Deckard by a heating of the powder and a thereby caused connection of the particles. As a result, the usable materials are more diverse and extend from plastics and polymers to metal, ceramic materials or even composites, i.e. composite materials made of several materials (in [230], Summary). The process of laser sintering differs from laser melting in that in sintering, by definition, the powder melts only superficially, causing the individual grains to bond tightly together as they solidify. In melting, on the other hand, the material actually goes into liquid melt. Whether the material is molten or not (and therefore whether it is laser sintered or laser melted) can be easily determined for metals by exceeding the melting point; for glass or thermoplastics, the "glass transition temperature" applies here, at which these materials change from a solid to a viscous state.

Laser melting was significantly developed by the Fraunhofer Institute for Laser Technology (ILT) in Aachen[153] in order to be able to realize not only the outer form but also the material properties in industrial prototype production, which enables direct testing of the

[149] In addition to all those already mentioned, the Frenchmen André, de Witte and Le Méhauté are described as particularly unfortunate, as they had developed a 3D printer shortly before Hull [227], but the development was not pursued by the applicant at the time (CILAS ALCATEL) [228].

[150] Whereby the name "Selective Laser Sintering" only appears in a later application (1989) by Beaman and Deckard [229].

[151] If, for example, a sphere is to be printed not only as a hemisphere (relief) but as a full sphere, then the rear shape (which the viewer does not see) would have to become smaller and smaller from the equator towards the back (i.e. it would have to be cut in, as it were).

[152] But again, not limited to the light from the laser, but described generally as a beam (e.g., electron beam, beam of energetic particles, X-ray or UV beam, and even a matter beam of reactive chemicals) see, e.g., in [231] claims 8–14.

[153] In 2011, this was rewarded with the Innovation Award of the State of NRW.

"printed" prototypes [232]. The process was optimized and popularized by the Lübeck-based company SLM Solutions Group under the name "Selective Laser Melting[154]". In order to introduce even higher energies into the powder and thus also to print materials with a high melting point (e.g. implants made of titanium), electron beams are also used for melting.[155] An electron beam gun is based on the principle of the Braun tube, but is optimized for high power.

All the above mentioned processes are rather complex and more suitable for the professional user than for the interested enthusiast and are more manufacturing machines than 3D printers. What is generally considered a 3D printer today houses a much simpler technology. American engineer Crum was looking for a simple, cheap and fast way to make prototypes in the late 1980s. His starting point, similar to Kodama's, was to observe a 2D plot. On a plotter, the pen moves across the drawing surface, leaving an ink trail on the paper. What if, Crump thought, melted plastic came out of the pen instead of ink? The implementation was quite simple. In the 80s hot glue guns were already quite common (in the USA probably preferably 3M's "Polygun") and so Crump also experimented with such a device [235]. His 1989 patent aims at a device that can be connected to an office computer and produce a 3D object in a reasonable time. In this process, the print head moves over a surface, printing the liquefied print material onto the surface in a defined amount. Crump's key innovation is not the displacement of the print head or the liquefaction of the material. It is the type of material and its delivery to the nozzle. The print substance is fusible material rolled up in a flexible strand on a spool. Crump's company, StrataSys, called the process "fused deposition modeling". Since the associated abbreviation "FDM" is a trademark registered by StrataSys, the abbreviation FFF has become established in the open source community for the same process. This stands for "Fused Filament Fabrication[156]". This designation was suggested by the members of the RepRap[157] project in order to avoid possible legal problems with StrataSys. In fact, the initiator Bowyer intended to use Neuman's self-replication for the benefit of society in the sense of democratizing the means of production. Bowyer himself writes (from [238]), "So the replicating rapid prototyping machine will allow the revolutionary ownership, by the proletariat, of the means of production. But it will do so without all that messy and dangerous revolution stuff, and even without all that messy and dangerous industrial stuff. Therefore I have decided to call this process Darwinian Marxism...".

[154]Selective Laser Melting, SLM is also a registered trademark of SLM Solutions Group.

[155]E.g. by the Swedish company Arcam, whose founder Larson applied for a patent in 1993, where the energy was still entered into the material in the form of unspecific "energy waves" [233], later it was the "electron gun" [234], Arcam also holds the trademark EBM (for "Electron Beam Melting").

[156]Melt-filament production, with filament (fibre) from Latin filamentum = thread work.

[157]RepRap is short for "replicating" and "rapid prototyping", a machine that can print its own parts and can be replicated by anyone. The open source project [236] was developed in 2004 by the British engineer and mathematician Bowyer [237].

The researchers from the "Three Dimensional Printing Laboratory" at the Massachusetts Institute of Technology took a different approach. This is meant not only technically, but also commercially, because MIT not only developed the process, but later licensed it to the 3D printing industry. In the process, which was developed at MIT as early as 1989, the powder is bonded with a binder that is applied by a print head [239]. Except for the powder stage, the process is most similar to that of normal printing. When 3D printing[158] is mentioned, this process is usually meant.

The basic processes for all technologies were patented in the 1980s. In the meantime, these property rights have expired, which explains the large number of market participants today.

Points of Light in the Space

3D printing and all other additive or generative manufacturing processes allow the realization of virtuality in a limited space. It is always additive manufacturing when something is added to the component during production; we speak of generative manufacturing when the component is "created" from the medium, so to speak. The term solidification, which has already been used several times, describes what happens in this process. From an undefined substrate, a powder or a liquid, objects are created in space. If one were to imagine this space divided into a three-dimensional spatial grid, each individual point in space could be addressed by its x-y-z coordinate. Viewed as a digital image, these points in space would then no longer be pixels, but voxels or "volume pixels".

From the beginning, the idea of 3D prototyping could not be separated from 3D display. Swainson had also connected three-dimensional moving images in the volume with his process and thus designed the basis for a three-dimensional color television system.

Although the idea was not yet implemented by Swainson, it was later applied in some volumetric displays. The basic idea behind it is the defined control of individual voxels in a room, which can then be seen from all sides at a certain position. This creates a new problem; you need some substance in the volume where you can activate light points. This can be gases, liquids or solids that can somehow be excited by energy radiation.[159] A simpler way is to project onto multiple planes. The simplest

(continued)

[158] Of course, the terms 3D print, binder jetting or inkjet 3D print are also used.

[159] There were some inventions for this purpose, but all of them did not get beyond the prototype stage; a comprehensive compilation can be found in [240].

liquid crystal displays can be used for this, which have no polarizing filters stuck on them and can switch from transparent to scattering. The American company Lightspace[160] had arranged several such panels one behind the other to create a layered image with multiple planes [242]. This static volume display also requires more and more projection planes when the space is finely staggered. However, the increase in the number of planes cannot be unlimited, since the LC layers cannot become completely transparent, but always scatter a little.

So you need the smallest possible number of levels with a large volume and a small distance between levels at the same time. Some young people from the VFN[161] in Stade had an exciting idea. They deformed a projection screen into a helix and projected spatial figures onto it with a laser while it rotated [243]. A very similar form of projection was chosen by the American company Actuality Systems, whose "Perspecta" system presented the projection in a small dome (approx. 30 cm in diameter) [244]. Despite the fact that the Perspecta system—at least for a while—had a certain resonance, such systems are still immature. The projection surface is usually small, the display of the graphics rather simple and the system expensive.

However, such systems have a decisive advantage. You do not need to print out the 3D file and can still view it from all sides as in reality.

Literature

1. The Palace Museum. Zhang Zeduan: life along the Bian River at the Pure Brightness Festival, Painting. The Palace Museum – Collections. 2019. https://en.dpm.org.cn/collections/collections/2009-10-19/990.html. Accessed on: 17. Aug. 2019.
2. Barker R. Specification of the Patent granted to Mr. Robert Barker, of the City of Edinburgh, Portrait Painter; for his Invention of an entire new Contrivance or Apparatus, called by him La Nature a Coup d'Œil. Repertory of Arts and Manufactures. London: Wilkie and Robinson; 1796. pp. 165–167.
3. University of Cambridge. A panoramic view I A damned serious business. Panorama, Leicester Square. https://exhibitions.lib.cam.ac.uk/waterloo/artifacts/pam-6-81-231-folding-plate-opposite-title-page/. Accessed on: 17. Aug. 2019.
4. Daguerre LJM. Das Daguerreotyp und das Diorama. Stuttgart: J. B. Metzlersche Buchhandlung; 1839.
5. le Gall G. Die Welt von Paris aus betrachtet. In: Wedekind G, Haug S, editors. Die Stadt und ihre Bildmedien. Paderborn: Wilhelm Fink; 2018.
6. Wood RD. The diorama in great Britain in the 1820s. Hist Photogr. 1993;17:284–95.

[160] Now resident in Latvia [241].

[161] Association for the promotion of young scientists, which has been working on the development of volumetric displays since the 1990s.

7. von Zedlitz L, editor. Neustes Conversations-Handbuch für Berlin und Potsdam zum täglichen Gebrauch der Einheimischen und Fremden aller Stände. Berlin: Eifersdorff; 1834.
8. Bertuch FJ. Porte-Feuille – Bertuch's Bilderbuch. Weimar: Verlag des Gr. H. S. priv. Landes-Industrie-Comptoir; 1830.
9. Milano A. Wunderbare Lichteffekte mit Papier- Tag- und Nachtbilder in den Guckkasten gezaubert. In: Brückner W, Vanja K, Lorenz D, Nagy S, editors. Arbeitskreis Bild Druck Papier Tagungsband Modena 2010 Il Gruppo di Studio Bild Druck Papier (Immagine Stampa Carta) a Modena. Münster: Waxmann; 2011.
10. Mannoni L. Le grand art de la lumière et de l'ombre: archéologie du cinéma. Paris: Nathan; 1995.
11. Menier & Cie, Pharmaciens-Droguistes. Maison Centrale de Droguerie, Prix Courant Général. Paris: Plon Frères; 1854.
12. Houdin R. Magie et physique amusante. Paris: Ancienne Maison Michel Lévvy Freres; 1877.
13. Evans HR. The old and the new magic. Chicago: Kegan Paul, Trench, Trübner & Co.; 1906.
14. Mannoni L. La lanterne magique du Boulevard du Crime. Henri Robin Fantas Magic 1895;1994 (16):5–26.
15. Brockhaus, Editor. Conversations-Lexicon der neuesten Zeit und Literatur. M bis R. Leipzig: F. A. Brockhaus; 1833.
16. Langhans CF, Kopisch A. Pleorama, erfunden und aufgestellt von C. F. Langhans. Erläuterungen der in dem Pleorama erscheinenden Gegenstände. Breslau: Eduard Philipp; 1831.
17. Waller F, Walker R. Motion picture theater. Patent, US 2,280,206. 1937.
18. von Hueck A. Von den Grenzen des Sehvermögens. Archiv für Anatomie, Physiologie und wissenschaftliche Medicin. 1840.
19. Seldes G, Garrison R. Your world of tommorow. Rogers-Kellogg-Stillson. 1939;
20. Martin RG. Mr. Cinerama. TRUE the Man's Magazine 1953;
21. Panorama on a Giant Screen. Popular Mechanics Magazine. 1949:110–1.
22. Boyle JW. And Now... CINERAMA. American Cinematographer. 1952;480–1:498–500.
23. Waller F, Huntington NY, Dresser WR. Method and apparatus for projecting motion pictures. Patent, US 2,392,440.
24. Rolfe JM, Staples KJ. Flight simulation. Repr. Cambridge: Cambridge University Press; 1997.
25. Hawker TM. Hawker VC: RFC ace: the life of Major Lanoe Hawker VC DSO 1890–1916. 2013th edition. Barnsley, South Yorkshire: Pen & Sword Aviation; 2013.
26. Broelmann J. Lehrschaukel zur Fliegekunst – Der Flugsimulator Franz Drexlers. Kultur & Technik. 1995;19:24–5.
27. Biermann K, Cielewicz E. Flugplatz Döberitz: Geburtsort der militärischen Luftfahrt in Deutschland. 1. Aufl. Berlin: Links; 2005.
28. Lender M, Heidelberg P. Appareil pour l'apprentissage sans danger de la conduite des avions. Patent, FR 503648A.
29. La Section de Cinématographie Technique de la Direction des Inventions. Le Piloteur. Frankreich: Direction des Inventions, des études et des expériences techniques. 1919. http://www.cnc-aff.fr/internet_cnc/Internet/ARemplir/parcours/EFG1914/pages_DE/4161.html.
30. Crossman EC. "Dry Shooting" for Airplane Gunners. Pop Sci Mon 1919;94:13–14.
31. Heilig ML. EL Cine del Futuro: the cinema of the future. Presence Teleoper Virtual Environ. 1992;1:279–94.
32. Heilig ML. The cinema of the future. (1955). In: Packer R, Jordan K, Editor. Multimedia: from Wagner to virtual reality. Expanded ed. New York: Norton; 2002. pp. 239–251.
33. Jütte R. Geschichte der Sinne: von der Antike bis zum Cyberspace. München: Beck; 2000.
34. Aristoteles. Aristoteles über die Seele. Aus dem Griechischen übersetzt und mit Anmerkungen begleitet. Frankfurt: Johann Herrl; 1794.

35. Kant I. Träume eines Geistersehers, erläutert durch Träume der Metaphysik. Königsberg: Johann Jacob Kanter; 1766.
36. Becker-Carus C, Wendt M. Wahrnehmung. Allgemeine Psychologie. Berlin: Springer; 2017. pp. 73–156.
37. Schopenhauer A. In: Bernel L, editor. Die Welt als Wille und Vorstellung. München: Georg Müller; 1912.
38. Schopenhauer A. In: Bernel L, editor. Die Welt als Wille und Vorstellung. München: Georg Müller; 1913.
39. Heilig M. Sensorama simulator. Patent, US 3,050,870. 1961.
40. Borgwald C. Abtauchen in die fünfte Dimension. OSTSEE-ZEITUNG Rostock; 13. Nov. 2011.
41. Heilig ML. Experience theather. Patent, US 3,469,837. 1966.
42. Heilig ML. Beginnings: Sensorama and the Telesphere Mask. In: Dodsworth C, editor. Digital illusion: entertaining the future with high technology. New York: ACM Press, Addison-Wesley; 1998. p. 343–51.
43. Heilig ML. Stereoscopic-television apparatus for individual use. Patent, US 2,955,156. 1957.
44. Montgomery C. Recalling Columbia's startling trip to the rose bowl. The New York Times. New York; 27. Dec. 1981;2.
45. Robertson P. The guiness book of film: fact & feats. London: Guinness Superlatives Limited; 1980.
46. Huxley A. Schöne neue Welt: ein Roman der Zukunft. 68. Aufl. Frankfurt a. M.: Fischer; 2012.
47. [S.N.]. Duftender Film Zürcher Illustrierte 1940;16.
48. R. S. Camera Eye: Reporter-At-Large. Business Screen Magazine 1940;3:4.
49. Burnstock T. Scent of Mystery (1960). Scented storytelling. https://scentedstorytelling.com/2017/02/25/scent-of-mystery-1960/. Accessed on: 25. Aug. 2019.
50. Hoffman FW, Bailey WG. Arts and entertainment fads. New York: The Hawort Press; 1990.
51. Weinbaum SG. Pygmalions spectacles Wonder Stories 1935;7:28–39.
52. Lem S. Summa technologiae. 7. Aufl. Frankfurt a. M.: Suhrkamp; 2016.
53. Sutherland IE. The ultimate display. Proceedings of the Congress of the Internation Federation of Information Processing (IFIP). 1965. pp. 506–508.
54. StarTrek.com Staff. Meet The Man Behind The Holodeck, Part 1. STAR TREK. 2014. https://intl.startrek.com/article/meet-the-man-behind-the-holodeck-part-1. Accessed on: 1. Sept. 2019.
55. Pike OJ, Mackenroth F, Hill EG, Rose SJ. A photon–photon collider in a vacuum hohlraum. Nat Photonics. 2014;8:434.
56. Breit G, Wheeler JA. Collision of two light quanta. Phys Rev. 1934;46:1087–91.
57. Wallach H, O'connell DN. The kinetic depth effect. J Exp Psychol. 1953;45(4):205–17.
58. Green JBF. Figure coherence in the kinetic depth effect. J Exp Psychol. 1961;62:272–82.
59. Sutherland IE. A head-mounted three dimensional display. Proceedings of the December 9–11, 1968, fall joint computer conference, part I on – AFIPS '68 (Fall, part I). San Francisco: ACM Press; 1968. p. 757.
60. Delaney B, Furness TA. Sex drugs and tesselation: the truth about virtual reality as revealed in the pages of CyberEdge Journal. Oakland: CyberEdge Information Service; 2014.
61. Cicero MT. Tusculanische Unterredungen. Stuttgart: J. B. Metzlersche Buchhandlung; 1827.
62. Broderick D. The Judas Mandala. New York: Pocket Books; 1982.
63. Artaud A. The theater and its double. New York: Grove Press; 1958.
64. Lanier J. Virtually there. Sci Am. 2001;284:66–75.
65. Langer SK. Feeling and form. New York: Charkles Scribner's Sons; 1953.
66. Evenden I. The history of virtual reality. Virtual reality – the complete guide. Bristol: Immediate Media Co; 2016.

67. Sutherland IE. Sketch pad a man-machine graphical communication system. Proceedings of the SHARE design automation workshop on – DAC '64. Not Known: ACM Press; 1964. pp. 6.329–6.346.
68. Davis MR, Ellis TO. The RAND tablet: a man-machine graphical communication device. Proceedings of the October 27–29, 1964, fall joint computer conference, part I on XX – AFIPS '64 (Fall, part I). San Francisco, California: ACM Press; 1964. p. 325.
69. Atkinson P. A bitter pill to swallow: the rise and fall of the tablet computer. Des Issues. 2008;24: 3–25.
70. Roberts LG. The Lincoln WAND. Proceedings of the November 7–10, 1966, fall joint computer conference on XX – AFIPS '66 (Fall). San Francisco, California: ACM Press; 1966. p. 223. http://portal.acm.org/citation.cfm?doid=1464291.1464317. Accessed on: 9. Sept. 2019.
71. Gallé K. Erziehungsalarm: Weckruf für Eltern und Bildungsverantwortliche. 1. Aufl. Wien: Braumüller Verlag; 2015.
72. Goldstein GD, Neumann AJ, Editor. TRANSAC models – PHILCO Digit Comp Newsl 1957;9: 7–8.
73. Comeau CP, Bryan JS. Headsight television system provides remote surveillance. Electronics. 1961;34:86–90.
74. Minsky M. Telepresence. OMNI 1980.
75. Fisher SS. Viewpoint dependent imaging: an interactive stereoscopic display. Massachusetts Institute of Technology. Cambridge, Massachusetts; 1981 Oct.
76. Clarke AC. The second century of the telephone. The view from Serendip. New York: Random House; 1977.
77. Heinlein RA. Waldo. Die Zeit der Hexenmeister (Waldo & Magic Inc). 4. Aufl. München: Heyne; 1980.
78. Fisher SS, McGreevy M, Humphries J, Robinett W. Virtual environment display system. Proceedings of the 1986 workshop on Interactive 3D graphics – SI3D '86. Chapel Hill, North Carolina, United States: ACM Press; 1987. pp. 77–87.
79. Krueger MW. Responsive environments. Proceedings of the June 13–16, 1977, national computer conference on – AFIPS '77. Dallas: ACM Press; 1977.
80. University of Minnesota PowerWall Team. University of Minnesota PowerWall. SC94 Press Release. 1994. https://www.lcse.umn.edu/research/powerwall/powerwall.html. Accessed on: 6. Sept. 2019.
81. Czernuszenko M, Pape D, Sandin D, DeFanti T, Dawe GL, Brown MD. The immersadesk and infinity wall projection-based virtual reality displays. SIGGRAPH Comput Graph 1997;31:46–49.
82. Sandstrom TA, Henze C, Levit C. The hyperwall. Proceedings International Conference on Coordinated and Multiple Views in Exploratory Visualization – CMV 2003. London: IEEE; 2003. pp. 124–133.
83. Miller GA. The magical number seven, plus or minus two: some limits on our capacity for processing information. Psychol Rev. 1956;63:81–97.
84. Cruz-Neira C, Sandin DJ, DeFanti TA, Kenyon RV, Hart JC. The CAVE: audio visual experience automatic virtual environment. Commun ACM. 1992;35:64–72.
85. DeFanti TA, Sandin DJ. Final Report to the National Endowment of the Arts. 1977. Report No.: US NEA R60-34-163.
86. evl staff. Sayre Glove (first wired data glove). evl/electronic visualization laboratory. 1977. https://www.evl.uic.edu/entry.php?id=2162. Accessed on: 6. Sept. 2019.
87. Grimes GJ. Digital data entry glove interface device. Patent, US4,414,537. 1981.
88. Schmitt OH. A thermionic trigger. J Sci Instrum. 1938;15:24–6.

89. Bonet JP. Reducción de las letras y arte para enseñar a hablar los mudos. Madrid: Francisco Abarca de Angulo; 1620.
90. Haggerty JJ. Spinoff 1990. Washington: NASA; 1990. Report No.: NASA-NP-138, NAS 1.83: 138.
91. Zimmermann TG. Optical flex sensor. Patent, US 4,542,291. 1982.
92. Zimmerman TG, Lanier J, Blanchard C, Bryson S, Harvill Y. A hand gesture interface device. Proceedings of the SIGCHI/GI conference on human factors in computing systems and graphics interface – CHI '87. Toronto: ACM Press; 1987. p. 189–92.
93. Harvill YL, Zimmerman TG, Grimeaud J-J. Motion sensor which produces an asymmetrical signal in response to symmetrical movement. Patent, US 5,097,252. 1989.
94. Gentile CT, Wallace M, Avalon TD, Goodman S, Hall T. Angular displacement sensors. Patent, US 5,086,785. 1989.
95. Saggio G, Riillo F, Sbernini L, Quitadamo LR. Resistive flex sensors: a survey. Smart Mater Struct. 2016;25:013001.
96. Ditlea S. Grand Illusion coming soon, to your home.... New York Mag. 1990;23:27–34.
97. Kuipers J. Tracking and determining orientation of object using coordinate transformation means, system and process. Patent, US3,983,474. 1975.
98. Hamilton WR. Lectures on quaternions: containing a systematic statement of a new mathematical method. Dublin: Hodges and Smith; 1853.
99. Kuipers JB. Quaternions and rotation sequences: a primer with applications to orbits, aerospace, and virtual reality. Princeton: Princeton University Press; 2002.
100. Bolt RA. "Put-that-there": voice and gesture at the graphics interface. SIGGRAPH Comput Graph. 1980;14:262–70.
101. Sturman DJ, Zeltzer D. A survey of glove-based input. IEEE Comput Grap Appl. 1994;14:30–9.
102. Anderson B. Case study: a look at video animation. Broadcast Eng. 1979;21:86–8.
103. Ginsberg CM. Human body motion as input to an animated graphical display. Massachusetts: Massachusetts Institute of Technology. Cambridge; 1983.
104. Maxwell DR. Graphical marionette: a modern-day Pinocchio. Massachusetts: Massachusetts Institute of Technology. Cambridge; 1983.
105. Woltring HJ. New possibilities for human motion studies by real-time light spot position measurement. Biotelemetry. 1974;1:132–46.
106. Woltring HJ. Calibration and measurement in 3-dimensional monitoring of human motion by optoelectronic means. I preliminaries and theoretical aspects Biotelemetry 1975;2:169–196.
107. Burton RP, Sutherland IE. Twinkle box: a three-dimensional computer input device. Proceedings of the May 6–10, 1974, national computer conference and exposition on – AFIPS '74. Chicago: ACM Press; 1974. p. 513.
108. Yost L. Yost Labs 3-Space sensor product family. Ohio: Yost Labs Inc; 2017.
109. Mann S. Steve Mann my augmediated life – IEEE spectrum. IEEE spectrum: technology, engineering, and science news. 2013. https://spectrum.ieee.org/geek-life/profiles/steve-mann-my-augmediated-life. Accessed on: 13. Sept. 2019.
110. Maffiuletti NA, Minetto MA, Farina D, Bottinelli R. Electrical stimulation for neuromuscular testing and training: state-of-the art and unresolved issues. Eur J Appl Physiol. 2011;111:2391–7.
111. Purchese R. Happy go luckey: meet the 20-year-old creator of Oculus Rift. Eurogamer.net. 2013. https://www.eurogamer.net/articles/2013-07-11-happy-go-luckey-meet-the-20-year-old-creator-of-oculus-rift. Accessed on: 13. Sept. 2019.
112. The Guardian Staff. Facebook closes its $2bn Oculus Rift acquisition. What next?/technology/The Guardian. 2014. https://www.theguardian.com/technology/2014/jul/22/facebook-oculus-rift-acquisition-virtual-reality. Accessed on: 13. Sept. 2019.

113. Stynes T, Walker J. Stryker to acquire mako surgical for about $1.65 Billion. Wall Street Journal. 25. September 2013. https://www.wsj.com/articles/stryker-to-acquire-mako-surgical-for-about-165-billion-1380114400. Accessed on: 15. Sept. 2019.
114. Quaid AE, Abovitz RA. Guidance system and method for surgical procedures with improved feedback. Patent, EP 1 680 007 B1. 2004.
115. Quaid A, Kang H, Moses D, Ferre M, Hajaj B, Roche M, et. al. Haptic guidance system and method. 2008.
116. D'Onfro J. GOOGLE'S $500 MILLION MAN: Meet the „Weird" guy trying to invent a new computing platform. Business Insider Deutschland https://www.businessinsider.de/rony-abovitz-magic-leap-2014-12. Accessed on: 15. Sept. 2019.
117. Webber M. In place of a new TED Talk from Magic Leap, we bring you . . . [Internet]. TED Blog 2015. https://blog.ted.com/a-mysterious-message-in-a-tedx-talk/. Accessed on: 15. Sept. 2019.
118. Rittenhouse D. An Optical Problem, proposed by Mr. Hopkinson, and solved by Mr. Rittenhouse Transactions of the American Pholosophical Society Philadelphia: Robert Aitken; 1786. pp. 201–206.
119. Fraunhofer J. Bestimmung des Brechungs- und des Farbenzerstreungs-Vermögens verschiedener Glasarten, in Bezug auf die Vervollkommnung achromatischer Fernröhre. Ann Phys. 1817;56:264–313.
120. Fraunhofer J. Neue Modifikation des Lichtes durch gegenseitige Einwirkung und Beugung der Strahlen, und Gesetze derselben. München: Lentner; 1820.
121. Haüy R-J. Essai d'une Théorie sur la Structure des Crystaux. Paris: Gogué & Née de La Rochelle; 1784.
122. Bravais A. Abhandlung über die Systeme von regelmässige auf einer Ebene im Raum vertheilten Punkten. Leipzig: Wilhelm Engelmann; 1897.
123. Bragg WH, Bragg WL. The reflection of X-rays by crystals. Proc Roy Soc Math Phy Eng Sci. 1913;88:428–38.
124. Levola T. Diffractive optics for virtual reality displays. J Soc Inf Disp. 2006;14:467.
125. Levola T, Saariko P. Device for expanding an exit pupil in two dimensions. Patent, US 8,160,411. 2006.
126. Bradski GR, Miller SA, Abovitz RA. Methods and systems for creating virtual and augmented reality. Patent, US 2016/0026253 Juni 13, 2015.
127. Watts M. Semiconductor engineering – molecular imprints becomes a virtual reality company. Semiconductor Engineering 2015. https://semiengineering.com/molecular-imprints-becomes-a-virtual-reality-company/. Accessed on: 15. Sept. 2019.
128. Macnamara JG. Three dimensional virtual and augmented reality display system. Patent, US 8,950,867 Nov 23, 2012.
129. Kolbow-Lehradt B. Microsoft HoloLens angetestet: Ich sehe was, was du nicht siehst – Netzpiloten.de. Netzpiloten Magazin. 2016. https://www.netzpiloten.de/microsoft-hololens-angetestet-augmented-reality-brille/. Accessed on: 15. Sept. 2019.
130. Janssen J-K, Czerulla HA. Microsoft HoloLens im Test: Tolle Software, schwaches Display. https://www.heise.de/ct/artikel/Microsoft-HoloLens-im-Test-Tolle-Software-schwaches-Display-3248670.html. Accessed on: 15. Sept. 2019.
131. Sauter M. Magic Leap One ausprobiert: Für 2 Milliarden US-Dollar hätten wir mehr erwartet – Golem.de. golem.de IT-News für Profis. 2018. https://www.golem.de/news/magic-leap-one-ausprobiert-fuer-2-milliarden-us-dollar-haetten-wir-mehr-erwartet-1812-137902.html. Accessed on: 15. Sept. 2019.
132. Bikoulis A. Magic Leap One im Hands-On-Test: Ist Augmented Reality bereit für die Wirklichkeit? PC Ganes Hardware. 2018. https://www.pcgameshardware.de/Virtual-Reality-

Hardware-258542/Specials/magic-leap-im-test-augmented-reality-1269329/. Accessed on: 15. Sept. 2019.
133. Spinger C. Magic Leap One im Test: Die derzeit beste Mixed-Reality-Brille [Internet]. VR·Nerds. 2018. https://www.vrnerds.de/magic-leap-one-im-test-die-derzeit-beste-mixed-reality-brille/. Accessed on: 15. Sept. 2019.
134. Fröhlich B, Plate J. The cubic mouse: a new device for three-dimensional input. Proceedings of the SIGCHI conference on human factors in computing systems – CHI '00. The Hague, The Netherlands: ACM Press; 2000. pp. 526–531.
135. Mosher RS. Handyman to Hardiman. 1967. https://www.sae.org/content/670088/. Accessed on: 21. Sept. 2019.
136. Fick BR, Makinson JB. Final Report on Hardiman I Prototype for Machine Augmentation of Human Strength and Endurance. Washington und Fort Belvoir: Engineering Psychology Programs Office of Naval Research, Naval Air Systems Command and Army Mobility Equipment Research and Development Center; Report No.: N00014-66-C0051.
137. Fukuda H, Morishita T, Ogata T, Saita K, Hyakutake K, Watanabe J, et. al. Tailor-made rehabilitation approach using multiple types of hybrid assistive limb robots for acute stroke patients: a pilot study. Assistive Technol 2016;28:53–56.
138. Sankai Y. HAL: Hybrid assistive limb based on cybernics. In: Kaneko M, Nakamura Y, editors. Robotics research. Berlin: Springer; 2010. p. 25–34.
139. Cyberdyne Inc. CYBERDYNE. http://www.cyberdyne.jp/. Accessed on: 21. Sept. 2019.
140. Iida F, Anastasiades B, Gunura K. Wearable posture assisting device. Patent, EP 2 842 527.
141. Noonee. noonee. Chairless chair. 2019. https://www.noonee.com/. Accessed on: 21. Sept. 2019.
142. Meech KJ, Weryk R, Micheli M, Kleyna JT, Hainaut OR, Jedicke R, et. al. A brief visit from a red and extremely elongated interstellar asteroid. Nature 2017;552:378–381.
143. Mahr K. Der Kampf um die Pyramiden. München: Moewig; 1965.
144. Francis HG. Stützpunkt Donnergott. München: Moewig; 1974.
145. Naish J. Why scientists say this object could be an alien spaceship. 2017. http://www.dailymail.co.uk/sciencetech/article-5173329/Why-scientists-say-object-alien-spaceship.html. Accessed on: 29. Dec. 2017.
146. Parsons J. Oumuamua might not be an alien spacecraft after all. Mirror 2017. http://www.mirror.co.uk/science/oumuamua-might-not-alien-spacecraft-11699149. Accessed on: 29. Dec. 2017.
147. Sample I. Is Oumuamua an alien spacecraft? Initial scans show no signs of technology. The Guardian. 15. Dec. 2017. http://www.theguardian.com/science/2017/dec/14/is-oumuamua-an-alien-spacecraft-first-scans-show-no-signs-of-technology. Accessed on: 29. Dec. 2017.
148. Online F. Stephen Hawking ist alarmiert: Liefert Oumuamua Beweis für außerirdisches Leben? – Video. FOCUS https://wwwfocusde/wissen/weltraum/ Accessed on: 29. Dec. 2017.
149. Fuge L. How to recognise an alien spaceship. Cosmos Magazine 2017. https://cosmosmagazine.com/space/how-to-recognise-an-alien-spaceship. Accessed on: 29. Dec. 2017.
150. von Neumann J. In: Burks AW, editor. Theory of self-reproducing automata. Urbana, USA: University of Illinois Press; 1967.
151. Williams GV. MPEC 2017-V17 : New designation scheme For interstellar objects. Minor planet center, Smithsonian Astrophysical Observatory. 2017. https://www.minorplanetcenter.net/mpec/K17/K17V17.html. Accessed on: 29. Dec. 2017.
152. Breakthrough Initiatives. Breakthrough listen to observe interstellar object 'Oumuamua. 2017. https://breakthroughinitiatives.org/news/14. Accessed on: 29. Dec. 2017.
153. Breakthrough Initiatives. Breakthrough listen releases initial results and data from observations of Oumuamua. 2017. https://breakthroughinitiatives.org/news/15. Accessed on: 29. Dec. 2017.
154. Pauls E. Zauberwesen und Hexenwahn am Niederhein. Beiträge zur Geschichte des Niederrheins. Düsseldorf: Ed Lintz; 1898. p. 134.

155. Bouginé CJ. Handbuch der allgemeinen Litterargeschichte. Zürich: Oreli, Geßner, Füßli und Comp; 1789.
156. de Caus S. Von Gewaltsamen bewegungen: Beschreibung etlicher, so wol nützlichen alß lustigen Machiner beneben Unterschiedlichen abriessen etlicher Höllen od. Grotten und lust Brunne Frankfurt: Abraham Pacquart; 1615.
157. Hänggi CE. Automatenkunst im 18. Jahrhundert Kunst + Architektur in der Schweiz 2012;63:4–9.
158. Descartes R, v. Kirchmann JH. René Descartes' philosophische Werke. Erste Abtheilung. Lebensbeschreibung. Abhandlung über die Methode. Zweite Abtheilung. Untersuchungen über die Grundlagen der Philosophie. Berlin: L. Heimann; 1870.
159. Vaucanson J. Le Mécanisme du fluteur automate présenté à messieurs de l'Académie royale des sciences. Paris: Jacques Guerin; 1738.
160. de Vaucanson J. Beschreibung eines mechanischen Kunst-Stucks und Automatischen Flöten-Spielers so denen Herren von der Königlichen Academie der Wissenschaften zu Paris durch den Herrn Vaucanson Erfinder dieser Maschine überreicht worden, samt Einer Description sowohl einer künstlich-gemachten Ente, die von sich selbst das Essen und Trincken hinein schluckt . . . Augsburg: Johann Andreas Erdmann Maschenbaur; 1748.
161. von Helmholtz H. Ueber die Wechselwirkung der Naturkräfte und die darauf bezüglichen neuesten Ermittlungen der Physik. Königsberg: Gräfe & Unzer; 1854.
162. Schneider B. Blumenweberei für Esel: Musterwebstühle und Androiden im 18. Jahrhundert Kultur Technik. 2007;31:26–9.
163. Babbage C. Passages from the life of a Philosopher. London: Longman, Roberts & Green; 1864.
164. Babbage C. On the application of machinery to the computation of astronomical and methematical tables. London: Richard Taylor; 1824.
165. Alex J. Zur Entstehung des Computers: von Alfred Tarski zu Konrad Zuse: zum Einfluss elementarer Sätze der mathematischen Logik bei Alfred Tarski auf die Entstehung der drei Computerkonzepte des Konrad Zuse. Düsseldorf: VDI; 2007.
166. Gautier T. Le Club Des Hachichins. Revue des Deux Mondes (1829–1971). 1846;13:520–535.
167. Moreau J-J. Du hachisch et de l'aliénation mentale: études psychologiques. Paris: S.E.M.P; 1845.
168. Gautier T. Photosculpture – Extrait du Moniteur universel. Paris: Paul Dupont; 1864.
169. Scheiner C. Pantographice, seu Ars delineandi res quaslibet per parallelogrammum lineare seu cavum, mechanicum, mobile. Rom: Ludouici Grignani; 1631.
170. Willème F. Photo-Sculpture. Patent, US 43,822. 1864.
171. Gautier T, Alisch J, Roes M. Avatar. 1. Aufl. Berlin: Matthes & Seitz; 2011.
172. Stolze F. Photosculptur. In: Eder JM, Editor. Jahrbuch für Photographie und Reproductionstechnik. Halle a. S.: Wilhelm Knapp; 1893. pp. 494–8.
173. o. V. Selkes Photoskulptur. Neuhaus R, Juhl E, Editor. Photograph Rundsch. 1899;13:364.
174. Pötschke H. Verfahren, Gegenstände mit Hülfe der Photographie plastisch abzubilden. 1891.
175. Selke W. Photosculpture apparatus. Patent, US 675,417. 1898.
176. o. V. Über Selkes Photoskulptur. Neuhaus R, Juhl E, Editor Photograph Rundsch 1900;14:59.
177. 3 Dcopysystems. Big Alice – Welcome to the wonderful world of Alice. 2019. https://3dcopysystems.com/big-alice/. Accessed on: 27. Sept. 2019.
178. Blanther JE. Manufacture of contour relief maps. Patent, US 473,901. 1890.
179. Mueller B, Kochan D. Laminated object manufacturing for rapid tooling and patternmaking in foundry industry. Comput Ind. 1999;39:47–53.
180. Feygin M, Hsieh B. Laminated Object Manufacturing (LOM): a simpler process. Solid freeform fabrication – conference. Austin, Texas; 1991. pp. 123–130.

181. Siemens Healthineers MedMuseum. Die Geschichte der Computertomographie bei Siemens Healthineers. Erlangen: Siemens Healthcare; 2018.
182. Cormack AM. Representation of a function by its line integrals, with some radiological applications. J Appl Phys. 1963;34:2722–7.
183. Cormack AM. Representation of a function by its line integrals, with some radiological applications. II J Appl Phys. 1964;35:2908–13.
184. Hounsfield GN. Computerized transverse axial scanning (tomography): Part 1. Descrip Syst BJR. 1973;46:1016–22.
185. Radon J. Über die Bestimmung von Funktionen durch ihre Integralwerte längs gewisser Mannigfaltigkeiten. Leipzig: B. G. Teubner; 1917. p. 262–77.
186. Baese C. Photographisches Verfahren zur Herstellung plastisch richtiger Bildwerke. Photographische Rundschau und Photographisches Centralblatt 1904;18. Jahrgang:127–9.
187. Baese C. Photoskulptur Photograph Korrespond 1904;41:391–389.
188. Kernreuter O, Räth K. Die Reliefphotographien. Photograph Korrespond. 1899:134–41.
189. Allgeyer J. Handbuch über das Lichtdruck-Verfahren. Leipzig: Karl Scholtze; 1881.
190. Namias R. Ueber eine neue Eigenschaft der Bichromatgelatine. Jahrb Photograph Reproduktionstechnik Jahr 1909;1909(23):89–92.
191. Baese C. Photographic Process for the Reproduction of Plastic Objects. Patent, US 774,549. May 17, 1904.
192. Pötschke H, Laric O. Bust of Johann Wolfgang von Goethe (Photosculpture)/Three D Scans. Three D Scans. http://threedscans.com/albertina/poetschke/. Accessed on: 1. Oct, 2019.
193. Berger K, Ruhl K, Schroeder Y, Bruemmer C, Scholz A, Magnor M. Markerless motion capture using multiple color-depth sensors. Vision 2011; Modeling:8 pages.
194. Nitzan D, Brain AE, Duda RO. The measurement and use of registered reflectance and range data in scene analysis. Proc IEEE. 1977;65:206–20.
195. Leica Geosystems. Leica ScanStation P50 – Technische Daten. Leica Geosystems AG; 2017.
196. Morioka I. Process for manufacturing a relief by the aid of photography. Patent, US 2,015,457. Feb 20, 1933.
197. Przybilla H-J. Streifenprojektion – Grundlagen, Systeme und Anwendungen. Schriftenreihe des DVW, Gesellschaft für Geodäsie, Geoinformation und Landmanagemen Augsvurg: Wißner; 2007. pp. 111–122.
198. Monteath FH. Photomechanical process for producing bas-reliefs. Patent, US 1,516,199. Sep 23, 1922.
199. Munz OJ. Photo-Glyph recording. Patent, US 2,775,758. May 25, 1951.
200. Munz OJ. Verfahren und Vorrichtung zur Herstellung von Photoreliefs. Patent, DE 1 148 859. Nov 16, 1956.
201. Swainson WK. Verfahren und Material zur Herstellung einer dreidimensionalen Figuren. Patent, DE 1 797 599. 1967.
202. Gabor D. Nobel Lecture – Holography [Internet]. Stockholm. 1971. https://www.nobelprize.org/prizes/physics/1971/gabor/lecture/. Accessed on: 3. Oct, 2019.
203. Young II. The Bakerian Lecture. On the theory of light and colours. Phil Trans R Soc Lon. 1802;92:12–48.
204. Fresnel A. Premier mémoire sur la diffraction de la lumière. In: de Senarmon H, Verdet E, Fresnel L, editors. Œuvres complètes d'Augustin Fresnel (1866–1870). Paris: Imprimerie Impériale; 1815.
205. Gabor D. A new microscopic principle. Nature. 1948;161:777–8.
206. Maiman TH. Stimulated optical radiation in ruby. Nature. 1960;187:493–4.
207. Maiman TH. Ruby laser systems. Patent, US 3,353,115.

208. Leith EN, Upatnieks J. Wavefront reconstruction with diffused illumination and three-dimensional objects. J Opt Soc Am. 1964;54:1295.
209. Leith EN, Upatnieks J. Reconstructed wavefronts and communication theory. J Opt Soc Am. 1962;52:1123.
210. Leith EN, Upatnieks J. Photography by laser. Sci Am. 1965;212:24–35.
211. Denisyuk YN. My way in holography. Leonardo. 1992;25:425.
212. Benton SA, Bove VM. Holographic imaging. Hoboken: Wiley-Interscience; 2008.
213. Cross LG, Cross C. HoloStories: reminiscences and a prognostication on holography. Leonardo. 1992;25:421.
214. Unterseher F, Hansen J, Schlesinger B. Handbuch der Holographie: Wie mache ich Hologramme selber? München: Popa; 1991.
215. Johnston SF. Holographic visions: a history of new science. Oxford: Oxford University Press; 2006.
216. Chwolson OD. Die Lehre von der strahlenden Energie. Braunschweig: Friedr. Vieweg & Sohn; 1922.
217. Swainson WK. Verfahren zur Erzeugung von Änderungen der Eigenschaften eines Speichermediums in drei Dimensionen. Patent, DE 26 33 036. Juli 22, 1976.
218. DiMatteo PL. Method of generating and constructing three-dimensional bodies. Patent, US 3,632,923. Oct 21, 1974.
219. Kodama H. Automatic method for fabricating a three-dimensional plastic model with photo-hardening polymer. Rev Sci Instrum. 1981;52:1770–3.
220. Tsuji A. 3D Printing and Nagoya University [Internet]. Meidai Watch. 2019. http://www.meidaiwatch.iech.provost.nagoya-u.ac.jp/en/2019/06/3d.html. Accessed on: 2. Oct, 2019.
221. Ponsford M, Glass N. "The night I invented 3D printing" [Internet]. CNN Business. 2014. https://www.cnn.com/2014/02/13/tech/innovation/the-night-i-invented-3d-printing-chuck-hall/index.html. Accessed on: 2. Oct, 2019.
222. Hull CW. Apparatus for production of three-dimensional objects by stereolithography. Patent, US. 1984;4:575,330.
223. Ward JE. Automatic Programming of Numerically Controlled Machine Tools. Cambridge: Electronic Systems Laboratory, Massachusetts Institute of Technology; 1960. Report No.: 6873-FR-3.
224. Ross DT. Origins of the APT language for automatically programmed tools. SIGPLAN Not. 1978;13:61–99.
225. Hull CW, Lewis CW. Stereolithographic supports. 1988.
226. Hull CW. Stereolithographic supports. 1989.
227. André J-C. Le Mehauté A. De Witte O Dispositif pour réaliser un modèle de pièce industrielle Juli. 1984;16
228. Mendoza HR. Alain Le Méhauté, The man who submitted patent for SLA 3D printing before Chuck Hull. 3DPrint.com/the voice of 3D printing/additive manufacturing. 2015. https://3dprint.com/65466/reflections-alain-le-mehaute/. Accessed on: 10. Oct, 2019.
229. Beaman JJ, Deckard CR. Selective laser sintering with assisted powder handling. Sep 5, 1989.
230. Deckard CR. Method and apparatus for producing parts by selective sintering. Oct 17, 1986.
231. Hull CW. Stereolithographische Vorrichtung und Verfahren zur Herstellung von dreidimensionalen Gegenständen. Aug 8, 1984.
232. Meiners W, Wissenbach K, Gasser A. Verfahren zur Herstellung eines Formkörpers, Patent, DE 196 49 865. Dec 2, 1996.
233. Larson R. Verfahren und Vorrichtung zur Herstellung dreidimensionaler Körper. May 12, 1993.
234. Andersson L-E, Larsson M. Device and arrangement for producing a three-dimensional object Apr 27, 2000.

235. Mantey D. A Man & His Glue Gun. manufacturing.net. 2012. https://www.manufacturing.net/blog/2012/06/man-his-glue-gun. Accessed on: 11. Oct, 2019.
236. Bowyer A, RepRap Community. Welcome to RepRap.org. RepRap. 2004. https://reprap.org/wiki/RepRap. Accessed on: 11. Oct, 2019.
237. Bowyer A. 3D Printing and humanity's first imperfect replicator. 3D printing and additive manufacturing. 2014;1:4–5.
238. Bowyer A. Wealth without money – The background to the Bath Replicating Rapid Prototyper Project. RepRap 2004. https://reprap.org/wiki/Wealth_Without_Money. Accessed on: 30. Jul. 2021.
239. Sachs EM, Haggerty JS, Cima MJ, Williams PA. Three-dimensional printing techniques. September, 1989.
240. Blundell B, Schwarz AJ. Volumetric three-dimensional display systems. New York: Wiley; 2000.
241. LightSpace Technologies. Gen3 Volumetric 3D image front view and surface displays – www.lightspace3d.com. https://www.lightspace3d.com/index.php/Products/gen3-volumetric-3d-image-displays/. Accessed on: 11. Oct, 2019.
242. Sullivan A. A solid-state multi-planar volumetric display. SID Symp Digest of Technical Papers 2003;34:1531.
243. Bahr D, Langhans K, Bezecny D, Homann D, Vogt C. FELIX: a volumetric 3D imaging technique. In: Proc. SPIE 3101 Refregier P, Ahlers R-J, Editor. 1997. pp. 202–210.
244. Favalora GE, Napoli J, Hall DM, Dorval RK, Giovinco M, Richmond MJ, et al. 100-million-voxel volumetric display. In: Proc. SPIE 4712 Hopper DG, Editor. 2002. pp. 300–312.

Concluding Remarks

6

The "Real" Virtual Reality

Summary

More than 50 years ago, the computer pioneer Sutherland had a vision of a virtual world, which he called the "ultimate display". At the end of our journey through times and virtual reality, things should be put in order once again. Finally, a look at what is possible today and a glimpse of what might be one day.

After the manifold descriptions of the peculiarities of virtual reality, we will finally try to describe a desirable, or even the "ultimate" VR display. However, knowing the actual possibilities, we can now focus on what would be feasible in the foreseeable future.

So for now, we'll ignore Dolgoff's idea of the factual materialization of holograms and Sutherland's dream of computerized control of matter. And since we're on the subject of Sutherland: Perhaps our ultimate display is not a display at all, but a pair of glasses?

But first things first. Our journey started with a visit to Plato's Cave, hinting at the discrepancy between reality and perception. What we perceive is not necessarily what corresponds to reality. It is only what we are led to believe through clever presentation, which we, in the absence of better knowledge, believe to be real. The light drawings with the camera obscura make the representation of the outer world possible even for those who are completely unsuited to painting. If one wants to manifest the fleeting images, the knowledge of perspective helps to create a realistic copy. Here too, through technical means such as photography, even the untalented can achieve a masterpiece. The work does not have to be duplicated by hand, but can be projected onto a flat wall with the successors of the magic lantern, whereby the flat wall may also be the screen itself.

If one wants to order the procedures and apparatuses for generating an artificial reality, one is quickly tempted to create a classification. Regardless of the deeper meaning of such an endeavor, new problems can arise, as the Argentinean writer Borges illustrates in his

A. Grasnick, *Basics of Virtual Reality*,
https://doi.org/10.1007/978-3-662-64201-6_6

essay “El idioma analítico de John Wilkins”.[1] He has the sinologist Franz Kuhn quote the classification of animals from the Chinese encyclopedia “Heavenly Treasury of Benevolent Knowledge” (“Himmlischer Warenschatz wohltätiger Erkenntnisse”) (from [4], pp. 15–16):

> a) belonging to the emperor, b) embalmed, c) tamed, d) milk pigs, e) sirens, f) mythical creatures, g) stray dogs, h) included in this classification, i) acting like mad, j) uncountable, k) drawn with the finest camel-hair brush, l) and so on, m) which broke the water jug, n) which look like flies from afar

In this description of the shortcomings of Wilkin’s language, Borges manifests doubts about the possibility of a universally valid classification that is not arbitrary. Borges, one of the fathers of “magic realism”[2], cleverly mixes the real with the fictional in his analysis, presenting real people with fictitious quotations from invented books. Here we again encounter literary art as virtual reality, which should not be surprising, since according to Langer art must inevitably create virtual realities.

Borges’ art is echoed in reality when the French philosopher Foucault, for his part, sets about an “Order of Things” [5]. In the preface to his book, Foucault mentions the pleasure, but also the uneasiness, that came over him when reading Borges’ text. The confused and absurd disorder in its categories makes one smile involuntarily, but also feeds the concern that things are difficult to order. The order of things, however, must somehow be able to be defined on the basis of segments, variations, similarities or differences (from [5], p. 22):

> Order is at the same time that which poses in things as their inner law, as their secret web, … which exists only through the grid of a gaze, an attention, a language.

In our case, there would be a wide field to sort. There are numerous proposals to classify the various ancient philosophies, styles of painting, procedures for three-dimensional imaging or generative technologies. Again, one can sort by nonsensical categories,[3] or one can classify only broadly. As stated in the preface to this book (and hopefully made clear), my main concern with the basics of virtual reality was to describe the possibilities of representation, of generating virtual reality. Now, some methods and devices have been described here, which are sometimes quite special. However, it is sufficient to distinguish between

---

[1] “The Analytical Language of John Wilkins” (in [1]), “Die analytische Sprache von John Wilkins” (in [2]) is primarily a critique language Wilkins, who in the seventeenth century tried to divide all things in the universe into only 40 categories [3].

[2] An artistic movement of the early twentieth century in painting and literature, in which the boundaries between reality and fantasy disappear.

[3] e.g. inventors ennobled by the emperor (Simon Ritter von Stampfer, Franz Freiherr von Uchatius or Joseph Ritter von Blanther), or assassins who made remarkable inventions (e.g. Eadweard Muybridge and again the Ritter von Blanther).

those methods that create a real spatial image and those that only convey an illusion of space. This simple distinction is sufficient to classify the processes. On the one hand we have the real spatial images, which are composed of real points in space. Since these can be not only solid materials but of course also points of light, all figurative representations, volumetric projections and finally holography fall under this category. Under the premise of this book, this is **real virtuality.** For this purpose, we leave the term **fictive virtuality** for the illusion-generating processes that do not provide a real spatial image (i.e., all monoscopic, stereoscopic and autostereoscopic technologies).[4] This is already a manageable distinction, but requires the generic term **virtuality,** which could be understood as the representation of the virtual space of virtual reality.

Virtuality In the spatial representation of virtual reality **(virtuality),** a distinction is made between **real virtuality,** in which substantial (real) objects are actually generated in space, and **fictive virtuality,** in which virtual (fictive) objects are apparently generated in space.

For the ultimate display, a screen that can display spatial images will definitely be used.

All fictional reality methods use some form of screen on which different perspectives are displayed simultaneously. In front of the screen there is generally some kind of 3D optics that extracts the correct perspectives for the respective eye from the mixed perspectives. In a 3D screen, the optics is usually a lenticular or parallax barrier. In stereoscopy, the 3D optics are glasses located far in front of the screen close to the eye. In VR glasses, the eyepieces take over the function of the 3D glasses thus leading to the familiar design. It is obvious to do without glasses and to equip a large rotunda with autostereoscopic monitors. However, we know about the problems of viewing angle and resolution reduction. In addition, there is the problem of fictitious points in space. If the room is large enough, the points can be far in front of the wall and require focusing on that point. This leads again to the well-known accommodation-convergence-problem. So it is a method of real virtuality after all. So we fill the room with Swainson's reactive solutions, Hull's polymer or Crump's powder, go inside and start shooting powerful beams of energy into the room to create the points of space around us.

Obviously, that doesn't seem like a good idea either. With that, we ruefully turn to Dolgoff's holographic projectors and want to start building a holodeck. Unfortunately, neither holographic projectors nor film-quality holographic screens are currently available. Holography requires a resolution that is still only achieved by photographic material on large screens. A hologram is usually static and thus does not meet the requirements of virtual reality.

The best candidate for our project would indeed be a modern form of Lippmann integral photography that combines almost all the advantages of real and fictitious virtuality. Hyperview technology would probably be quite suitable for this. Although such a

[4] Not to say "virtual virtuality".

hyperview dome has never been built, it would at least be possible with sufficient financial support.

In order to get in touch with the space and to fill Lanier's vision of social interaction in virtual space with life, one would need cameras, microphones and sensors to record one's own position and register the actions performed. So data gloves, suit and helmet...

But if you're already wearing Minsky's jacket, Waldo's gloves and Sutherland's hat, why not Pygmalion's glasses or Lem's counter-eye?

And this brings us to the previously proclaimed thesis: the truly ultimate display may not be a display at all—but a complete 3D suit.

This would of course contain 3D glasses with which the 3D world could be viewed and through which one could also do without the display rotunda. Thus, the interaction can take place in any room and is not limited to large installations that can only be performed by a few.

The VR participant must wear a suit that provides haptic feedback. Thus, the wearer can not only feel temperature and humidity,[5] but also experience a counter-pressure during virtual mechanical work or cyber-game. For this purpose, devices would have to be attached to the joints of the suit, the fingers, etc., which would be controlled by the computer. This would allow the player, if attached to the floor with magnetic shoes, to then sit back on a virtual chair and read a virtual book as a reminiscence to the analog past.

The virtual space should not be limited to the real space of the VR friend. For a virtual bike ride or a leisurely rowboat tour, you can use devices and treadmills in which these movements can be performed. A car ride or a flight can be accomplished with the appropriate simulator. It becomes more difficult when a walker or runner is supposed to move casually in a virtual world. But even for this there are already possibilities such as the "Cyberwalk" from the Max Planck Institute for Biological Cybernetics.[6] If you tilt the entire treadmill, you could even use it to simulate mountain hikes.

In order to be seen in the virtual world, the suit would also have to be reasonably chic. But wait, does it really have to? In the virtual world we don't have to present a real image of ourselves, but can transfer our movements and interactions to an avatar which, as a phantom of ourselves, doesn't even have to bear any resemblance to the person controlling it. The telepresence of our phantom allows interaction with other phantoms in the virtual world and thus an extended form of social contact.

Goodbye in virtual space!

---

[5] If desired as in Heiligs Sensorama also wind and smell.

[6] An omni-directional treadmill (which can move in all directions on the plane of the floor) [6].

Literature

1. Borges JL. Otras inquisiciones. Other inquisitions, 1937–1952 . . . Translated by Ruth L.C. Simms. Introduction by James E. Irby. Austin: Zbuniversity of Texas Press; 1964.
2. Borges JL. The one and the many – essays on literature. Munich: Hanser; 1966.
3. Wilkins J. An essay towards a real character, and a philosophical language. London: Gellibrand & John Martyn; 1668.
4. Borges JL. The analytic language of John Wilkins. In: Hsu M, editor. f.fwd – reader. Zurich: ETH Zurich, Department of Architecture, Chair of CAAD; 2002. pp. 13–8.
5. Foucault M. The order of things: an archaeology of the human sciences. 24th ed. Frankfurt a. M.: Suhrkamp; 2017.
6. Campos J, Bülthoff H. Multimodal integration during self-motion in virtual reality. The neural bases of multisensory processes. New York: CRC Press; 2011. p. 603–28.

Register of Persons

Abovitz, Rony

Rony A. Abovitz (born 1971), American engineer, founder of Mako Surgical and Magic Leap

Adams, William

William Grylls Adams (1836–1915), British physicist, Professor of Physics at Kings College in London, discovered the photoelectric effect with Day

Adelson, Edward

Edward H. Adelson, (born 1952), American neuroscientist, Professor of Vision Science at the Massachusetts Institute of Technology

Aëtios

Aëtios of Amida (503–575), Byzantine physician in Alexandria and medico of the Roman emperor Justinian in Constantinople

Agatharchus

Agatharchus (c. 500 B.C.), Greek painter, mentioned by Vitruvius as the author of a treatise on skenographia, regarded as a "rapid painter"

Aglaophon

Aglaophon "the Younger" (c. 500 BC), son of Aristophon, grandson of Aglaophon "the Elder", praised by Pliny and Cicero

Aguilonius, Franciscus

Franciscus Aguilonius, also François d'Aguilon (1567–1617), Belgian Jesuit and mathematician

Aeschylus

Aeschylus (525 B.C.–456 B.C.), Greek poet, his tragedies (especially "The Persians" [Die Perser] and "The Oresteia" [Die Orestie]) were also translated into German and are still performed today

Akasaki, Isamu

Isamu Akasaki (born 1929), Japanese engineering scientist, General Manager of Matsushita Research Laboratory, Professor of Electronics at Nagoya University and Meijo University, Nobel Prize in Physics 2014 (together with Amano and Nakamura)

A. Grasnick, *Basics of Virtual Reality*,
https://doi.org/10.1007/978-3-662-64201-6

Alberti, Leon

Leon Battista Alberti (1404–1472), Italian art and architecture theorist, especially known for his theoretical writings on painting and sculpture

Alhazen

Abu Ali al-Hasan ibn al-Haitham, also Latin Alhazen or Alhacen (c. 965–1040), Arab optician and astronomer

Amano, Hiroshi

Hiroshi Amano (born 1960), Japanese physicist, professor at Nagoya University, Nobel Prize in Physics 2014 (jointly with Akasaki and Nakamura)

Ames, Adelbert

Adelbert Ames Jr. (1880–1955), American physiologist, professor of physiological optics at Dartmouth College, founder of the Dartmouth Eye Institute, known for the Ames Room

Amidror, Isaac

Isaac (Itzhak) Amdidror (born 1953), Isreali-Swiss computer scientist, moiré researcher at the Ecole Polytechnique Fédérale de Lausanne (EPFL)

Amosis

Ahmose I. (1560 B.C.–1525 B.C.), Pharaoh of the 18th Dynasty

Anaxagoras

Anaxagoras (499 B.C.–428 B.C.), Greek philosopher, teacher of Pericles and Euripides

Anaximander

Anaximander, also Anaximandros (610 B.C.–547 B.C.), Greek philosopher, student of Thales

Anaximenes

Anaximenes (585 B.C.–524 B.C.), Greek philosopher and observer of nature, considered a disciple of Anaximander

Anderton, John

John Anderton (1843–1905), English optician, especially active in the field of polarization

Anschütz, Ottomar

Ottomar Anschütz (1846–1907), German photographer, inventor of chronophotography and the electrotachyscope

Antiphilos

Antiphilos (c. 330 BC), Greek painter, contemporary and rival of Apelles

Apelles

Apelles (c. 375 B.C.–300 B.C.), Greek painter, pupil of Pamphilos, considered a masterful painter by the writers of his time

Archimedes

Archimedes of Syracuse (287 B.C.–212 B.C.), Greek mathematician and natural scientist, known among other things for the calculation of the number Pi, the laws of leverage or the static buoyancy of bodies

Ardenne, Manfred

Manfred Baron von Ardenne (1907–1997), German inventor and scientist, professor at the TU Dresden, founder of the Ardenne Institute, developed (with Loewe) the radio tube and electronic television, involved in the development of the Soviet atomic bomb

Argand, Aimé

François Pierre Aimé Argand (1750–1803), chemist and distiller from Geneva, invented the Argand oil lamp

Aristarchos

Aristarchus of Samos (310 B.C.–230 B.C.), Greek astronomer, pupil of Straton of Lampsakos, after Archimedes representative of a heliocentric world view (with the sun in the center)

Aristotle

Aristotle (384 B.C.–322 B.C.), Greek philosopher and naturalist, student of Plato; teacher of Alexander the Great, later king of Macedonia, influential polymath and theorist in the fields of logic, ethics, rhetoric, natural history, and epistemology

Arndt, Ernst Moritz

Ernst Moritz Arndt (1769–1860), German writer, historian and patriot, professor in Greifswald and Bonn, founding member of the "Urburschenschaft" in the "Grüne Tanne" in Jena

Artaud, Antonin

Antoine Marie Joseph Paul Artaud (1896–1948), French actor, writer and director, theorist of theatre, founder of the "Theatre of Cruelty"

Auer, Carl

Carl Auer von Welsbach (1858–1929), Austrian chemist and manufacturer, inventor and producer of the Auer incandescent mantle, created the OSRAM brand name

Babbage, Charles

Charles Babbage (1791–1871), British mathematician, professor of mathematics at Cambridge University, holder of the Lucasian Chair, inventor of the "Difference Engine No. 1" calculating machine

Baccei, Tom

Thomas J. Baccei (born 1944), American computer scientist and entrepreneur, founder of Pentica Systems, author of the bestseller "Magic Eye"

Bach, Ludwig

Ludwig Bach (1865–1912), German ophthalmologist, professor of ophthalmology in Marburg

Bacon, Roger

Roger Bacon (c. 1220–1292), English Franciscan and natural philosopher, authored writings on astronomy and optics

Baer, Karl

Karl Ernst Ritter von Baer (1792–1876), German-Baltic physician, professor in Königsberg

Baese, Carlo

Carlo Baese di Castelvecchio (1877–1943), Italian engineer, inventor of a relief photograph

Baird, John

John Logie Baird (1888–1946), Scottish inventor, famous developer of television

Baker, Henry

Henry Baker, pseudonym "Henry Stonecastle of Northumberland" (1698–1774), English naturalist, endowed the "Bakerian Lecture" of the Royal Society

Balthasar, Theodor

Theodor Balthasar (c. 1677–after 1721), German physician and mathematician, professor at the Knights' Academy (Ritterakademie) of Erlangen; city physicist of Erlangen

Balzac, Honoré

Honoré de Balzac (1799–1850), French writer, best known for his cycle of novels "The Human Comedy" (La Comédie humaine)

Bartholin, Erasmus

Erasmus Bartholin, also Bartholinus (1625–1698), Danish scientist, professor of mathematics in Copenhagen, studied birefringence

Beaman, Joe

Joseph J. Beaman (born c. 1950), American engineer, professor at the University of Texas, co-founder of DTM

Beard, Terry

Terry D. Beard (born 1946), American sound and video engineer, inventor of the Nuoptix process

Benton, Steve

Stephen Anthony Benton (1941–2003), American physicist, developer at Polaroid, professor at the Massachusetts Institute of Technology, researched in particular in the field of holography, inventor of the rainbow hologram

Bergen, James

James R. Bergen (born c. 1950), American mathematician, psychologist, and biophysicist, researcher at Bell Labs, RCA, and Sarnoff

Bernanose, André

André Bernanose (1912–2002), French chemist and physicist, professor at the University of Nancy

Bernauer, Ferdinand

Ferdinand Bernauer (1892–195) German mineralogist and geologist, professor of mineralogy in Berlin, inventor of the Zeiss "Bernotar" polarizing filter

Bernier, Robert

Robert V. Bernier (1911–1976), American officer, developer of stereoscopic projection techniques

Berthier, Auguste

Auguste Berthier (c. 1850–1915) French photographer, described the idea of the parallax barrier

Berthon, Rodolphe

Rodolphe Berthon (probably born c. 1880, died after 1937), French optician and astronomer, inventor of the lenticular process in colour photography

Bertuch, Friedrich

Friedrich Johann Justin Bertuch (1747–1822), German translator and publisher, secretary to the Duke of Saxe-Weimar-Eisenach, initiator of the princely free drawing school in Weimar, famous for his richly illustrated picture book for children

Bettini, Mario

Mario Bettini, lat. Marius Bettinus (1582–1657), Italian mathematician

Biard, James R.

James R. "Bob" Biard (born 1931), American engineer, researcher at Texas Instruments, Chief Scientist at Honeywell, Professor at Texas A&M University

Bitzer, Donald

Donald L. Bitzer (born 1934), American engineer, inventor of the plasma screen (together with Slottow and Wilson), professor at North Carolina University

Blanther, Joseph

Joseph E. Ritter von Blanther (1859–1898), Austrian officer, inventor of a layer method for 3D maps, charged with murder

Boccacio, Giovanni

Giovanni Boccaccio (1313–1375), Italian writer, author of the "Decamerone"

Bohr, Niels

Niels Bohr (1885–1962), Danish physicist, professor at the University of Copenhagen, developed the "Bohr model", Nobel Prize in Physics 1922

Bolas, Mark

Mark Bolas (b. 1960), American computer scientist, professor of interactive media at the University of Southern California (USC), co-founder of Fakespace Labs

Bolt, Richard

Richard A. Bolt (born c. 1940) American computer scientist, professor at the Massachusetts Institute of Technology, founder of the Human Interface Lab

Bolyai, Farkas

Farkas Wolfgang Bolyai (1775–1858), Hungarian mathematician, father of János Bolyai, professor of mathematics at the Reformed College (today Liceul Vocational Reformat) in Neumarkt am Mieresch (today Targu Mures), friend of Gauss, researched on Euclid's axiom of parallels

Bolyai, János

János (Johann) Bolyai (1802–1860), Hungarian mathematician, son of Farkas Bolyai, developed from Euclid's axiom of parallels a non-Euclidean (hyperbolic) space geometry

Bonet, Juan

Juan Pablo Bonet (1573–1633), Spanish civil servant, secretary and administrator of the Velasco family, secretary to King Felipe IV

Borchardt, Ludwig

Ludwig Borchardt (1863–1938), German Egyptologist and architect, director of the excavations of the German Orient Society 1911–1914

Borges, Jorge Luis

Jorge Francisco Isidoro Luis Borges Acevedo (1899–1986), Argentine writer and philosopher, director of the National Library of Argentina, one of the creators of magical realism

Born, Max

Max Born (1882–1970), physicist, professor of physics in Frankfurt, Berlin and Göttingen, friend of Einstein, Nobel Prize in Physics 1954

Botticelli

Sandro Botticelli, also Sandro di Mariano di Vanni Filipepi, called Botticelli (1445–1510), Italian Renaissance painter, his motifs are often Greek mythology or biblical history

Bouton, Charles

Charles Marie Bouton (1781–1853), French painter, together with Daguerre inventor of the diorama

Bower, Joseph

Joseph L. Bower (born 1938), American organization theorist, professor at Harvard

Bowyer, Adrian

Adrian Bowyer (born 1952), British engineer and mathematician, lecturer at the University of Bath, developer of the open-source printer RepRap

Bradley, Milton

Milton Bradley (1836–1911), American game designer and entrepreneur, founder of the Milton Bradley Company (now part of the Hasbro group)

Bragg, William Henry

Sir William Henry Bragg (1862–1942), British physicist, professor at the University of Adelaide (Australia), Leeds and London, received the Nobel Prize in Physics in 1915 together with his son William Lawrence Bragg

Bragg, William Lawrence

Sir William Lawrence Bragg (1890–1971), British physicist, professor of physics at the universities of Manchester and Cambridge, received the Nobel Prize in Physics in 1915 together with his father William Henry Bragg

Brahe, Tycho

Tycho Brahe, actually Tyge Ottesen Brahe (1546–1601), Danish astronomer, court mathematician to Emperor Rudolph II in Prague, described e.g. the celestial phenomena supernova and comet, left extensive observation records, which were evaluated after his death by his assistant Kepler

Brandenburg-Schwedt, Friederike Charlotte

Friederike Charlotte von Brandenburg-Schwedt (1745–1808), Princess of Prussia, prince abbess of Herford Abbey, known especially for her correspondence with Euler

Brandt, Willy

Willy Brandt, real name Herbert Frahm (1913–1992), German politician, SPD party leader and fourth German Chancellor

Braun, Ferdinand

Karl Ferdinand Braun (1850–1918), German physicist, teacher of physics in Leipzig, professor of physics in Marburg, Karlsruhe, Tübingen and Strasbourg, inventor of the Braun tube, Nobel Prize in Physics 1909 (together with Marconi)

Braunstein, Rubin

Rubin Braunstein (1922–2018), American physicist, researcher at RCA, professor of physics at the University of California, pioneer of LED technology

Bravais, Auguste

Auguste Bravais (1811–1863), French officer and naturalist, professor of astronomy in Lyon, professor of physics in Paris, defined crystals via lattice systems

Brewster, David

Sir David Brewster (1781–1868), British physicist, professor of physics at the University of St. Andrews, principal of the University of Edinburgh, discoverer of the Brewster angle, inventor of the kaleidoscope and the lens stereoscope, Newton's biographer

Broderick, Damien

Damien Francis Broderick (born 1944), Australian literary scholar and writer, science fiction author

Brown, Glenn

Glenn Halstead Brown, (1915–1995), American chemist, professor of chemistry at Kent State University, founder of the "Liquid Crystal Institute"

Brunelleschi, Fillipo

Filippo Brunelleschi (1377–1446), Italian architect and sculptor, architect of the dome of the cathedral in Florence, known for the geometric construction of perspective and its application in architecture and painting

Buffon, Georges-Louis

Georges-Louis Leclerc de Buffon, proper name Georges-Louis Leclerc (1707–1788), French naturalist and botanist, director of the Royal Botanical Gardens in Paris, author of several works on natural history

Burton, Robert

Robert Preece Burton (born c. 1946), American physicist, professor at Brigham Young University, founder of the Hyperdimensional Research Group

Buti, Ludovico

Ludovico Buti, also Lodovico di Tommaso Buti (c. 1550–1613), Italian painter, pupil of Santi di Tito (Titi)

Carmack, John

John D. Carmack II (born 1970), American game developer, co-founder of id software, best known for the games Wolfenstein and Doom, CTO of Oculus

Celsus, Aulus Cornelius

Aulus Cornelius Celsus (c. 25 BC–c. 50 AD), Roman medical writer

Chairephon

Chairephon of Sphettos (470 B.C.–399 B.C.), Greek philosopher, pupil of Socrates

Chartres, Bernard

Bernard of Chartres, lat. Bernardus Carnotensis (c. 1075–1126), French philosopher, chancellor of the cathedral school of Chartres

Childe, Henry

Henry Langdon Childe (1781–1874), English stage performer, known for his shows with cross-fades through several projectors

Christensen, Clayton

Clayton M. Christensen (born 1952), American economist, professor at Harvard

Christison, Robert

Sir Robert Christison (1797–1882), Scottish physician and toxicologist, professor of medical law and medicine in Edinburgh, president of the Royal Society of Edinburgh

Cicero

Marcus Tullius Cicero (106 B.C.–63 B.C.), Roman philosopher, politician and famous orator, considered the stylistic model of the classical Latin language

Cimabue

Cenni di Pepo, nickname Cimabue (1240–1302), Italian painter, possibly the teacher of Giotto

Clarke, Arthur

Sir Arthur Charles Clarke (1917–2008), British physicist and writer, mastermind of geostationary satellites, science fiction author

Clarke, Edward

Edward Marmaduke Clarke (1830–1850), Irish instrument maker, co-founder of the London Electrical Society, creator of the Royal Panopticon of Science and Art

Clarke, John

John Mason Clarke (1857–1925), American geologist and paleontologist, professor of geology and mineralogy at Rensselaer Polytechnic Institute (New York), president of the Geological Society of America

Clarkson, Euan

Euan Neilson Kerr Clarkson (born 1937), British palaeontologist, professor of palaeontology at the University of Edinburgh

Claudet, Antoine

Antoine François Jean Claudet (1797–1867), photographer and daguerreotypist, photographed many famous personalities of his time, inventor of the stereomonoscope

Clement of Alexandria

Titus Flavius Clemens or Clemens Alexandrinus (150–215), Greek theologian and writer, author of the “Stromateis”

Clottes, Jean

Jean Clottes (born 1933), French archaeologist, professor and Inspector General of Archaeology as well as scientific advisor on prehistoric rock art at the French Ministry of Culture, studied the Chauvet Cave, among other sites

Colbert, Jean-Baptiste

Jean-Baptiste Colbert, Marquis de Seignelay (1619–1683), French politician, finance minister under Louis XIV, initiator of the French Academy of Sciences

Colt, Samuel

Samuel Colt (1817–1862), American gun manufacturer, invented a pistol with a revolving cylinder, named "Colt" after him

Commandino, Federico

Federico Commandino, Latin Federicus Commandinus (1506–1575), Italian translator of ancient mathematicians, private secretary to Pope Clement VII

Conard, Nicholas

Nicholas John Conard (born 1961), American archaeologist, professor of prehistory and early history in Tübingen, is considered the discoverer of the Venus of "Hohle Fels" and the mammoth from Vogelherd

Constantinus

Constantinus Africanus, also Constantine the African (1020–1087), North African Benedictine, physician and translator

Cormack, Allan

Allan MacLeod Cormack (1924–1998), South African-American physicist, theorist of X-ray tomography, professor at Tufts University, Nobel Prize for Medicine 1979 together with Godfrey Hounsfield

Cross, Lloyd

Lloyd Gerald Cross (1935–2015), American physicist and holographic artist, inventor of multiplex holograms, co-founder of the Multiplex Company, co-founder "School of Holography" (both with Fred Unterseher), best known work is his multiplex hologram "The Kiss"

Crump, S. Scott

Steven Scott Crump (born 1953), American engineer, founder of StrataSys, inventor of FDM printing

Cuff, John

John Cuff (1708–1772), English optician and instrument maker, especially known for his microscopes

Curtze, Maximilian

Ernst Ludwig Wilhelm Maximilian Curtze (1837–1903), German mathematics historian and Copernicus researcher, high school professor in Thorn (Copernicus' birthplace)

Czermak, Johann

Johann Nepomuk Czermak (1828–1873), Austrian physiologist, professor of physiology in Graz, pupil of Purkinje

d'Arcy, Patrick

Patrick d'Arcy (1725–1779), Irish-French mathematician and officer, president of the Académie Royale des Sciences, Marechal des camps in the royal army, discoverer of angular momentum as well as the afterimage effect

da Fabriano, Gentile

Gentile da Fabriano (1370–1427), Italian painter, known for realistic painting, e.g. in the picture "The Adoration of the Magi"

da Vinci, Leonardo

Leonardo da Vinci (1452–1519), Italian painter and scholar, his numerous inventions and extraordinary works of art earned him the nickname "universal genius"

Daguerre, Louis

Louis Jacques Mandé Daguerre (1787–1851), French painter, pioneer of photography

d'Almeida, Joseph-Charles

Joseph-Charles d'Almeida (1822–1880), French physicist, professor of physics at the Lyceum of Algiers, founder of the Journal of Physics, is considered, together with Rollmann, to be the discoverer of the anaglyph process

Damianos

Damianos of Larissa (c. 400 A.D.), Greek author of a treatise on optics, possibly son or pupil of Heliodorus

Dante

Dante Alighieri (1265–1321), Italian poet, wrote the "Divine Comedy"

Darwin, Charles

Charles Robert Darwin (1809–1882), English naturalist and explorer, his main work "On the Origin of Species" is considered the beginning of the scientific theory of evolution

Day, Richard

Richard Evan Day (1848–1929), English physicist, lecturer at Kings College in London, discovered the photoelectric effect with Adams

de Bois-Clair, Gaspar

Gaspar Antoine de Bois-Clair (1654 1704), French painter

de Caus, Salomon

Salomon de Caus, also de Caux et al. (1576–1626), French architect and engineer, master builder of Frederick of the Palatinate in Heidelberg, builder of parks and inventor of various automata

de Fontana, Johannes

Johannes de Fontana, also Giovanni da Fontana (1395–1455), Italian physician and author, city doctor of Udine

de Monconys, Baltasar

Baltasar de Monconys (1611–1665), French diplomat, wrote travelogues

de Montebello, Roger

André Roger Lannes, Duke of Montebello (1908–1986), French photographer and inventor

Dechales, Claude

Claude François Milliet Dechales (1621–1678), French Jesuit and mathematician, professor of mathematics in Lyon

Deckard, Carl

Carl R. Deckard (born 1961), American engineer and businessman, inventor of 3D laser sintering, co-founder of DTM (acquired by 3D Systems), co-founder of Structured Polymers (acquired by Evonik)

Dee, John

John Dee (1527–1608), English mathematician and astrologer, advisor to Queen Elizabeth I

DeFanti, Tom

Thomas Albert DeFanti (born 1948), American computer scientist, professor at the University of Illinois at Chicago, co-founder with Sandin of the Electronic Visualization Laboratory

della Francesca, Piero

Piero della Francesca, also Petrus Pictor Burgensis or just Piero (c. 1410–1492) Italian painter and art theorist, teacher of Luca Pacioli (a teacher of Da Vinci)

della Porta, Giovanni

Giovanni (also Giambattista) Battista della Porta, also Johann Baptista Porta (1535–1615), Italian physician and scholar, wrote the "Magia naturalis" an extensive work on magic and the secrets of nature

Democritus

Democritus of Abdera, also Democritus (459 B.C.–371 B.C.), Greek philosopher, pupil of Leucippus, with Leucippus founder of atomism

Denisyuk, Yuri

Yuri Nikolaevich Denisyuk (1927–2006), Russian physicist, professor and head of the holography laboratory at the Vavilov Institute of Optics, inventor of white-light holograms

Descartes, René

René Descartes, lat. Renatus Cartesius (1596–1650), French philosopher and mathematician, the Cartesian coordinate system is named after him

Destriau, Georges

Georges Destriau (1903–1960), French physicist, professor at the Sorbonne in Paris, discoverer of electro-photoluminescence

di Banco, Maso

Maso di Banco (c. 1300–1348), Italian painter, famous for his fresco on the life of St. Sylvester (Pope Sylvester I)

Dibutades

Dibutades of Sicyon, also Butades (perhaps 600 BC), Greek potter and sculptor, inventor of shadow projection

Dickson, William

William Kennedy Laurie Dickson (1860–1935), Scottish-American engineer, assistant to Edison, inventor of the kinetoscope, film producer

DiMatteo, Paul

Paul L. DiMatteo (born 1928), American engineer and inventor, manager of Dynell Electronics, Robotic Vision, Vivax Medical, and Nextek

Diogenes

Diogenes of Apollonia (c. 500 B.C.), Greek physician and philosopher, wrote a treatise on nature in which air is a divine element endowed with spiritual power

Dircks, Henry

Henry Dircks (1806–1873) British engineer, inventor of the "Dircks' Fantasmagoria" (later Pepper's ghost)

Döbler, Ludwig

Leopold Ludwig Döbler (1801–1864), Austrian physicist and magician, inventor of the phantascope

Dolgoff, Gene

Eugene Dolgoff (born 1950), American engineer and inventor, developer of LC projection solutions

Donatello

Donatello, Donato di Niccolò di Betto Bardi (1386–1466), Italian sculptor, first assistant to Ghiberti, later known as a sculptor for the realism of his figures

Dörpfeld, Wilhelm

Wilhelm Dörpfeld (1853–1940), German architect and archaeologist, professor at the University of Jena, excavations in Olympia, on the Acropolis and with Schliemann in Troy

Dove, Heinrich

Heinrich Wilhelm Dove (1803–1879), German physicist and meteorologist, professor of physics in Königsberg and Berlin, inventor of the Dove prism

Drexler, Franz

Franz Drexler (b. c. 1880–d. unknown), German engineer and airplane pilot, pioneer of automatic flight controls

Drummond, Thomas

Thomas Drummond (1797–1840), British officer and surveyor, British Under-Secretary for Ireland, inventor of the lime light

du Hauron, Louis

Louis Arthur Ducos du Hauron (1837–1920), French explorer and inventor, pioneer of colour photography

Duboscq, Jules

Jules Duboscq (1817–1886), French optician and photographer, especially stereophotographer

Dudley, Leslie

Leslie Peter Dudley (1905–1998), British-American author and inventor, developer of numerous devices for stereoscopic cinema

Dudnikov, Yuri

Yuri Anatoljewitsch Dudnikov, (unknown, probably born around 1930) Soviet scientist, author of writings on autostereoscopy, often together with Boris Rozhkov

Dumas, Alexandre

Alexandre Dumas the Elder, also Alexandre Dumas Davy de la Pailleterie (1802–1870), French writer, best known for his novels "The Three Musketeers" (Les Trois Mousquetaires) and "The Count of Monte Cristo" (Le Comte de Monte-Cristo)

Dürer, Albrecht

Albrecht Dürer (1471–1528), German painter and art theorist, with his lifelike graphics and paintings he belongs to the outstanding Renaissance painters, remarkable is also his mathematics book "Underweysung der Messung mit dem Zirckel und Richtscheyt"

Dvořák, Max

Max Dvořák (1874–1921), Czech art historian, professor of art history at the University of Vienna

Eames, Charles and Ray

Charles Ormond Eames, Jr. (1907–1978) and Bernice Alexandra "Ray" Kaiser Eames (1912–1988), American designers, designed the famous Eames Lounge Chair

Ebers, Georg

Georg Moritz Ebers (1837–1898), German Egyptologist, professor of Egyptology in Jena and Leipzig, author of historical novels and editor of illustrated popular science picture books

Eddington, Arthur

Sir Arthur Stanley Eddington (1882–1944), British astronomer and physicist, researched and published on relativity, professor of astronomy at Cambridge, director of Cambridge Observatory

Edison, Thomas

Thomas Alva Edison (1847–1931), American inventor and businessman, was involved in more than 1000 inventions, including the phonograph, the light bulb and the kinetoscope

Ehrenberg, Christian

Christian Gottfried Ehrenberg (1795–1876), professor at the Friedrich Wilhelm University in Berlin (today Humboldt University) and protozoologist, travel companion of Alexander Humboldt's Russian expedition (1829)

Ehrenberger, Bonifacius

Bonifacius Heinrich Ehrenberger (1681–1759), professor of mathematics and logic at the Gymnasium Hildburghausen and professor of mathematics and metaphysics at the Gymnasium Coburg

Eichenlaub, Jesse

Jesse B. Eichenlaub, (born 1955) American physicist and inventor, 3D display developer, CTO at Dimension Technologies in Rochester New York

Einstein, Albert

Albert Einstein (1879–1955), German physicist, teacher of mathematics and physics, technical expert at the Swiss Patent Office, professor of physics in Zurich, Prague and Princeton, Nobel Prize in Physics 1921, known for the theory of relativity

Eisler, Paul

Paul Eisler (1907–1992), Austrian engineer, inventor of the printed circuit board

Empedocles

Empedocles (495 BC–435 BC), Greek natural philosopher and orator, considered by Aristotle to be the inventor of rhetoric

Epicurus

Epicurus, also Epikouros (341 B.C.–271 B.C.), Greek philosopher, disciple of Pamphiles, teacher of Hermarchus, founder of the Epicurean School

Erasistratos

Erasistratos (305 B.C.–250 B.C.), Greek physician and anatomist, student of Praxagoras

Estanave, Eugène

Eugène Pierre Estanave (1867–c. 1936) French mathematician and inventor, vice-secretary of the "Société Mathématique de France", collaborator in the laboratory of v. Lippmann at the Sorbonne, developer of stereoscopic X-rays

Eugenius

Eugenius of Palermo, also Eugenius Amiratus or Eugenio Panormitano (1130–1203), Greek translator and writer, ammiratus (admiral or emir) of the kingdom of Sicily

Euclid

Euclid of Alexandria, Eukleides, Lat. Euclides (c. 300 BC), Greek. Mathematician, famous for his works on mathematics and optics

Euler, Leonhard

Leonhard Euler (1707–1783), Swiss mathematician and physicist, professor of physics and mathematics in Petersburg, eponym of Euler's number **e**

Exner, Sigmund

Siegmund Ritter Exner von Ewarten (1846–1926), Austrian physiologist, professor at the Physiological Institute in Vienna

Faraday, Michael

Michael Faraday (1791–1867), British natural scientist, experimental physicist, professor of chemistry at the Royal Institution, known for his numerous discoveries in the field of electricity, the unit of electrical capacity is called "Farad" after him

Faris, Sadeg

Sadeg M. Faris (born 1945), American engineer and entrepreneur, founder of Reveo

Fergason, James

James Fergason (1934–2008) American physicist and inventor, researched liquid crystals at Westinghouse, founder of LC manufacturer ILIXCO (now LXD)

Fisher, Scott

Scott Fisher (born 1951), American engineer, head of NASA's "Virtual Environment Workstation Project" (VIEW), professor of interactive media at the University of Southern California, (USC)

Fitton, William

William Henry Fitton (1780–1861), British physician and geologist, member of the Royal Society and president of the Geological Society of London

Flaubert, Gustave

Gustave Flaubert (1821–1880), French writer, best known for his novel "Madame Bovary"

Foucault, Michel

Michel Foucault (1926–1984), French philosopher and psychologist, professor of the history of thought systems at the Collège de France

Frank, Charles

Sir Frederick Charles Frank (1911–1998), British physicist, professor at the University of Bristol, particularly active in the field of crystal research

Fraunhofer, Joseph

Joseph Ritter von Fraunhofer (1787–1826), German optician, professor of the Physical Cabinet of the Bavarian Academy of Sciences and Humanities, eponym of the Fraunhofer-Gesellschaft

Fresnel, Augustin

Augustin Jean Fresnel (1788–1827), French physicist and optician, researched the wave nature of light, developer of the Fresnel lens named after him

Frisch, Karl

Karl Ritter von Frisch (1886–1982), German-Austrian zoologist, professor of zoology at the University of Munich, Nobel Prize in Physiology or Medicine in 1973 (together with Konrad Lorenz and Nikolaas Tinbergen) for the discoveries of social behaviour patterns in bees

Frisius, Gemma

Gemma R. Frisius, real name Jemme Reinerszoon (1508–1555), Frisian mathematician, astronomer and instrument maker, professor of medicine and mathematics at the University of Leuven

Fröhlich, Bernd

Bernd Fröhlich (born 1962), German computer scientist, professor at the Bauhaus University Weimar

Gaarder, Jostein

Jostein Gaarder (born 1952), Norwegian writer, best known for the philosophical book for young people "Sophie's World"

Gábor, Dennis

Dennis Gábor (1900–1979), Hungarian-British engineer, developer at Siemens & Halske in Berlin and Thomson-Houston in Rugby, professor of physics at Imperial College in London, Nobel Prize in Physics 1971

Gaddi, Taddeo

Taddeo Gaddi, also Taddeo Gherardo di Zanobi (1290–1366), Italian painter, son of Gaddo Gaddi, godson of Giotto

Galen

Galenos of Pergamon, lat. Aelius Galenus or Galienus (128–199), Greek physician and anatomist, personal physician of Commodus

Galilei, Galileo

Galilei Galileo (1564–1641), Italian astronomer and mathematician, professor of mathematics in Padua, court mathematician to the Medicis in Florence, proponent of the heliocentric view of the world, improved the telescope for observing the heavens

Gauß, Carl Friedrich

Johann Carl Friedrich Gauß (1777–1855), German mathematician and physicist, professor at the University of Göttingen known for mathematical solutions and descriptions of physical phenomena, non-Euclidean geometry and geodesy

Gautier, Théophile

Théophile Gautier (1811–1872), French writer, wrote "Avatar", a book about soul-swapping

Gershon, Levi

Levi ben Gershon, also Gersonides, Leo de Balneolis, Ralbag (as an abbreviation of Rabbi Levi Ben Gershon), or Leo Hebraeus (1288–1344), Jewish mathematician and religious scholar

Gershun, Andrej

Andrej Aleksandrovich Gershun (1903–1952), Russian physicist, professor at the Leningrad Technical School of Precision Mechanics and Optics (now Saint Petersburg ITMO State University of Information Technologies, Mechanics and Optics)

Ghiberti, Lorenzo

Lorenzo Ghiberti (1378–1455), an Italian relief sculptor, architect and writer, whose masterpiece is his Gate of Paradise at the Baptistery of Florence Cathedral, wrote a book on art, the "Commentarii"

Giltsch, Adolf

Adolf Giltsch (1852–1911), lithographer in Jena

Ginsberg, Carol

Carol Marsha Ginsberg Brown (born 1951), American engineer and designer

Giotto

Giotto di Bondone (1267–1337), Italian painter and architect, possibly a pupil of Cimabue, master builder at Florence Cathedral, famous Renaissance painter

Glaucon

Glaucon of Athens (428 B.C.–382 B.C.), Greek writer, brother of Plato, known primarily from Plato's dialogues

Gleichen-Rußwurm, Wilhelm Friedrich

Wilhelm Friedrich Freiherr von Gleichen-Rußwurm, also called "Rußwurm" (1717–1783), German officer and naturalist, microscopist

Gmelin, Philipp

Philipp Friedrich Gmelin (1721–1768), German physician and botanist, city physicist of Tübingen, professor of botany and chemistry

Goethe, Johann Wolfgang

Johann Wolfgang von Goethe (1749–1832), German poet, with his extensive works is considered one of the most important German writers, also wrote some contributions to natural science (e.g. on the metamorphosis of plants and animals or the theory of colors)

Goldstein, Bernard

Bernard R. Goldstein (born 1938), American historian of science, professor and director of Jewish Studies at the University of Pittsburgh

Gortler, Steven

Steven S. Gortler (born 1959), American computer scientist, professor of computer science at Harvard

Gould, Gordon

Gordon Gould (1920–2005), American physicist, professor at the Polytechnic University of New York, founder of Optelecom, co-inventor and eponym of the laser

Gould, Stephen

Stephen Jay Gould (1941–2002), American paleontologist and evolutionary biologist, professor at Harvard University

Govi, Gilberto

Gilberto Govi (1826–1889), Italian physicist, professor of physics in Turin, representative of Italy in the International Metre Commission

Gray, George

George William Gray (1926–2013), British chemist, professor of chemistry at the University of Hull, influential researcher in the field of liquid crystals

Griendel, Johann

Johann Franz Griendel (von Ach auf Wankhausen), also Gründel or Grindelius (1631–1687), German optician, Nuremberg manufacturer of projectors and microscopes

Grimes, Gary

Gary J. Grimes (born 1948), American physicist, researcher at Bell Labs, professor of telecommunications at the University of Alabama at Birmingham (UAB)

Gropius, Karl

Karl Wilhelm Gropius (1793–1870), German painter, pupil of Schinkel, landscape, decorative, diorama and theatre painter

Gullstrand, Allvar

Allvar Gullstrand (1862–1930), Swedish ophthalmologist, professor of physical and physiological optics at Uppsala University, Nobel Prize for Medicine 1911

Gyges

Gyges (before 680 B.C.–644 B.C.), king of Lydia, legendary founder of a royal dynasty that perished with Croesus, used an invisible ring to seduce the queen and seize power

Haeckel, Ernst
Ernst Heinrich Philipp August Haeckel (1834–1919), German physician and zoologist, professor of zoology at the University of Jena
Hales, Thomas
Thomas Callister Hales (born 1958), American mathematician, professor at the University of Michigan, best known for his extensive computer proof of Kepler's conjecture following Hsiang's attempt at a computer proof
Hamilton, William
Sir William Rowan Hamilton (1805–1865), Irish mathematician, professor of astronomy at Trinity College in Dublin, inventor of quaternions
Hammond, Laurens
Laurens Hammond (1895–1973), American inventor, known for the Hammond organ
Hanrahan, Pat
Patrick "Pat" Hanrahan (born 1955), American biophysicist and computer scientist, co-founder of Tableau Research (now part of Salesforce), professor of computer science at Stanford, won a total of 3 Oscars for his work in rendering
Harrison, Lee
Lee Harrison III (1929–1998) American engineer, founder of Computer Image Corporation, Emmy Award for Scanimate technology
Harsdörffer, Georg
Georg Philipp Harsdörffer (1607–1658), German jurist and poet, still known today for his "Germanizations" of foreign-language terms
Hauslab, Franz
Franz Ritter von Hauslab (1798–1883), Austrian officer and cartographer, General and Director of Artillery at the Ministry of War, Professor of Situation Drawing at the Imperial and Royal Academy of Engineering (k. k. Ingenieur-Akademie), teacher of Emperor Franz Joseph I
Haüy, René-Just
René-Just Haüy (1743–1822), French naturalist, priest, teacher and botanist, professor of mineralogy at the École des mines de Paris, studied the structure of crystals
Hawker, Lanoe
Lanoe George Hawker (1890–1916), British officer and pilot, died in air combat with Manfred von Richthofen, developer of the "Rocking Nacelle" flight simulator
Heilig, Morton
Morton Leonard Heilig (1926–1997), American filmmaker, pioneer of virtual reality
Heilmeier, George
George Harry Heilmeier (1936–2014), American engineer and manager, director of DARPA, CTO of Texas Instruments
Heinlein, Robert
Robert Anson Heinlein (1907–1988), American author, science fiction writer

Hecataeus

Hecataeus of Miletus (560 B.C.–480 B.C.), Greek explorer and historian, described the known world in his travelogue, improved Anaximander's world map, and wrote the first Greek history book

Helfrich, Wolfgang

Wolfgang Helfrich (born 1932), German physicist, professor of physics at the Free University of Berlin, researcher at RCA and La Roche in the field of liquid crystals

Helmholtz, Hermann

Hermann Ludwig Ferdinand von Helmholtz (1821–1894), German physiologist and physicist, professor of physiology in Berlin, Königsberg and Heidelberg, lastly professor of physics at the Friedrich-Wilhelms-Universität Berlin (today Humboldt-Universität)

Heraclitus

Heraclitus of Ephesus (520 B.C.–460 B.C.) Greek philosopher and historian, his philosophy of change (of the coming into being and passing away of things) is often summarized in the saying "panta rhei" (everything flows)

Herapath, William

William Bird Herapath (1820–1868), British physician and chemist, discoverer of artificial tourmaline

Herder, Johann Gottfried

Johann Gottfried von Herder (1744–1803), poet of the "Weimar Classicism", which is understood today as the time of the work of Herder, Goethe, Schiller and Wieland

Hering, Ewald

Karl Ewald Konstantin Hering (1834–1918), German physiologist, professor of physiology in Vienna and Prague, researched in particular on the sense of sight

Herophilos

Herophilos of Chalcedon (325 B.C.–255 B.C.), disciple of Praxagoras

Herschel, John

Sir John Frederick William Herschel (1792–1871), British astronomer, co-founder of the Royal Astronomical Society, son of Wilhelm Herschel, the discoverer of Uranus

Hesiod

Hesiod, also Hesiodios (c. 700 B.C.), Greek poet, wrote in the "Theogony" about the origin of the gods and the world (one of the oldest sources of Greek mythology) and describes in the poem "Works and Days", among other things, the Pandora's Box and the legend of Prometheus

Hess, Walter

Walter Rudolf Hess (1881–1973), Swiss ophthalmologist and physiologist, professor at the University of Zurich, head of the Physiological Institute, Nobel Prize for Medicine 1949

Hillebrand, Franz

Franz Hillebrand (1863–1926), Austrian philosopher and psychologist, professor at the University of Innsbruck, founder of the Institute for Experimental Psychology

Hirschberg, Julius

Julius Hirschberg (1843–1925), German ophthalmologist, professor at the University of Berlin, the association for the history of ophthalmology "Julius-Hirschberg-Gesellschaft" is named after him

Holbein, Hans

Hans Holbein the Younger (1497–1543), German painter

Holmes, Richard

Richard "Dick" B. Holmes (born c. 1940), American engineer, founder of Electronic Imaging Systems

Holonyak, Nick

Nick Holonyak (born 1928), American engineer, researcher at Bell Telephone Laboratories and General Electric, professor at the University of Illinois, is considered the inventor of the red light-emitting diode

Homer

Homer (after 850 B.C.–before 750 B.C.), Greek poet, creator of the Iliad and the Odyssey

Hooke, Robert

Robert Hooke (1635–1703), English scholar, professor of geometry at Gresham College in London, Royal Society experimenter, astronomer, optician, and microscopist

Hopkinson, Francis

Francis Hopkinson (1737–1791), American jurist and politician, writer and composer, judge, signer of the Declaration of Independence

Horner, William George

William George Horner (1768–1837), English mathematician and teacher, created a method for solving equations (Horner scheme), developed a description of the Daedalum

Hounsfield, Godfrey

Sir Godfrey Newbold Hounsfield (1919–2004), English engineer, developer at EMI Laboratories, inventor of computer tomography, Nobel Prize for Medicine in 1979 together with Allan Cormack

Hsiang, Wu-Yi

Wu-Yi Hsiang (born 1937), Chinese (Taiwanese) mathematician, professor at the University of California, known for his attempt at a computer proof of Kepler's conjecture

Hueck, Alexander

Alexander Friedrich von Hueck (1802–1842), German-Baltic physician, professor of medicine at the University of Dorpat (Estonia)

Hull, Chuck

Charles W. Hull (born 1939), American engineer, inventor and entrepreneur, founder and president of 3D-Systems

Hunain

Abu Zaid Hunain ibn Ishaq al-Ibadi, also Hunain ibn Ishaq or lat Johannitius (808–873), Christian-Arab scholar and translator, head of the House of Wisdom in Baghdad

Huxley, Aldous

Aldous Leonard Huxley (1894–1963), British author, wrote the bestseller "Brave New World"

Huygens, Christiaan

Christiaan Huygens (1629–1695), Dutch optician, physicist and astronomer, formulated a wave theory of light

Ivanov, Semyon

Semyon Pavlovich Ivanov (1906–1972) Russian engineer, researcher at the National Institute of Photographic Cinematography (NIKFI), inventor of the Moscow raster screen

Ives, Frederic

Frederic Eugene Ives (1856–1937), American photographer and inventor, pioneer of color photography, father of Herbert E. Ives

Ives, Herbert

Herbert Eugene Ives (1882–1953), American physicist and engineer, director of electro-optical research at Bell Labs, pioneer of autostereoscopy, known for conducting the Ives-Stilwell experiment to prove time dilation

Jacobsen, Holger

Holger Jacobsen, lat. Oligerus Jacobaeus (1650–1701), Danish physician, professor at the University of Copenhagen, compiled a catalogue of the royal curiosity collection

Janssen, Pierre

Pierre Jules César Janssen (1824–1907), French astronomer, director of the astrophysical institute in Meudon, discoverer of helium (at the same time as Lockyer), developed the "Janssen Revolver" for planetary observation

Jaquet-Droz, Pierre

Pierre Jaquet-Droz (1721–1790), Swiss watchmaker, designer of automata

Jell, Peter

Peter A. Jell, Professor in the School of Earth and Environmental Sciences at the University of Queensland, Australia

Johnston, Sean

Sean F. Johnston (born c. 1955), British physicist and historian, professor of Science, Technology and Society at the University of Glasgow

Joly, John

John Joly (1857–1933), Irish engineer and geologist, professor of geology at Trinity College, Dublin, inventor of the line grid method for colour photography, developed a method for determining the age of the earth from radioactive decay

Jordan, Max

Max Jordan (1837–1906), German art historian, lecturer at the University of Leipzig, director of the Royal National Gallery in Berlin

Julesz, Béla

Béla Julesz (1928–2003), Hungarian electrical engineer and psychologist, scientist at Bell Labs, professor of phsychology at Rutgers University New Jersey, creator of random-dot stereograms

Juncker, Christian

Christian Juncker (1668–1714), German teacher and writer

Kanolt, Clarence

Clarence Whitney Kanolt (1880–1964), American physicist and inventor, head of the cryogenic laboratory of the National Bureau of Standards and US Bureau of Mines, employee of Farrand Optical

Kant, Immanuel

Immanuel Kant (1724–1804), German philosopher, professor of logic and metaphysics in Königsberg

Kaplan, Sam

Sam H. Kaplan (1915–2000), American chemist and inventor, known for his theory of the parallax barrier and TV patents

Charles III

Charles III of Lorraine (1543–1608), Duke of Lorraine

Kawamoto, Hirohisa

Hirohisa "Hiro" Kawamoto (born 1938), Japanese engineer, researcher at RCA, General Manager Sony Consumer Laboratory, Vice President Sharp

Keller-Dorian, Albert

Albert Keller-Dorian (1856–1924), French engineer and industrialist, co-founder of lenticular colour photography (with Berthon)

Kepler, Johannes

Johannes Kepler, also Keppler, lat. Ioannes Keplerus, (1571–1630), German mathematician and astronomer, described the "Kepler's laws" of planetary motion, named after him today

Kerr, John

John Kerr (1824–1907), Scottish physicist, mathematics teacher at the Free Church Training College in Glasgow, discoverer of the Kerr effect

King, John

John King, jun. (1797–1871), John King & Son, manufacturers of mathematical and optical instruments in Clare Street, Bristol

Kircher, Athanasius

Athanasius Kircher, Latin Athanasius Kircherus Fuldensis (1602–1680), German Jesuit and scholar, professor of mathematics in Würzburg and at the Collegium Romanum in Rome, court mathematician of Emperor Ferdinand II in Vienna, polymath, explorer and writer

Knebel, Karl Ludwig

Karl Ludwig von Knebel (1744–1834), German poet and translator, teacher and court master at the Weimar court, friend of Goethe

Kodama, Hideo

Hideo Kodama (born 1950), Japanese physicist and patent attorney, researcher at the Nagoya Municipal Industrial Research Institute, Visiting Professor at Nagoya University, patent attorney in the firm of Kai-U

Koelbing, Huldrych

Huldrych Martin Koelbing (1923–2007), Swiss ophthalmologist and medical historian, professor of the History of Medicine at the University of Zurich

Kohlhans, Johann

Johann Christoph Kohlhans (1604–1677), German mathematician, professor of mathematics at the Gymnasium Casimirianum in Coburg

Krueger, Myron

Myron W. Krueger (born 1942), American computer artist, creator of interactive video installations

Kubrick, Stanley

Stanley Kubrick (1928–1999), American photographer, director and screenwriter, known for the films "Spartacus" and "Full Metal Jacket", Oscar for the film "2001: A Space Odyssey"

Kuhn, Franz

Franz Walther Kuhn (1884–1961) German jurist and sinologist, translator of the erotic novel Jin Ping Mei

Kuipers, Jack

Jack B. Kuipers (1921–2016), American engineer, professor of mathematics at Calvin College, inventor of a position sensor

Kusch, Polykarp

Polykarp Kusch (1911–1993), German-American physicist, developer at Westinghouse and Bell, professor of physics at Columbia University and University of Texas, research on the magnetic moment of the electron, Nobel Prize in Physics 1955

Laertios, Diogenes

Diogenes Laertios, Latin Laertius (presumably third century AD), historian of philosophy, author of the 10-volume work "On the Lives and Teachings of Famous Philosophers"

Lagrange, Joseph-Louis

Joseph-Louis Lagrange, real name Giuseppe Ludovico Lagrangia (1736–1813), Italian mathematician and physicist, professor of mathematics in Turin, director of the Royal Prussian Academy of Sciences in Berlin

Lambert, Johann

Johann Heinrich Lambert (1728–1777), German mathematician and philosopher, founder of photometry

Land, Edwin

Edwin Herbert Land (1909–1991), American scientist and entrepreneur, invented the plastic polarizer, founder of Polaroid

Langer, Susanne

Susanne Katherina Langer (1895–1985) American philosopher, philosophy lecturer at various American universities

Langhans, Carl

Carl Ferdinand Langhans (1781–1869), German architect, senior building official in Berlin, inventor of the pleorama

Langmuir, Irving

Irving Langmuir (1981–1957), American chemist and physicist, scientist at General Electric, researched among other things on the emission of gases and the theory of thin films, invented the name "plasma", Nobel Prize for Chemistry 1932

Lanier, Jaron

Jaron Lanier (born 1960), American computer artist, developer of the data glove, pioneer of virtual reality

Larson, Ralf

Ralf Larson (born 1946), Swedish engineer and entrepreneur, inventor of electron beam melting, founder of Arcom

Laube, Hans

Hans E. Laube (1900–1976), Swiss technician, inventor of the olfactory cinema

Laue, Max

Max von Laue (1879–1960), German physicist, professor of physics in Frankfurt and Berlin, researched, among other things, the wave nature of X-rays, Nobel Prize in Physics in 1914

Lebon, Philippe

Philippe Lebon d'Humbersin (1767–1804), French engineer, professor of mechanics at the École Nationale des Ponts et Chaussée, inventor of the gas lamp

Lechner, Bernard

Bernard J. Lechner (1934–2014), American engineer, engineer at RCA, invented the "active matrix display"

Ledermüller, Martin

Martin Frobenius Ledermüller (1719–1769), German lawyer, notary at Nuremberg, judicial councillor at Bayreuth, inspector of the Bayreuth Natural History Cabinet, microscopist and author of popular scientific microscopy books

Lefort, Pierre

Pierre Henri Armand Lefort (1804–1880), French photographer, inventor of the polyorama panòptic

Lehmann, Otto

Otto Lehmann (1855–1922), professor of Physics and Electrical Engineering at the Technical University of Karlsruhe (now Karlsruhe Institute of Technology)

Leibnitz, Gottfried Wilhelm

Gottfried Wilhelm Leibniz (1646–1716), German mathematician and philosopher, polymath, known for his contributions to differential and integral calculus and his extensive correspondence with the scholars of his time

Leith, Emmett

Emmett Norman Leith (1927–2005), American physicist, professor of electrical engineering at the University of Michigan, together with Juris Upatnieks developer of practical holography

Lem, Stanisław

Stanisław Herman Lem (1921–2006), Polish writer and philosopher, successful science fiction author, some of his works have been made into films

Leroi-Gourhan, André

André Leroi-Gourhan (1911–1986) French archaeologist and paleontologist, professor at the Sorbonne and at the Collège de France, classified art styles of cave paintings according to technique and age

Leucippus

Leucippus, also Leukippos (fifth century B.C.), Greek philosopher, pupil of Parmenides, teacher of Democritus, with Democritus founder of atomism

Levoy, Marc

Marc Levoy (born 1953), American architect and computer scientist, professor of computer science at Stanford

Lieberkühn, Johann

Johann Nathanael Lieberkühn (1711–1756); German physician and microscopist, member of the Royal Society and Leopoldina, improved and manufactured own microscopes

Lincoln, William

William Ensign Lincoln (1847–1944), American writer, son and biographer of John Larkin Lincoln, professor of Latin and literature

Link, Ed

Edwin Albert Link (1904–1981), American inventor, pilot and entrepreneur, inventor of the flight simulator "Link Trainers", founder of Link Aviation Devices (today L3 Commercial Training)

Linné, Carl

Carl von Linné, also Linnaeus (1707–1778), Swedish naturalist, professor at the University of Uppsala, author of an encyclopaedia on the systematics and terminology of nature

Lippmann, Gabriel

Jonas Ferdinand Gabriel Lippmann (1845–1921), French physicist, professor of physics at the Sorbonne in Paris, Nobel Prize for Physics 1908, inventor of integral photography

Lipton, Lenny

Lenny Lipton (born 1940), American stereographer, author and entrepreneur, founder of StereoGraphics (now RealD)

Listing, Johann

Johann Benedict Listing (1808–1882), German mathematician and physicist, professor of physics and mathematics in Göttingen, introduced the term topology to mathematics

Lobachevsky, Nikolai

Nikolai Ivanovich Lobachevsky (1792–1856), Russian mathematician, professor at the University of Kazan, researched on non-Euclidean geometry

Lorenzetti

Ambrogio Lorenzetti (1290–1348) and Pietro Lorenzetti (1280–1348), Italian painters

Lossew, Oleg

Oleg Vladimirovich Lossew (1903–1942), Russian radio engineer, researched at the Nizhny Novgorod Lenin Radio Laboratory in the field of semiconductor technology and electroluminescence

Lubschez, Ben

Ben Jehudah Lubschez (1881–1963), Ukrainian-American architect and photographer, author of books on subjects of representation

Luckey, Palmer

Palmer Freeman Luckey (born 1992), American entrepreneur, founder of Oculus

Louis XIV

Louis XIV (1638–1715), French nobleman, son of Louis XIII and Anne of Austria, already formally the king of France at the age of 4, real king at 22

Lucian

Lucian of Samosata (120–180), Greek writer, satirist, influential writer until modern times

Lucretius

Titus Lucretius Carus (94 B.C.–53 B.C.), Roman philosopher, student of Philodemos, considered one of the most important textual sources on Epicurus, atomist and materialist, "proved" the mortality of the soul

Lumière, Auguste

Auguste Marie Louis Nicolas Lumière, (1862–1954), French photographic entrepreneur, founder of Laboratoires Lumière, inventor of the cinematograph together with his brother Louis

Lumière, Louis

Louis Jean Lumière (1864–1948), French photography entrepreneur, inventor of the cinematograph together with his brother Auguste

MacCrellish, Frederick

Frederick MacCrellish (1828–1882), American editor and newspaper publisher, partner in the daily newspaper "Alta California", San Francisco

Mach, Ernst

Ernst Waldfried Josef Wenzel Mach (1838–1916), Austrian physicist, physiologist and philosopher, professor in Graz, Prague and Vienna

Magnus, Albertus

Albertus Magnus, also Albert the Great, Albert the German or Albert of Lauingen, lat. Albertus Theutonicus or Albertus Coloniensis (c. 1200–1280), German ecclesiastical scholar, Dominican, bishop of Regensburg

Magnus, Eduard
Leopold Eduard Samuel Magnus (1799–1872), German painter, professor at the Berlin Art Academy (Berliner Kunstakademie)

Magnus, Hugo
Hugo Magnus (1842–1907), German ophthalmologist and medical historian, professor in Breslau

Maiman, Theodore
Theodore Harold Maiman (1927–2007), American physicist, researcher at Hughes Research, founder of Korad Corporation, inventor of the laser

Malus, Louis
Étienne Louis Malus (1775–1812), French officer and engineer, the law of Malus is named after him

Manetti, Antonio
Antonio Manetti (1423–1497), Italian architect, biographer of Brunelleschi

Mann, Steve
Steve Mann (born 1962), Canadian computer scientist, professor at the University of Toronto, specialist in wearable computing

Marcks, Gerhard
Gerhard Marcks (1889–1981), German sculptor, professor at the Burg Giebichenstein art school, directed the Bauhaus pottery in Dornburg, his brother Dietrich Marcks was a member of Borchardt's excavation campaign

Marconi, Guglielmo
Guglielmo Marchese von Marconi (1874–1937), Italian radio engineer, founder of the Wireless Telegraph and Signal Co.; Nobel Prize in Physics 1909 (together with Braun)

Marey, Jules
Étienne-Jules Marey (1830–1904), French physiologist and engineer, professor of natural history at the Collège de France, president of the Académie des Science, inventor of medical and photographic equipment, founder of the Institut des Sciences du Mouvement (today named after him)

Marg, Elwin
Elwin Marg (1918–2010), American ophthalmologist, professor of physiological optics and optometry at the University of Berkeley, founder of the Minerva Foundation (together with his wife)

Martini, Martino
Martino Martini (1614–1661), Italian Jesuit and historian, worked as a missionary and scholar in China, is best known for his historical and geographical works on China

Marx, Karl
Karl Marx (1818–1883), German philosopher, theorist of socialism and communism, co-founder of the Communist League (Bund der Kommunisten), author of the "Manifesto of the Communist Party" ("Manifest der kommunistischen Partei") and "Das Kapital", member of the Royal Society of Arts

Masaccio

Tommaso di Ser Giovanni di Mone Cassai, called Masaccio (1401–1428), Italian painter, one of the founders of realistic painting in the early Renaissance period

Mau, August

August Mau (1840–1909), German archaeologist, researcher of Pompeii's wall paintings, which he divided into stylistic periods

Maxwell, Delle

Delle Rae Maxwell (born 1953), American computer scientist, daughter of the geophysicist Arthur E. Maxwell, married to Pat Hanrahan, co-founder of the Maxwell-Hanrahan Foundation

Maxwell, James

James Clerk Maxwell (1831–1879), Scottish physicist, professor of physics at Kings College, London, and Cambridge, famous for Maxwell's equations

May, Brian

Brian Harold May (born 1947), musician and physicist, guitarist (Queen), director of the London Stereoscopic Company

Mehmke, Rudolf

Rudolf Mehmke (1857–1944), German mathematician, professor at the Technical Universities of Darmstadt and Stuttgart

Meisenbach, Georg

Georg Meisenbach (1841–1912), German engraver and entrepreneur, inventor of a machine for the production of line screens

Melzi, Francesco

Francesco Melzi (1491–1570), Italian painter, pupil and heir of Leonardo da Vinci

Meyerhof, Max

Max Meyerhof (1874–1945), German physician and ophthalmologist in Cairo

Minkowski, Hermann

Hermann Minkowski (1864–1909), German mathematician and physicist, professor in Göttingen, friend of Hilbert, developed the four-dimensional spacetime (Minkowski space)

Molyneux, William

William Molyneux (1656–1698), Irish natural philosopher and jurist, translator of Descartes and Leibnitz, member of the Royal Society

Monteath, Frederick

Frederick Hutchinson Monteath (1866–1938), Australian photographer, with Monteath Photo Sculpture, Sydney, maker of photographic reliefs

Moon, Parry

Parry Hiram Moon (1898–1988), American engineer, scientific author, professor at the Massachusetts Institute of Technology

Moreau, Jacques-Joseph

Jacques-Joseph Moreau (1804–1884) French psychiatrist, researched the effects of intoxicants on humans

Morgan, William

William Morgan, (born pres. before 1820 - unknown) British publisher and lithographer, editor of “Dioramic Views”

Morioka, Isao

Isao Morioka (1894–1985), Japanese artist and sculptor, founder of the Rittaishashinzo Studios

Moser, Ludwig

Ludwig Moser (1805–1880), German physicist, professor of physics in Königsberg

Mosher, Ralph

Ralph S. Mosher (1920–2008), American engineer, developer of walking machines and exoskeletons by General Electrics, founder of Robotics Inc.

Müller, Johannes

Johannes Peter Müller (1801–1858), German physician and physiologist, professor of physiology in Bonn and Berlin, author of the “Handbuch der Physiologie des Menschen” (Manual of Human Physiology)

Munz, Otto

Otto John Munz, also Münz (1907–1980), American lawyer and inventor

Muth, Heinrich

Heinrich Ludwig Muth (1673–1754), German optician, optician and court mathematician in Kassel

Muybridge, Eadweard

Eadweard Muybridge, real name Edward James Muggeridge (1830–1904), British photographer and inventor, his studies of movement led to the moving image

Nakamura, Shuji

Shuji Nakamura (born 1954), Japanese-American electronics engineer, development manager at Nichia, professor at the University of California and at the University of Bremen, Nobel Prize in Physics 2014 (together with Akasaki and Amano)

Negroponte, Nicholas

Nicholas Negroponte (born 1943), American computer scientist, professor at the Massachusetts Institute of Technology, honorary professor at the Berlin University of the Arts, co-founder of the MIT Media Lab

Neumann, John

John von Neumann, also Johann von Neumann, real name János Lajos Neumann von Margitta (1903–1957), Hungarian-American mathematician, pioneer of computer technology

Newton, Isaac

Sir Isaac Newton (1643–1727), English physicist and mathematician, holder of the Lucasian Chair of Mathematics at Trinity College, Cambridge

Ng, Ren

Ren Ng (born 1979), American computer scientist, professor of computer science at Stanford, founder of Lytro

Nicolai, Ernst

Ernst Anton Nicolai (1722–1802), German physician, professor of medicine in Halle and Jena, already authored medical writings in German

Niépce, Nicéphore

Joseph "Nicéphore" Niépce (1765–1833) French inventor, pioneer of photography, inventor of heliography

Nietzsche, Friedrich

Friedrich Wilhelm Nietzsche (1844–1900), German philologist and philosopher, professor of classical philology in Basel, best known for his work "Also sprach Zarathustra"

Nipkow, Paul

Paul Julius Gottlieb Nipkow (1860–1940), German inventor, invented the Nipkow disc

Noaillon, Edmond

Edmond Henri Victor Noaillon (unknown, presumably born before 1910), Belgian engineer and inventor

Nefertiti

Nefertiti (c. 1370–1330 BC), wife of Akhenaten and mother-in-law of Tutankhamun

Norling, John

John A. Norling (1895–1957), American film engineer and entrepreneur, co-founder of Loucks-Norling Studios

North, Alex

Alex North (1910–1991), American composer, created numerous film scores

Obama, Barack

Barack Hussein Obama (born 1961), American politician, 44th President of the USA (2009–2017)

Oseen, Carl

Carl Wilhelm Oseen (1879–1944), Swedish physicist, professor of mathematical physics at Uppsala University, director of the Nobel Institute

Ovid

Publius Ovidius Naso (43 B.C.–17 A.D.), Roman poet, widely read writer until modern times, especially known for the "Metamorphoses"

Pacioli, Luca

Luca Pacioli, lat. Lucas de Burgo S. Sepulchri (1445–1514), Italian Franciscan and mathematician, teacher of da Vinci

Pamphilos

Pamphilos, lat. Pamphilus (390 B.C.–350 B.C.), Greek painter, teacher of Apelles

Panofsky, Erwin

Erwin Panofsky (1892–1968), German art historian, professor in Hamburg and Princeton, also studied perspective in ancient wall painting and Renaissance paintings, among other things

Panum, Peter

Peter Ludvig Panum (1820–1885), Danish physiologist, professor of physiology in Kiel and Copenhagen

Paris, John

John Ayrton Paris (1785–1856), British physician, president of the Royal College of Physicians, member of the Royal Society, is considered the inventor of the thaumatrope

Parrhasios

Parrhasios of Ephesos (ca 440 B.C.–385 B.C.), Greek painter, praised by contemporary witnesses for the delicacy and realism of his painting

Paullini, Christian

Christian Franz Paullini (1643–1712), Thuringian physician and writer, personal physician of the bishop of Münster and city physician of Eisenach, author of curious writings

Peckham, Johannes

Johannes Peckham, lat. Also Ioannes de Pecham (c. 1220–1292), English clergyman, author of writings on mathematics and optics, rector of Paris University, later archbishop of Canterbury

Pepper, John

John Henry Pepper (1821–1900), British teacher and scientific presenter, professor of chemistry and director at the Royal Polytechnic Institution in London

Pericles

Pericles (490 B.C.–429 B.C.), Greek politician, disciple of Anaxagoras, strategist (general) of the Attic democracy

Petit, Pierre

Pierre Petit (1594–1677), French mathematician, astronomer and instrument maker, influential engineer (general superintendent of fortifications) at the court of Louis XIV, involved with Pascal in the vacuum experiment, known today mainly as a pen pal of Huygens

Pettitt, Paul

Paul Barry Pettitt, British archaeologist, professor of Archaeology at Durham University, UK, specialising in Palaeolithic art

Phylidor

Phylidor, also Philidor, Philidore, Phillidore or Paul de Philipsthal in different variations (probably second half of the eighteenth century - 1829), Dutch-French (Brabantine) magician, known for his stage performances with ghostly apparitions

Pisanello

Antonio Pisanello, also Antonio di Puccio Pisano (1395–1455), Italian painter and draughtsman, known for his realistic animal and plant drawings

Pittman, Gary

Gary E. Pittman, (born c. 1930), American engineer, Texas Instruments engineer, statistician

Planck, Max

Max Karl Ernst Ludwig Planck (1858–1947), German physicist, professor of physics in Kiel and Berlin Göttingen, Nobel Prize in Physics 1918 for the discovery of the energy quantum

Plateau, Joseph

Joseph Antoine Ferdinand Plateau (1801–1883), Belgian physicist, professor in Ghent, researched in the field of afterimage effects, inventor of the phenakistiscope, the mathematical Plateau problem is named after him

Plato

Plato (427 B.C.–348 B.C.), Greek philosopher, pupil of Socrates

Platter, Felix

Felix Platter (1536–1614), Swiss physician and anatomist, professor of practical medicine and rector at the University of Basel, conducted public dissections, wrote influential works on human anatomy, clinical medicine and mental illnesses

Pliny

Pliny the Elder, Gaius Pliny Secundus (23 A.D.–79 A.D.), Roman scholar and writer

Plutarch

Plutarch, also Plutarchos, lat. Plutarchus (45–125), Greek writer, author of numerous biographies, but also philosophical writings

Polhemus, Bill

William L. "Bill" Polhemus (1924–1998), American officer and engineer, founder of Polhemus Navigation, president of the Institute of Navigation (ION), Virginia

Polyhistor, Alexander

Alexander Polyhistor, also Lucius Cornelius Alexander Polyhistor (c. 100 B.C.–40 B.C.), Greek scholar

Pope, Martin

Martin Pope (born 1918), American chemist, professor of chemistry at New York University, researched the fundamentals of organic light-emitting diodes

Pöppel, Ernst

Ernst Pöppel (born 1940), German psychologist, professor of psychology in Munich

Portlock, Joseph

Joseph Ellison Portlock (1794–1864) British geologist and soldier, member of the Royal Society and General of the Royal Engineers

Pötschke, H.

Hermann? Pötschke (worked between 1890 a. 1917), German master mason and photosculptor

Ptolemaios I

Ptolemaios I. Soter, lat. Ptolemaeus (366 B.C.–283 B.C.), general of Alexander the Great and later regent in Alexandria

Ptolemy, Claudius

Claudius Ptolemy (c. 100–160), Greek astronomer and philosopher, developed the geocentric world view (with the earth in the center), wrote works on optics and epistemology

Pulfrich, Carl

Carl Pulfrich (1858–1927), German physicist and optician, employee at Carl Zeiss, worked especially in the field of stereoscopy

Purkinje, Jan

Jan Evangelista Purkyně or John Evangelista Knight of Purkinje (1787–1869), Czech physiologist, professor of physiology in Prague, researched in the field of sensory perception, the Purkinje effect is named after him

Pythagoras

Pythagoras of Samos (570 BC–510 BC), Greek philosopher and mathematician, founder of the Pythagorean philosophical school, the Pythagorean theorem is attributed to him

Qin

Qin Shihuangdi, actually Ying Zheng (259 B.C.–210 B.C.), King of Qin, first Emperor of China, founder of the Qin dynasty, known for his tyranny and large-scale projects (especially the Great Wall of China and his mausoleum)

Quantz, Johann

Johann Joachim Quantz (1697–1773), German flautist, town piper of Pirna, flautist at the Electoral Saxon Chapel in Dresden, court composer of Frederick the Great

Quinquet, Antoine

Antoine-Arnoult Quinquet (1745–1803), French pharmacist, worked with Argand on the balloon for Montgolfier, improved the Argand lamp

Radon, Johann

Johann Karl August Radon (1887–1956), Austrian mathematician, professor of mathematics in Hamburg, Greifswald, Breslau, Erlangen and Vienna, president of the Austrian Mathematical Society, especially known for the Radon transformation

Raphael

Raphael da Urbino, also Raffaello Sanzio da Urbino (1483–1520), Italian painter and architect, among other things painter for the murals of the Apostolic Palace, construction supervisor for St. Peter's Basilica in Rome

Ramses II

Ramses II (1303 BC–1213 BC), pharaoh of the 19th dynasty, son of Sethos I

Ramses III.

Ramses III. (1221 B.C.–1156 B.C.), Pharaoh of the 20th Dynasty, son of Sethnacht

Ranke, Hermann

Hermann Ranke (1878–1953), German Egyptologist, professor of Egyptology at the University of Heidelberg, director of the Egyptological Institute there, participated in several excavations in Egypt, was involved with Borchardt in the discovery of the bust of Nefertiti

Rankine, William

William John Macquorn Rankine (1820–1872), Scottish physicist and civil engineer, professor of civil engineering in Glasgow, influential thermodynamicist, founder of the Institution of Civil Engineers

Reinitzer, Friedrich

Friedrich Richard Kornelius Reinitzer (1857–1927), professor of botany at the German Technical University in Prague

Reisch, Emil

Emil Reisch (1863–1933), Austrian archaeologist, professor of archaeology at the University of Vienna, Director of the Austrian Archaeological Institute

Rembrandt

Rembrandt van Rijn (1606–1669), Dutch painter

Reynaud, Émile

Charles-Émile Reynaud (1844–1918), French mechanic and photographer, inventor of the Praxinoscope and the Théâtre Optique (film projection)

Riemann, Bernhard

Georg Friedrich Bernhard Riemann (1826–1866), German mathematician, professor of mathematics in Göttingen, provided foundations for the theory of higher-dimensional space

Risner, Friedrich

Friedrich Risner (1533–1580), German mathematician and translator, known especially for his Latin translation and edition of Alhazen and Witelos Optik

Rittenhouse, David

David Rittenhouse (1732–1796), American astronomer and mathematician, professor of astronomy at the State University of Pennsylvania, president of the American Philosophical Society

Roberts, Lawrence

Lawrence Gilman "Larry" Roberts (1937–2018), American engineer, co-developer of the ARPANET

Robertson, Stephan

Stephan Kaspar Robertson or d'Étienne-Gaspard Robertson, proper name Étienne-Gaspard Robert, pen name (1763–1837), Belgian magician and balloonist

Robin, Henri

Henri Robin, real name Henri Joseph Donckèle (1811–1874), French magician

Roddenberry, Gene

Eugene Wesley Roddenberry (1921–1991), American television and film producer, screenwriter, author and producer of the "Star Trek" series

Roget, Peter

Peter Mark Roget (1779–1869), British physician and physiologist, professor of physiology at the Royal Institution, author of "Roget's Thesaurus"

Rohr, Moritz

Moritz von Rohr (1868–1940) German optician, assistant to Abbe at Carl Zeiss, professor of optics at the University of Jena, developer of various optical devices

Rollmann, Wilhelm

Wilhelm Rollmann (1821–1890), German teacher, teacher of physics and mathematics at the Gymnasium Stargard, is considered together with d'Almeida to be the discoverer of the anaglyph process

Röntgen, Wilhelm

Wilhelm Conrad Röntgen (1845–1923), German physicist, professor of physics in Strasbourg, Hohenheim, Giessen, Würzburg and Munich, discoverer of "X-rays", Nobel Prize in Physics 1901

Rosenbach, Ottomar

Ottomar Ernst Felix Rosenbach (1851–1907), German physician, psychotherapist, chief physician of the All Saints Hospital in Breslau, professor at the University of Breslau

Ross, Douglas

Douglas Taylor Ross (1929–2007), American computer scientist, co-developer of the computer language APT, founder of SofTech

Rothafel, Roxy

Samuel Lionel "Roxy" Rothafel, real name Rothapfel (1882–1936), American cinema manager, founder of the Roxy Theatre

Round, Henry

Henry Joseph Round (1881–1966), English engineer, chief engineer at Marconi, discoverer of the Round effect

Rubens

Peter Paul Rubens (1577–1640), Flemish painter

Salisbury, Johannes

John of Salisbury, lat. Joannes Salisberiensis (c. 1115–1180), English theologian, Bishop of Chartres

Sandin, Daniel

Daniel J. Sandin (born 1942), American computer scientist, professor at the University of Illinois at Chicago, co-founder with DeFanti of the Electronic Visualization Laboratory

Sandrart, Joachim

Joachim von Sandrart (1606–1688), German painter and art historian, author of important writings on art history

Sanka, Yoshiyuki

Yoshiyuki Sanka (born 1958), Japanese engineer and entrepreneur, professor at the University of Tsukuba, developer of the HAL robotic suit, founder of the company Cyberdyne

Schadt, Martin

Martin Schadt (born 1938), Swiss physicist, developer of liquid crystal displays, founder of the company Rolic

Scheiner, Christoph

Christoph Scheiner (1573–1650), German Jesuit, physicist, optician and astronomer, professor for physics and astronomy in Ingolstadt, later rector of the monastery school St. Anna (today the lyceum "Carolinum") in Silesian Neisse (today Polish Nysa)

Scherer, Alexander

Alexander Nicolaus Scherer (1772–1824), German-Russian chemist, professor of chemistry in Dorpat (Estonia) and St. Petersburg

Schopenhauer, Arthur

Arthur Schopenhauer (1788–1860), German philosopher, author of influential works

Schott, Caspar

Caspar Schott, also Kaspar Schott (1608–1660), German Jesuit, professor of mathematics and physics in Würzburg, pupil of Athanasius Kircher

Schumpeter, Joseph

Joseph Alois Schumpeter (1883–1950), Austrian economist and politician, Harvard professor

Schwenter, Daniel

Daniel Schwenter, (1585–1636), German mathematician, professor at the Nuremberg University in Altdorf "Altdorfina" (dissolved 1809)

Scott, Leader

Leader Scott, pseudonym for Lucy Baxter (1837–1902), English writer, daughter of the poet William Barnes, authored several writings on Italian art and architecture

Sedgwick, Adam

Adam Sedgwick (1785–1873), English geologist and clergyman, professor of geology at Cambridge, critic of Darwin's theory of evolution

Seeber, Ludwig

Ludwig August Seeber (1793–1855), professor of physics in Freiburg and Karlsruhe

Séguin, Pierre

Pierre Séguin (unknown–1884), French glass painter, inventor of the polyorama and polyoscope

Selke, Willy

Willy Alfred Carl Selke (worked c. 1895–1915, life data unknown), German photographer and inventor, photosculptor

Sethos I

Sethos I (1323 BC–1279 BC), pharaoh of the 19th dynasty, father of Ramses II

Shakespeare, William

William Shakespeare (1564–1616), English writer and actor

Shaw, Wilfrid

Wilfrid "Bud" Shaw, (born 1930), American chemist, conducted research with Brown on the use of liquid crystals

Smith, Robert

Robert Smith (1689–1768), British mathematician, professor of astronomy at Cambridge

Smith, Willoughby

Willoughby Smith (1828–1891), English engineer, discovered the photosensitivity of selenium

Snow, Dean

Dean R. Snow (born 1940), American anthropologist, professor of anthropology, Pennsylvania State University, USA, from 2007–2009 president of the Society for American Archaeology

Socrates

Socrates (469 B.C.–399 B.C.), Greek philosopher, disciple of Archelaus (who was a disciple of Anaxagoras), teacher of Plato and Xenophon, practitioner of philosophical disputation, sentenced to death as an old man and executed

Solon

Solon (640 B.C.–560 B.C.), Greek politician and philosopher, the Solonian order is considered the beginning of democracy

Sombrowsky, Reiner

Reiner Sombrowsky (born 1950), German mathematician, developer of the Holotron monitor

Sophocles

Sophocles (497 B.C.–406 B.C.), Greek poet, his tragedies (especially "Antigone" and "Oedipus the King") were also translated into German and are still performed today

Spencer, Domina

Domina Eberle Spencer (born 1920), American mathematician, professor of mathematics at the University of Connecticut, wife of Parry Moon

Spooner, William

William Matthias Spooner (1796–1882), British publisher, inventor of the "Protean Views"

Sprigg, Reginald

Reginald Claude Sprigg (1919–1994), Australian geologist and entrepreneur, discoverer of the Ediacara fauna (macrofossils)

Stampfer, Simon

Simon Ritter von Stampfer (1790–1864), Austrian mathematician, professor at the Polytechnic Institute in Vienna (today TU Vienna)

Stanford, Leland

Amasa Leland Stanford (1824–1893), American entrepreneur and politician, co-founder and president of the Central Pacific Railroad, governor of California, founder of Stanford University

Steenblik, Richard

Richard A. Steenblik (born 1957), American engineer, inventor of the ChromaDepth process

Steiner, Johann

Johann Ludwig Steiner (1711–1779), Swiss optician and watchmaker, co-founder of the Zurich Natural Research Society, translator of optical writings

Stierlin, Henri

Henri Stierlin (born 1928), Swiss art historian; publisher of illustrated art books, argued that Borchardt had forged the bust of Nefertiti

Stumpf, Felix

Fleix Stumpf (1885–1970), German physicist, private lecturer at the Technical University of Berlin, son of the psychologist and music researcher Carl Stumpf, after his death also editor of the "Erkenntnislehre"

Sturm, Johannes

Johannes Christoph Sturm (1635–1703), German mathematician and physicist, professor of mathematics and physics at the University of Altdorf

Sutherland, Ivan

Ivan Edward Sutherland (born 1938), American computer scientist, professor at Harvard, University of Utah and the California Institute of Technology, developer of graphical input and output systems

Swainson, Wyn

Wyn Kelly Swainson (born 1939), Danish-American chemist and engineer, developer of early 3D printing applications, founder of Formigraphic (later Omtec Replication)

Swedenborg, Emanuel

Emanuel Swedenborg, proper Swedberg (1688–1772), Swedish scholar and mystic, founder of the New Church (Swedenborgian)

Symons, Walter

Walter Symons (probably 1855–1905), British publishing manager, inventor of the Motograph pictures

't Hooft, Gerard

Gerard or Gerardus 't Hooft (born 1946), Dutch physicist, professor at the University of Utrecht, Nobel Prize for Physics 1999 together with Martinus Veltman

Tacquet, André

André Tacquet, also Andrea Tachenius (1612–1660), Belgian Jesuit and mathematician. Teacher at the Jesuit College in Louvain, author of mathematical textbooks

Talbot, Henry

William Henry Fox Talbot (1800–1877), English naturalist, pioneer of photography, known for the discovery of the self-imaging effect (now called the "Talbot effect")

Tang, Ching

Ching Wan Tan (born 1947), Chinese-American chemist, researcher at Eastman Kodak, professor of chemistry at the University of Rochester, considered the inventor of the OLED (together with van Slyke)

Thales

Thales of Miletus (c. 623 BC–548 BC), Greek philosopher, one of the Seven Sages

Theodotos

Theodotos of Aetolia, also Theodotas (third cent. B.C.), general of Ptolemaios IV, known for his betrayal of Ptolemaios

Theophrast

Theophrastus, also Theophrastos of Eresos (371 B.C.–287 B.C.), Greek philosopher and botanist, pupil of Aristotle

Thomson, William

William Thomson, Lord Kelvin (1924–1907), British physicist, professor of natural philosophy in Glasgow, researched in the fields of electricity and thermodynamics, the unit Kelvin is named after him

Thutmosis

Thusmosis, also Thustmose (fourteenth century BC), Egyptian sculptor at the time of Pharaoh Akhenaten, known through the discovery of his studio by Borchardt

Timoshenko, Gregory

Gregory Stephen Timoshenko (1904–2000), Ukrainian-American physicist, professor of electrical engineering at the University of Connecticut, son of mechanics pioneer Stephen Timoshenko

Tiraboschi, Girolamo

Girolamo Tiraboschi (1731–1794), Italian Jesuit and historian, head of the library at the court of Modena

Tóth, László

László Fejes Tóth (1915–2005), Hungarian mathematician, professor at the Alfred Renyi Institute in Budapest

Trentsensky, Mathias

Mathias Trentsensky (1790–1868), Austrian lithographer and publisher, made picture sheets, produced and published the picture series for Stamper's stroboscopic discs in the art shop Trentsensky und Vieweg

Tutankhamun

Tutankhamun (1342 B.C.–1323 B.C.), pharaoh of the 18th dynasty, especially known for the discovery of his richly furnished tomb in the Valley of the Kings in the twentieth century

Tyler, Christopher

Christopher W. Tyler (born 1943), English neuroscientist, professor at the City University of London

Uchatius, Franz

Franz Freiherr von Uchatius (1811–1881), Austrian officer, last in rank of field marshal lieutenant, invented the steel bronze and a moving picture projector

Unterseher, Fred

Fred Unterseher (born 1946), American artist, inventor of the holographic kinetic figures, co-founder of the Multiplex Company, co-founder "School of Holography" (both with Llloyd Cross)

Upatnieks, Juris

Juris Upatnieks (born 1936), Latvian-American physicist and inventor, professor of electrical engineering at the University of Michigan, with Emmett Leith developer of practical holography

Valius, Nikolai

Nikolai Adamovich Valius (unknown, probably born around 1920), Soviet engineer, professor at the National Institute of Photographic Cinematography (NIKFI), author on scanning optics

van Berkel, Cees

Cees van Berkel (born 1958), Dutch physicist, display specialist, and data scientist

Van Slyke, Steven

Steven Van Slyke (born 1956), American engineer, researcher at Eastman Kodak, CTO at Kateeva, considered the inventor of the OLED (together with Tang)

Vasari, Giorgio

Giorgio Vasari (1511–1574), Italian architect, painter and art historian, author of an extensive work of artist biographies (more than 300 artists), coined the terms Gothic and Renaissance

Vaucanson, Jacques

Jacques de Vaucanson (1709–1782), French engineer, chief inspector of silk manufactories, designer of automatic machines

Vermeer

Jan Vermeer van Delft (1632–1675), Dutch painter

Vesalius, Andreas

Andreas Vesalius, real name Andreas Witinck (1514–1564), Flemish physician and anatomist, professor of surgery and anatomy in Venice and Padua

Vieth, Gerhard

Gerhard Ulrich Anton Vieth, also G. U. A. Vieth (1763–1836), German teacher, professor for mathematics in Dessau, wrote some mathematical treatises, is counted among "gymnastics fathers" like Guts Muths or Jahn

Vignola, Giacomo

Giacomo Barozzi da Vignola (1507–1573), Italian architect, painter and perspective draughtsman

Vitruvius

Vitruvius, rarely also Marcus Vitruvius Polli (c. 80 BC–c. 15 BC), Roman architect and engineer, his main work is the "Ten Books on Architecture"

Voigt, Woldemar

Woldemar Voigt (1850–1919), German physicist, professor of physics in Königsberg and Göttingen, known especially for his work in electrooptics and crystal physics

Voltaire

Voltaire, real name François-Marie Arouet (1694–1778), French philosopher and social critic

Wackenfels, Johannes

Johann Matthäus Wacker von Wackenfels (1550–1619), German jurist and diplomat, friend of Kepler

Walch, Johann

Johann Ernst Immanuel Walch (1725–1778), philologist and geologist, professor of philosophy, logic and rhetoric at the University of Jena

Walgenstein, Thomas

Thomas Walgenstein, proper name Thomas Rasmussen Walgenste(e)n, also Thomas Erasmi Gotlandus or Walgenstenius (1627–1681), Danish mathematician, is regarded as the inventor of the magic lantern

Waller, Fred

Frederic Waller (1886–1954), American inventor and filmmaker, employee at Paramount, inventor of the Viterama panorama cinema, Oscar 1953 for the Cinerama

Weigel, Erhard

Erhard Weigel (1625–1699), German mathematician, professor of mathematics and for a time rector of the University of Jena, famous for the "Weigel House" with elevator and wine tapping system - one of the "seven wonders" of Jena

Weinbaum, Stanley

Stanley G. Weinbaum (1902–1935), American writer, science fiction author

Wheatstone, Charles

Sir Charles Wheatstone (1802–1875), British physicist, professor of experimental physics at King's College, London, known for the development of Wheatstone's measuring bridge and the stereoscope

White, James

James White (1824–1884), Scottish optician, instrument maker for the University of Glasgow, notably for William Thomson's laboratory

Wiener, Christian

Ludwig Christian Wiener (1826–1896) German mathematician, professor of geometry in Karlsruhe, author of a textbook on descriptive geometry

Wilkins, John

John Wilkins (1614–1672), English clergyman and scholar, head of Wadham College, Oxford, master of Trinity College, Cambridge, Bishop of Chester, founding member of the Royal Society, defender of Copernicus, developer of a universal philosophical language

Willème, François

François Willème (1830–1905), French artist, developer of the photo-sculpture copy

Williams, Richard

Richard Williams (born 1927), American chemist, researched liquid crystals at RCA

Winnek, Douglas

Douglas Fredwill Winnek Coffey (1907–1999), American inventor, son of bigamist and murderer William Noah Coffey

Witelo

Witelo, also Witelo de Viconia or Vittelio (c. 1237–1285), Thuringian-Silesian monk and scientific author in the field of optics

Wittmann, Marc

Marc Christoph Wittmann (born 1966), German psychologist, researches in particular on the psychology of time

Wolf, Max

Maximilian Franz Joseph Cornelius Wolf (1863–1932), German astronomer, professor of astronomy in Heidelberg, director of the Heidelberg Observatory

Wollaston, William

William Hyde Wollaston (1766–1828), English physician and chemist, private scholar, member and president of the Royal Society, developed his own instruments for observing nature, discoverer of palladium and rhodium

Woltersdorff, Johann

Johann Lucas Woltersdorff (1721–1772), preacher of the St. Gertrud church in Berlin and mineralogist

Woltring, Herman

Herman Jozef Woltring (1943–1992) Dutch psychologist and mathematician, biomechanist, professor at the University of Nijmegen

Xenophanes

Xenophanes of Colophon (570 B.C.–475 B.C.), Greek philosopher, critic of Homer's and Hesoid's world of gods, according to him men created the gods in their image

Xenophon

Xenophon (430 B.C.–425 B.C.), Greek philosopher and politician, pupil of Socrates, created with "Oikonomikos" a work on economics, with "Hipparchikos" the first book on horsemanship

Young, Thomas

Thomas Young (1773–1829), British physician and physicist, researched the accommodation of the eye and the wave theory of light, decipherer of Egyptian hieroglyphs

Zahn, Johann

Johann(es) Zahn (1641–1707), German mathematician and optician, professor of mathematics in Würzburg, canon in the order of the Premonstratensians

Zanoni, Louis

Louis A. Zanoni (1933–2018), American engineer, Hellmeier's employee at RCA, founder co-founder of the digital watch company Optel and founder of Zantech (service for digital watches)

Zeduan, Zhang

Zhang Zeduan or Chang Tse-tuan (1085–1145), Chinese painter, creator of the Qingming scroll

Zeuxis

Zeuxis of Herakleia (c. 500 B.C.), Greek painter, painted the palace of Archelaos, famous for his naturalistic paintings

Zimmerman, Tom

Thomas G. "Tom" Zimmerman (born c. 1957), American engineer, developer at VPL, researcher at the MIT Media Lab and at IBM Research

Zöllner, Frank

Frank Zöllner (born 1956), German art historian, professor of art history at the University of Leipzig

Zuse, Konrad

Konrad Ernst Otto Zuse (1910–1995), German engineer and inventor, developer of the first computer Z3, developer of the programming language “Plankalkül”

Index

A. Grasnick, *Basics of Virtual Reality*,
https://doi.org/10.1007/978-3-662-64201-6

MIX
Papier aus verantwortungsvollen Quellen
Paper from responsible sources
FSC® C105338

If you have any concerns about our products,
you can contact us on
ProductSafety@springernature.com

In case Publisher is established outside the EU,
the EU authorized representative is:
Springer Nature Customer Service Center GmbH
Europaplatz 3, 69115 Heidelberg, Germany

Printed by Libri Plureos GmbH
in Hamburg, Germany